Y0-CUN-553

Adhesion and microorganism pathogenicity

The Ciba Foundation is an international scientific and educational charity. It was established in 1947 by the Swiss Chemical and Pharmaceutical company of CIBA Limited – now CIBA-GEIGY Limited. The foundation operates independently in London under English trust law.

The Ciba Foundation exists to promote international cooperation in medical and chemical research. It organizes international multidisciplinary meetings on topics that seem ready for discussion by a small group of research workers. The papers and discussions are published in the Ciba Foundation series.

The Foundation organizes many other meetings, maintains a library which is open to graduates in science or medicine who are visiting or working in London, and provides an information service for scientists. The Ciba Foundation also functions as a centre where scientists from any part of the world may stay during working visits to London.

Adhesion and microorganism pathogenicity

Ciba Foundation symposium 80

1981

Pitman Medical

© Ciba Foundation 1981

ISBN 0-272-79615-8

Published in February 1981 by Pitman Medical Ltd, London. Distributed in North America by CIBA Pharmaceutical Company (Medical Education Administration), Summit, NJ 07901, USA.

Suggested series entry for library catalogues:
Ciba Foundation symposia.

Ciba Foundation symposium 80
x + 346 pages, 59 figures, 58 tables

British Library Cataloguing in publication data:

Adhesion and microorganism pathogenicity. –
(Ciba Foundation. Symposia; 80).
1. Micro-organisms – Physiology – Congresses
2. Adhesion – Congresses
3. Medical microbiology – Congresses
I. O'Connor, Maeve
II. Whelan, Julie III. Elliott, Katherine
IV. Series
576'.11'8 QR84

Set in 10pt Press Roman by Freeman Graphic, Tonbridge
Printed and bound in Great Britain at the Pitman Press, Bath

Contents

Symposium on Adhesion and microorganism pathogenicity held at the Ciba Foundation, London, 13-15 May 1980
Editors: Katherine Elliott (Organizer), Maeve O'Connor and Julie Whelan

Participants

E. H. BEACHEY Veterans' Administration Medical Center, 1030 Jefferson Avenue, Memphis, Tennessee 38104, USA

W. BREDT Institut für Allgemeine Hygiene und Bakteriologie, Zentrum für Hygiene, Universität Freiburg, Hermann-Herder-Strasse 11, D-7800 Freiburg, FRG

D. C. A. CANDY Institute of Child Health, The Nuffield Building, Birmingham Children's Hospital, Ladywood, Birmingham B16 8ET, UK

P. W. CHOPPIN Department of Virology and Medicine, The Rockefeller University, 1230 York Avenue, New York, NY 10021, USA

A. D. ELBEIN Department of Biochemistry, The University of Texas Health Science Center, 7703 Floyd Curl Drive, San Antonio, Texas 78284, USA

T. FEIZI Division of Communicable Diseases, MRC Clinical Research Centre, Northwick Park, Watford Road, Harrow, Middlesex HA1 3UJ, UK

R. FRETER Department of Microbiology and Immunology, 6643 Medical Science Building II, The University of Michigan Medical School, Ann Arbor, Michigan 48109, USA

J. FRIEND Department of Plant Biology, University of Hull, Hull HU6 7RX, UK

A. HELENIUS European Molecular Biology Laboratory, Postfach 10.2209, Meyerhofstrasse 1, D-6900 Heidelberg, FRG

R. J. HOWARD Malaria Section, Laboratory of Parasitic Diseases, National Institute of Allergy and Infectious Diseases, National Institutes of Health, Bethesda, Maryland 20205, USA

R. C. HUGHES Department of Biochemistry, MRC National Institute for Medical Research, The Ridgeway, Mill Hill, London NW7 1AA, UK

C. LEBEN Department of Plant Pathology, Ohio Agricultural Research and Development Center, Wooster, Ohio 44691, USA

M. M. LEVINE Center for Vaccine Development, Division of Infectious Diseases, University of Maryland School of Medicine, 29 South Greene Street, Baltimore, Maryland 21201, USA

D. MIRELMAN Department of Biophysics, The Weizmann Institute of Science, Rehovot, Israel

P. C. NEWELL Department of Biochemistry, University of Oxford, South Parks Road, Oxford OX1 3QU, UK

J. H. PEARCE Department of Microbiology, South West Campus, University of Birmingham, PO Box 363, Birmingham B15 2TT, UK

S. RAZIN Department of Membranes and Ultrastructure, The Hebrew University–Hadassah Medical School, PO Box 1172, Jerusalem, Israel

M. H. RICHMOND Department of Bacteriology, University of Bristol Medical School, University Walk, Bristol BS8 1TD, UK

J. M. RUTTER ARC Institute for Research on Animal Diseases, Compton, Near Newbury, Berkshire, RG16 0NN, UK

N. SHARON Department of Biophysics, The Weizmann Institute of Science, Rehovot, Israel

F. J. SILVERBLATT Veterans Administration Medical Center, Sepulveda, California 91343, USA

M. SUSSMAN Department of Microbiology, University of Newcastle upon Tyne Medical School, Newcastle upon Tyne NE1 7RU, UK

C. SVANBORG EDEN Department of Clinical Immunology, Institute of Medical Microbiology, University of Göteborg, Guldhedsgatan 10, S-413 46 Göteborg, Sweden

D. TAYLOR-ROBINSON Division of Communicable Diseases, MRC Clinical Research Centre, Northwick Park, Watford Road, Harrow, Middlesex HA1 3UJ UK

E. C. TRAMONT Division of Infectious Diseases, Department of Bacterial Diseases, Walter Reed Army Institute of Research, Walter Reed Army Medical Center, Washington, DC 20012, USA

K. VOSBECK K-125/212, CIBA-GEIGY Limited, CH-4002 Basle, Switzerland

D. F. H. WALLACH Radiobiology Division, Tufts University School of Medicine –New England Medical Center, 171 Harrison Avenue, Boston, Massachusetts 02111, USA

J. W. WATTS Department of Ultrastructural Studies, The John Innes Institute, Colney Lane, Norwich NR4 7UH, UK

Introduction

D. TAYLOR-ROBINSON

Division of Communicable Diseases, MRC Clinical Research Centre and Northwick Park Hospital, Watford Road, Harrow, Middlesex HA1 3UJ, UK

When Katherine Elliott and I discussed the form this symposium might take, a number of possible ways of approaching it were evident. The rather formal systematic approach would have been to start with the largest microorganisms and proceed to the smallest ones, the viruses, or vice versa. Another possible approach would have been to consider the anatomical location of microorganisms, discussing the gastrointestinal tract, then the urinary tract, and so on. This, however, would have involved a great deal of microorganism overlap and, furthermore, would have made it difficult to fit in some aspects of the subject under consideration. There certainly seemed to be merit in organizing the symposium into quite specific topics, for example mechanisms of attachment, ways of preventing attachment, and the like. In the end the form has been conditioned by inviting those we felt were doing good work and asking them to discuss it. In so doing I think we shall cover the main groups of microorganisms and many of the interesting and contentious areas.

The first part of the programme may look a hotchpotch but we deliberately decided to have a mix of diverse microorganisms and so create a situation which will stimulate discussion. Eventually, of course, we shall consider aspects which are more clinical, and there will be a progression towards discussing factors which prevent adhesion, or what can be done to prevent adhesion. Although, as I have already said, we shall cover the main groups of microorganisms, there is certainly no possibility of dealing with every microorganism or all aspects of this vast and growing topic. I hope, however, that if there appears to be a serious omission those who feel able to make a contribution will feel free to do so in the discussion.

I believe our discussions should have some direction. We should obviously think first about the mechanisms of adherence: do these differ for different microorganisms or are there similarities between different microorganisms? Secondly, the

1981 Adhesion and microorganism pathogenicity. Pitman Medical, Tunbridge Wells (Ciba Foundation symposium 80) p 1-2

relationship of microbial adherence to pathogenicity is what this meeting is fundamentally about but we should reflect on whether adherence is a dominant factor in pathogenicity or only a minor one. We might also give some thought to the subsequent events that are stimulated by adherence. After all, adherence is only the initiating factor in a chain of events that culminates in disease. And of course we must discuss ways in which adherence may be thwarted as a means of preventing disease.

In the final general discussion we might come back to these various points and ask ourselves whether we have covered them adequately. I am sure there will be some gaps and, in an effort to fill these and to summarize our thoughts, I believe we should draw up a table indicating what is known about adherence mechanisms for each of the microorganisms that we discuss and the bearing that they have on pathogenicity.

Attachment of mycoplasmas to inert surfaces

W. BREDT, J. FELDNER and I. KAHANE*

*Institute for General Hygiene and Bacteriology, Centre for Hygiene, University of Freiburg, D-7800 Freiburg, West Germany, and *Department of Membranes and Ultrastructure, Hebrew University–Hadassah Medical School, Jerusalem, Israel*

Abstract As well as adhering to a variety of animal cells *Mycoplasma pneumoniae* attaches firmly to inert surfaces such as glass and plastic. This property is the basis for the gliding motility of this species and provides a useful experimental model. The mechanism by which the organism attaches to glass appears to be influenced by the presence of protein in the medium. In buffer without protein the attachment is pH-dependent and seems to be determined mainly by electrostatic forces. Addition of bovine serum albumin (BSA) reduces attachment by about 90% but in serum-containing growth medium the cells adhered to the glass in great numbers. Experiments in BSA-containing buffer with added glucose (0.25 mg/ml) showed a 10-fold increase in attachment. This effect was dose-dependent, with the intermediates pyruvate and phosphoenolpyruvate decreasing attachment again at higher concentrations. Other metabolizable sugars caused a similar increase in adherence. 3-*O*-Methylglucopyranoside, 2-deoxyglucose, iodoacetate, fluoride, arsenite, carbonylcyanide-*m*-chlorophenylhydrazone and dicyclohexylcarbodiimide reduced the glucose-stimulated adherence significantly. The ionophore valinomycin improved adherence by about 20%. The results suggested that successful attachment of *M. pneumoniae* to glass in a protein-containing environment may require energy. In contrast, adherence to eukaryotic cells was not influenced by protein and seemed to be less sensitive to changes in energy metabolism. The exact mechanism of attachment of mycoplasmas to inert surfaces is so far unknown. There are certain differences from the mechanism(s) mediating adherence to animal cells. It is suggested that this particular type of adherence is involved in the pathogenesis of *M. pneumoniae* infection in a way as yet unknown.

Adherence of microorganisms to eukaryotic cells seems to have little in common with adherence to inert surfaces. However, in *Mycoplasma pneumoniae* we have a proven pathogen which colonizes only the mucous surfaces of the human respiratory tract yet is also capable of attaching firmly to inert materials such as glass or plastic

1981 Adhesion and microorganism pathogenicity. Pitman Medical, Tunbridge Wells (Ciba Foundation symposium 80) p 3-16

and of gliding along these surfaces with considerable speed (Bredt 1979, Radestock & Bredt 1977). Since *M. pneumoniae* exists only in humans and a few experimental animal hosts, this particular property must have some biological value for survival of the microorganism *in vivo.* It therefore seems reasonable to discuss its adherence to glass in the context of pathogenicity, and we shall try to summarize some of our results here.

The phenomenon

Somerson et al (1967) were the first to describe the adherence of *M. pneumoniae* to the glass surface of Povitsky bottles; Taylor-Robinson & Manchee (1967) then broadened the spectrum of species and in addition observed the same phenomenon on plastic surfaces. Other authors (Purcell et al 1971) examined the conditions of adherence of several mycoplasma species and found improved attachment when the serum content of the medium was low. However, these data were considered mainly in relation to antigen production and other purposes for which the mycoplasmas had to be washed. Another phenomenon that depends heavily on the ability of the cells to adhere to inert surfaces is the gliding movement of some motile *Mycoplasma* species (Bredt 1979). It was in this context that in addition to cell adherence we started to investigate the adherence of *M. pneumoniae* and *Mycoplasma gallisepticum* to glass (Feldner et al 1979b, Kahane et al 1979). A third motile species, *Mycoplasma pulmonis,* also showed firm attachment to glass. Other species vary in this respect. Laboratory strains of several species maintained for a long time *in vitro* seem to be less able to grow on the glass surface of coverslip chambers than freshly isolated strains.

Experimental systems

For quantitative studies *M. pneumoniae* cells (strain FH) labelled with $[^3H]$palmitate were allowed to settle for 3 h on circular glass coverslips in the wells of plastic multi-well trays (Feldner et al 1979b). Shortening of the attachment phase by centrifugation produced difficulties with suspension liquids containing protein and was therefore discontinued. Adherence of *M. gallisepticum* was studied in a slightly different system using test tubes (Kahane et al 1979), and in the preliminary qualitative experiments with *M. pneumoniae* we used protein-coated latex spheres (Gorski & Bredt 1977).

Properties of inert surfaces

Little precise information is available on the properties of the inert surfaces used in the experiments. It is known that mycoplasmas grow on glass as well as on plastic

(Taylor-Robinson & Manchee 1967). Pyrex glass was found to be superior to soda glass. In our experiments *M. pneumoniae* also grew on carbon-coated glass and on Formvarfilms coating platinum or steel grids. In all cases the morphology was similar, with the cells spreading irregularly (Bredt 1979), indicating that the conditions for attachment were similar. Silicon treatment of the glass resulted in the same growth pattern and quantitative results as seen on untreated surfaces. Growth in a serum-containing medium (10–20%), however, is inevitably accompanied by the development of a protein layer on the surface and this layer may at least help the cells to colonize the various substrata. This factor was certainly also involved when adherence to inert materials was tested with serum-coated polystyrene particles (Gorski & Bredt 1977). In the glass assay, pretreatment of the coverslips with growth medium apparently abolished the attachment-reducing effect of bovine serum albumin (BSA) (see below) (Feldner et al 1979b).

Properties of the mycoplasma surface

The results of treatments acting on the mycoplasma surface were somewhat contradictory. Trypsin treatment of *M. pneumoniae* (up to 50 μg/mg of mycoplasma protein) had only a limited effect on glass adherence (about 50% decrease) in contrast to its effect on cell adherence. This certainly seems to exclude proteins as the principal mediating structures. On the other hand after 3 h of settling time the test system may allow the cells to replace some of the lost surface peptides (Gorski & Bredt 1977). An involvement of protein is also suggested by the fact that adherence to glass is reduced by 50% by 10^{-3} M-phenylglyoxal (binding to arginine) and by 70% by 10^{-3} M-*N*-α-*p*-tosyl-L-lysine chloromethyl ketone HCl (TLCK, interacting with histidine) (J. Feldner, unpublished results). Whether this indicates a direct involvement of the respective amino acids or peptides remains unknown. Addition of amino acids to the buffer did not interfere with the mycoplasma–glass interaction. Incubation with anti-*M. pneumoniae* antiserum significantly reduced the attachment.

An observation on *Mycoplasma hominis* suggests that at least in this species the attachment-mediating structure is not vital: after numerous passages in our laboratory the PG21 strain irreversibly lost its ability to adhere to glass surfaces but showed no other signs of degeneration.

Influence of suspension liquid

Experiments with *M. pneumoniae* in Tris buffer pH 7.2 showed an influence of pH with an optimum at pH 5.6, a distinct temperature dependence, and an inhibiting effect of higher salt concentrations. However the attachment of both species investigated, *M. pneumoniae* and *M. gallisepticum,* was reduced by about 80% after addition of low concentrations (1 mg/ml) of BSA (Feldner et al 1979b, Kahane et al 1979). When attachment was tested in growth medium (containing 20% horse serum) at 37 °C the number of adhering cells began to increase after about 3 h and reached extremely high values after 6 h (Feldner et al 1979b). Apparently the growth medium enabled the cells to attach more firmly to the glass by providing either a coating substance on the glass surface or substrates for active metabolism, or by affecting membrane fluidity. A purely physical factor such as protein coating the glass was rather unlikely because of the slow onset of the attachment. Furthermore, chloramphenicol (10 μg/ml) significantly reduced the effect of growth medium, indicating that there must be a considerable metabolic component. The observations suggested two types of attachment: one in a protein-free environment mediated mainly by electrostatic forces and perhaps salt bridges, and the other in protein-containing growth medium, requiring active efforts by the living cell itself. The reducing effect of BSA added to buffer was perhaps due to interference with the first type of attachment, while at the same time the mycoplasma cell could not use the second or 'metabolic' type because of the substrate-free buffer. The role of metabolism was therefore investigated in further experiments.

The role of sugar metabolism

The experiments were done in Tris buffer containing 10 mg BSA/ml (BSA buffer). Addition of glucose to this buffer resulted in a significant increase in attachment (18-fold) with an optimum at 0.5 mg glucose/ml, whereas several non-metabolizable sugars were without effect (Feldner et al 1979b). Two other sugars metabolized by *M. pneumoniae,* namely fructose and mannose, also increased adherence. These two had a distinct optimum at 0.25 mg/ml (Fig. 1). Two products of glucose catabolism, pyruvate and phosphoenolpyruvate, also increased attachment, pyruvate in concentrations comparable to glucose (Fig. 1) and phosphoenolpyruvate in concentrations about 100-fold lower than pyruvate. In preliminary experiments the influence of these substances on the ATP content of the mycoplasmas was tested using the luciferin–luciferase system. The addition of glucose and pyruvate resulted in a significant increase in ATP within the mycoplasmas.

Several possible ways of inhibiting sugar metabolism were therefore tested. The first substances tested were the glucose analogues. The effects of glucose, fructose and mannose were reduced by 3-*O*-methylglucopyranoside (3-*O*-MG) to 18, 34 and 40% respectively. The analogues 2-deoxyglucose and 6-deoxyglucose, when tested with glucose, were also effective, but to a lesser extent (37 and 46% respectively). The effect of 3-*O*-MG was overcome by adding pyruvate.

The effect of other inhibitors tested is shown in Fig. 2. The SH reagents iodoacetate and *p*-hydroxymercuribenzoate (PHMB) and the enolase-inhibiting sodium

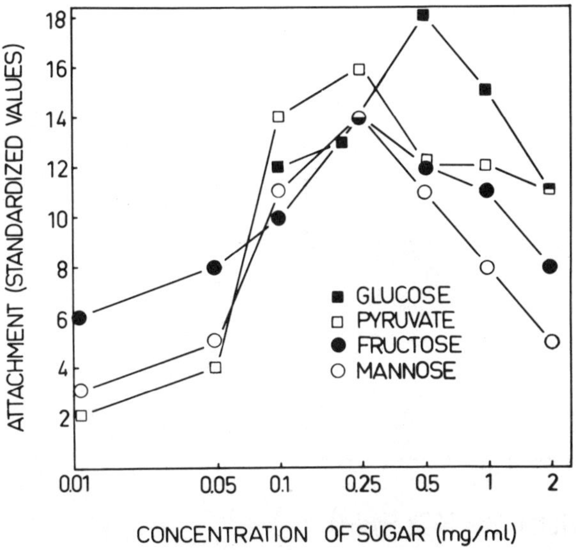

FIG. 1. Effect of various sugars on attachment of *M. pneumoniae* to glass. Control suspension in BSA buffer without sugar. Settling time, 3 h.

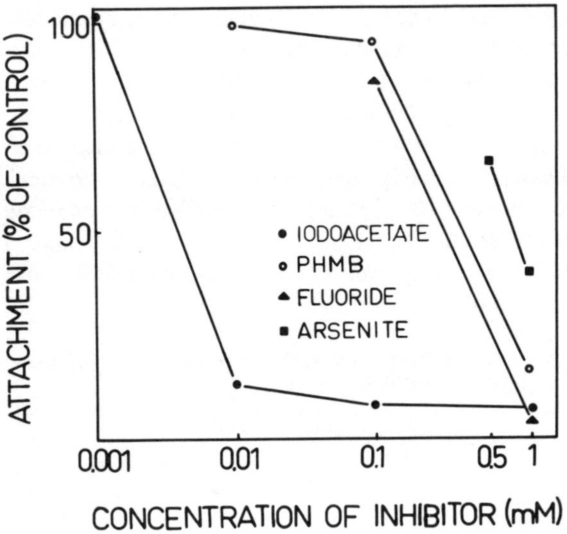

FIG. 2. Effect of inhibitors on attachment of *M. pneumoniae* to glass in the presence of BSA buffer with glucose.

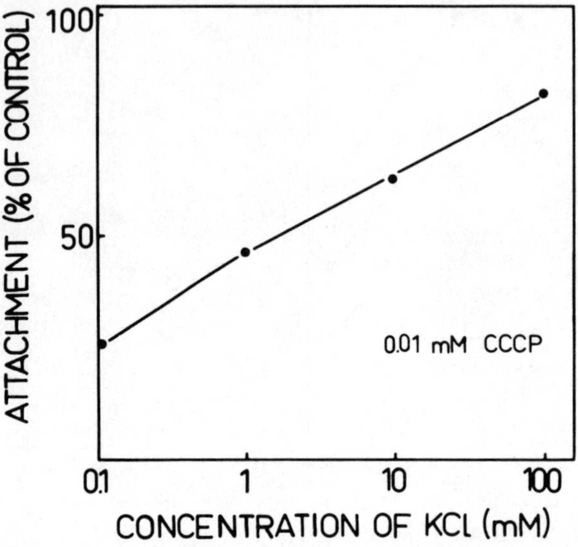

FIG. 3. Effect of increasing KCl concentrations on the inhibitory action of carbonylcyanide-*m*-chlorophenylhydrazone (CCCP) (0.1 mM). BSA buffer with 0.5 mg glucose/ml.

fluoride, as well as sodium arsenite, which inhibits the pyruvate dehydrogenase complex, effectively inhibited the attachment-enhancing effect of glucose. Substances that influence the electrochemical gradient, such as the ionophore valinomycin or the uncouplers 2,4-dinitrophenol (DNP) and carbonylcyanide-*m*-chlorophenylhydrazone (CCCP), did not react uniformly. DNP and CCCP, which both probably influence the proton-motive gradient across the membrane, reduced adherence significantly. On the other hand, addition of the ionophore valinomycin, which increases membrane permeability for K^+, resulted in a moderate increase in adherence. The effect of CCCP was partly abolished by increasing KCl concentrations (Fig. 3). The blocking of MG^{2+}-dependent ATPase by dicyclohexylcarbodiimide (DCCD) strongly inhibited the glucose effect.

When the glucose effect was tested on coverslips pretreated with 20% horse serum, adherence improved by about 40% (Table 1).

TABLE 1 Adherence of *M. pneumoniae* to coverslips, untreated or pretreated with 20% horse serum. Settling (3 h) in BSA buffer or BSA buffer with 0.5 mg glucose/ml

	Attachment (counts/min) to	
Mycoplasmas suspended in	*Untreated coverslips*	*Serum-treated coverslips*
BSA buffer	280	4400
BSA buffer with glucose	3550	6170

The results of the experiments with sugars and inhibiting substances all suggest that an intact sugar or energy metabolism is required for the attachment of *M. pneumoniae* cells to glass in protein-containing buffer.

Relation to cell adherence

The adherence of mycoplasmas to eukaryotic cells, discussed in detail by Razin et al (this volume), shows some striking differences from adherence to glass. Cell adherence is not improved by glucose, it is inhibited by pretreatment with neuraminic acid, which is not effective with glass, and inhibitors have only a minor effect. The available data suggest that in mycoplasma–cell adherence there is a more specific interaction between a mycoplasma binding site and a target cell receptor (Feldner et al 1979a), but qualitative results with protein-coated latex spheres did not show a definite difference. So at present we are unable to distinguish definitively between the two adherence phenomena.

Possible mechanism and biological role of adherence to inert material

Two questions have to be discussed. First, for which structure or mechanism is the energy or the sugar metabolism required? Second, what is the biological role of this type of attachment?

Several hypothetical answers to the first question are available: (1) permanent synthesis and secretion of a particular substance is required, or (2) the membrane surface has to undergo structural modifications.

The first possibility seems unlikely since attachment, at least in medium, is often followed by movement, which would require a large amount of the hypothetical substance. However an effect of a minor synthetic component cannot be excluded. The results with serum-treated glass suggest that a surface-coating substance, perhaps in part produced by the mycoplasmas themselves, may play a considerable role. For the second possible mechanism (changes in membrane structure) we may assume that proteins are involved in spite of the disappointing results of the trypsin experiments. This mechanism could operate in several ways. One would be changes in membrane fluidity, but in this case it is difficult to find a relationship to glucose metabolism. Another more plausible explanation is that modifications in the arrangement of surface protein occur that may also be involved in adherence to eukaryotic cells. The vertical disposition of membrane proteins regulated by the electrochemical gradient could be of importance. The increased adherence in the presence of valinomycin, which can also be observed in adherence to eukaryotic cells, suggests that there is a direct effect of membrane potential, since in buffer with a low potassium concentration valinomycin may produce hyperpolarization and therefore possibly increases the amount of protein exposed on the membrane surface. On the other hand, the horizontal arrangement may also be important for attachment. A clustering of membrane components may be required, which is triggered by contact with another surface. Such a clustering process could be regulated by contractile elements which require energy for their action. However, biochemical and ultrastructural information about these processes is not yet available.

So what is the possible role of this type of attachment – and of motility – *in vivo*? So far we can only speculate about this. Before *M. pneumoniae* cells reach the host cell surface they have to go through a layer of mucus, and on most parts of the bronchial epithelium they have to withstand the repelling action of the ciliary beat. Therefore they need a mechanism for sticking to and penetrating the mucus and for gliding down – possibly along the cilia – to the cellular surface. For technical reasons movement on the cell surface itself has not yet been observed. The finding that three pathogens of the respiratory tract, namely *M. pneumoniae, M. gallisepticum* and *M. pulmonis* (Bredt 1979), are motile and adhere strongly to glass supports the idea that this property is of biological importance in infection.

Conclusions

The attachment of *M. pneumoniae* in a protein-containing environment to glass surfaces is possibly an active process which requires energy or intact sugar metabolism. The exact nature of this mechanism is so far unknown. As microorganisms that lack walls, mycoplasmas have an opportunity unique to prokaryotes: they can rearrange their surface structure in direct response to an outside stimulus. Therefore adherence to inert surfaces as well as to eukaryotic cells may involve active changes in the patterns of membrane surface structures. The processes that facilitate adherence apparently also play a role in motility. So far it is only possible to speculate on the apparent biological importance of both properties in the mycoplasma-host relationship.

Acknowledgements

This work was supported by grant Br 296/13 from the Deutsche Forschungsgemeinschaft.

REFERENCES

Bredt W 1979 Motility. In: Barile MF, Razin S (eds) The mycoplasmas. Academic Press, New York, vol 1:141-155

Feldner J, Bredt W, Kahane I 1979a Adherence of erythrocytes to *Mycoplasma pneumoniae.* Infect Immun 25:60-67

Feldner, J, Bredt W, Razin S 1979b Adherence of *Mycoplasma pneumoniae* to glass surfaces. Infect Immun 26:70-75
Gorski F, Bredt W 1977 Studies on the adherence mechanism of *Mycoplasma pneumoniae.* FEMS (Fed Eur Microbiol Soc) Lett 1:265-268
Kahane I, Gat O, Banai M, Bredt W, Razin S 1979 Adherence of *Mycoplasma gallisepticum* to glass. J Gen Microbiol 111:217-222
Purcell RH, Valdesuso JR, Cline WL, James WD, Chanock RM 1971 Cultivation of mycoplasmas on glass. Appl Microbiol 21:288-294
Radestock U, Bredt W 1977 Motility of *Mycoplasma pneumoniae.* J Bacteriol 129:1495-1501
Razin S, Kahane I, Banai M, Bredt W 1981 Mycoplasma adhesion to eukaryotic cells. In this volume, p 98-113
Somerson NL, James WD, Walls BE, Chanock RM 1967 Growth of *Mycoplasma pneumoniae* on a glass surface. Ann N Y Acad Sci 143:384-389
Taylor-Robinson D, Manchee RJ 1967 Adherence of mycoplasmas to glass and plastic. J Bacteriol 94:1781-1782

DISCUSSION

Sharon: I understand that you and Professor Razin have isolated a lectin from mycoplasma membranes that binds to glycophorin. Does this affect adherence of the organisms to glass?

Bredt: We have not tested it yet. We would have to test it in the glucose system because the fraction is protein and it might interfere with the effect of bovine serum albumin.

Razin: We have some relevant observations on the adherence of *Mycoplasma gallisepticum* to glass (Kahane et al 1979). We tested glycophorin because it interferes with the adherence of *M. gallisepticum* to red blood cells. Glycophorin decreased adherence to glass but since glass does not contain sialic acid this effect appeared to be non-specific. Inhibition of attachment of *M. gallisepticum* to glass by bovine serum albumin supports this suggestion.

Treatment of mycoplasmas with trypsin abolishes or markedly reduces adherence to eukaryotic cells, but attachment to glass is only slightly affected (Kahane et al 1979) or not affected by trypsinization (Feldner et al 1979). Hence, different components of the mycoplasma membrane are apparently involved in attachment to glass and to eukaryotic cells.

Sharon: Have you tested the effects of changing the lipid composition of the medium on the adherence of mycoplasma to glass?

Bredt: Unfortunately such experiments can only be done with *Acholeplasma,* which can be grown without cholesterol or serum. *M. pneumoniae* needs at least 10% serum in the medium and it is very difficult to change the fatty acid composition or other components.

Razin: It is true that with *Acholeplasma laidlawii* it is much easier to change the lipid composition with respect to fatty acids or cholesterol than with *Mycoplasma* species, but it is still possible. We have not investigated this yet because we have been busy testing the influence of changes in the electrochemical-ion gradient of the mycoplasma membrane on attachment. Some time ago we tried to see whether the exposure of proteins on the mycoplasma cell surface is influenced by membrane fluidity. We changed membrane fluidity by various means and checked the effects of these changes on the disposition of membrane proteins. We failed to find any consistent relationship between changes in membrane fluidity and the degree of exposure of membrane proteins on the cell surface, as measured by the lacto-peroxidase-mediated iodination technique (Amar et al 1979). We have found, however, that if the electrochemical-ion gradient across the membrane is dissipated by ionophores there is a marked decrease in the exposure of proteins on the cell surface (Amar et al 1978). We are now trying to find out whether ionophores and uncouplers also decrease adherence by decreasing the exposure of the binding sites on the mycoplasma surface. Our preliminary results do not allow any definite conclusions. Although it appears that the electrochemical-ion gradient may influence the attachment of mycoplasma to glass, we still do not have conclusive answers about the effect of ionophores or uncouplers on mycoplasma attachment to erythrocytes.

Taylor-Robinson: Is there any possibility that the increased attachment rate seen with serum in the medium could be due to the serum stimulating multiplication of the organisms and making more of them available for attachment to glass?

Bredt: Adherence was quantified by the amount of radioactivity that attached to the glass. The adherence test itself was done in unlabelled growth medium. Therefore the total amount of label in the cells could not increase. The counts found on the glass were always representative of the sticking portion of the suspension, regardless of the number of cells.

Vosbeck: What happens to the label? If it is metabolized or secreted into the medium, your assay would have a different outcome. Does the palmitate label remain firmly attached to the cell during the whole assay period?

Bredt: The label is apparently an integral part of the membrane lipids. We found only a little in the supernatant. Only about 1% can be removed if we trypsinize the surface, trying to get rid of the surface lipoproteins, so we are quite sure that palmitate is a very stable label for mycoplasmas.

Hughes: I am interested in the serum factor which apparently helps to mediate attachment of mycoplasmas. A lot of work has recently been done with mammalian cells on the identification of the serum factors which mediate the attachment and spreading of trypsinized fibroblasts to a growth surface (see Grinnell 1978, Hughes et al 1980). One of the most important serum factors is the adhesive glycoprotein, fibronectin. Fibronectin very rapidly coats the inert plastic surface and provides a suitable substratum to which the cells stick and spread. Bovine serum albumin and a number of other serum proteins prevent cell attachment. Have you looked at the attachment of mycoplasma to inert surfaces in the presence of gelatin-treated serum, i.e. serum lacking fibronectin?

Bredt: We are aware of the work on mammalian cells but so far we haven't done anything on this problem. As a first step we want to fractionate horse serum to identify the fraction which apparently increases adherence.

Hughes: If you precoat glass with serum, how quickly do the mycoplasmas attach?

Bredt: So far we have no data on the kinetics of attachment to a precoated surface. The mycoplasmas may stick much more quickly to the protein-covered surface than to the uncoated glass.

Hughes: Perhaps the kinetics are very similar to the rate of attachment and spreading of trypsin-treated fibroblasts to serum-coated glass. It takes place very quickly, certainly within 3 h.

Bredt: We shall have to look into that.

Wallach: Is attachment to glass temperature-dependent, Professor Bredt? And is the serum albumin de-fatted? Is there any possibility of lipid exchange with albumin in the medium?

Bredt: Temperature is involved in each type of mycoplasma adherence. At 4 °C we get very little attachment. The optimum temperature for attachment of erythrocytes is 28-30 °C. Glass adherence is much better at 37 °C than at 4 °C, but we did not check intermediate temperatures.

We used normal not de-fatted BSA but we tested other protein-containing preparations such as glycophorin. Our trypsin/trypsin-inhibitor systems also inhibited non-specific adherence. I cannot exclude the possibility of an exchange of fatty acids between mycoplasmas and albumin but I think that the fatty acids are fairly well anchored in the membrane.

Wallach: Is there a monotonic increase of adherence with temperature up to an optimum?

Bredt: It seems that 37 °C is an optimum for glutaraldehyde-treated erythrocytes but we have not repeated these tests with glass adherence.

Elbein: Have you tried to compare the binding of these cells at different stages of growth? Can you bind mycoplasma membranes to glass and determine whether there is a lag in attachment, and whether at some point something is formed or secreted which results in increased attachment?

Bredt: This is a very peculiar system. Apparently the microorganisms have to be harvested at the logarithmic stage of growth. At pH values below 7 at harvest you cannot run a good adherence assay. Ageing colonies don't attach very well.

Elbein: When you referred to the effect of serum it looked as if there was a lag in attachment.

Bredt: We first put mycoplasmas in buffer and then cultured them in growth medium. They are apparently very sensitive to changes. There is probably a lag phase in metabolism and growth.

Elbein: Is something secreted by the cells during that lag period that might aid in their attachment? Or is there some change in the membrane preparations?

Bredt: I don't know. So far we have no evidence for secretion.

Razin: I would like to speculate on the role of albumin and energy in mycoplasma attachment to glass. Mycoplasmas usually die rather rapidly when suspended in buffer at 37 °C (Butler & Knight 1960). The addition of albumin protects the cells (D. Shinar & S. Rotem, unpublished work). Thus, albumin keeps the cells viable and metabolizing.

Mycoplasmas, unlike bacteria, can change their shape and, in addition, *M. pneumoniae* and *M. gallisepticum* possess attachment tips (see Razin et al, this volume). The presence of a cytoskeleton made of fibres forming a bundle at the attachment tip has been indicated recently (U. Goebel & W. Bredt, unpublished work). Energy is obviously required for keeping the cytoskeleton at the right conformation with the attachment tip exposed. Attachment through the tip is apparently the natural way for these organisms to attach to eukaryotic cells and probably also to inert surfaces. I would suggest that there are two types of attachment: one of viable mycoplasma cells and the other of dead cells in buffer. The pH optimum for mycoplasma attachment to glass in buffer is 5.5, which fits in with the electrostatic attachment that occurs with bacteria. With albumin the optimum pH value for attachment is 7.2, which is probably the optimum pH for sugar metabolism by the organism.

Bredt: The attachment increase in BSA buffer with glucose is apparently only a fraction of that in growth medium. Also we only see 40% inhibition with metabolic inhibitors. One can never suppress adherence as much in growth medium as in glucose buffer. So besides glucose energy there must be something else in the medium which enables the cells to attach even better.

Mirelman: Amoebae also stick very well to glass and this organism leaves a sort of footprint on the glass. These are specific particles of the membrane that remain attached and cannot be easily removed from the glass. Did you see footprints after the mycoplasmas attached?

Bredt: So far we have not seen footprints. It might just be a matter of dimension because each of those footprints you saw might be as large as a mycoplasma. We would detect such material only biochemically but this has not been done yet.

Taylor-Robinson: Valentine & Allison (1959) reported that viruses stick to glass and they discussed the kinetics of the interaction. It is a question of electrostatic forces, isn't it?

Choppin: Certainly some viruses stick very well to glass. But in the few instances where we know exactly which proteins on the virus are involved in binding and which proteins are involved on the surface of the cell, as we do for example with the myxoviruses and paramyxoviruses, the binding to the cell appears to be distinct from the binding to glass. One is purely electrostatic and the other has specific stereochemical requirements, and there is a temperature-independent initial episode.

Howard: One can demonstrate saturable binding of molecules such as lectins to glass and specific competition by sugars and oligosaccharides, yet clearly the lectin is not binding to sugars on the glass. Reactions of mycoplasmas with glass might therefore be misleading if you intend to extrapolate to the biological situation.

Hughes: I disagree entirely. For example let's look at the role of fibronectin when fibroblasts stick to glass. Fibronectin is not only a serum component; it is present not only in growth media but also on cell surfaces. It is expressed at the contact site between adherent cells. There is clearly a close analogy between two cells sticking together and cells sticking to an inert surface such as plastic or glass.

Howard: I am not sure that you can generalize from the situation with fibroblasts to other cell–glass and cell–cell interactions.

Sharon: Is adherence of fibroblasts inhibited by precoating the glass with fibronectin?

Hughes: Yes. It is just the normal competition of two proteins, fibronectin and albumin, for the available surface.

Howard: Professor Bredt, you demonstrated that trypsinization of mycoplasmas affects their binding to cells but not to glass. Can you explain this difference?

Bredt: Binding to glass was in fact affected by trypsin but to a lesser extent than cell adherence. We cannot test adherence immediately after trypsinization. We have done studies with cells which showed that adherence was restored in growth medium within 3 h. The material left in the mycoplasma might be sufficient to replace some of the binding site on the surface during the three hours of the test. We must find a way to test within half an hour.

Howard: Does trypsinization inhibit protein synthesis?

Bredt: Yes, but with glass adherence this is difficult to test. We can inhibit the restoration of adherence to cells, and we have to try again with glass. But we were really frustrated by the trypsin experiments.

Richmond: I was interested in the valinomycin effect. Have you looked at other peptides of that general type that do not have ionophore activity to see whether they give the same effect? This may be a useful way of studying the phenomenon because there are enormous numbers of peptides of known structure and you might be able to use them to analyse the effect. I am not convinced that valinomycin is exerting its action because of its involvement in metabolism. It may just be acting as a peptide.

Bredt: So far we have tested only valinomycin and at the moment the only explanation is the hyperpolarization of the membrane.

Richmond: The ionophore antibiotics are notorious for stacking themselves in membranes. A lot is known about the molecular variations which alter the stacking. You may have a precise way of studying this phenomenon by using different antibiotics of known structure but with variable properties.

Bredt: That would be an interesting thing to look at.

REFERENCES

Amar A, Rottem S, Razin S 1978 Disposition of membrane proteins as affected by changes in the electrochemical ion gradient across mycoplasma membranes. Biochem Biophys Res Commun 84:306-312

Amar A, Rottem S, Razin S 1979 Is the vertical disposition of mycoplasma membrane proteins affected by membrane fluidity? Biochim Biophys Acta 55:457-467

Butler M, Knight BCJG 1960 The survival of washed suspensions of *Mycoplasma.* J Gen Microbiol 22:470-477

Feldner J, Bredt W, Razin S 1979 Adherence of *Mycoplasma pneumoniae* to glass surfaces. Infect Immun 26:70-75

Grinnell F 1978 Cellular adhesiveness and extracellular substrata. Int Rev Cytol 53:65-144

Hughes RC, Pena SDJ, Vischer P 1980 Cell surface glycoproteins in fibroblast adhesion. In: Curtis A, Pitts JD (eds) Cell adhesion and motility. Cambridge University Press, Cambridge, p 329-356

Kahane I, Gat O, Banai M, Bredt W, Razin S 1979 Adherence of *Mycoplasma gallisepticum* to glass. J Gen Microbiol 111:217-222

Razin S, Kahane I, Banai M, Bredt W 1981 Mycoplasma adhesion to eukaryotic cells. In this volume, p 98-113

Valentine RC, Allison AC 1959 Virus particle adsorption I. Theory of adsorption and experiments on the attachment of particles to non-biological surfaces. Biochim Biophys Acta 34:10-23

Adhesion properties of *Entamoeba histolytica*

DAVID MIRELMAN and DAVID KOBILER

Department of Biophysics and Unit for Molecular Biology of Parasitic Diseases, Weizmann Institute of Science, Rehovoth, Israel

Abstract Trophozoites of *Entamoeba histolytica* adhere to and phagocytize red blood cells and bacteria. Furthermore, in the initial step of the amoebic infectious process the parasite attaches to intestinal epithelial cells. A lectin (carbohydrate-binding protein) which apparently has a role in the attachment of the parasite to host cells was found in trophozoites of *E. histolytica.* When amoeba cells were disrupted by freeze-thawing, the lectin activity, as determined by haemagglutination of human erythrocytes, remained associated with the sedimented membrane fraction. This activity was pH-dependent and heat- and oxidation-sensitive, and was destroyed by proteolysis and on autoincubation. Moreover, the lectin activity was inhibited by a variety of *N*-acetylglucosamine-containing compounds such as chitin and chitin oligosaccharides, bacterial peptidoglycan, rabbit colonic mucus, bovine and human serum, an IgA fraction isolated from human colostrum, and IgG from sera of amoebiasis patients. These glycoconjugates also interfered with the adherence of intact radiolabelled amoeba trophozoites to human intestinal epithelial cells as well as with their attachment to red blood cells. Although the lectin activity and the toxin-like activity previously found in *E. histolytica* seem to be two separate substances, they share a number of properties which suggest that they are related and may have a function in pathogenicity.

The ability of many pathogenic microorganisms to attach to particular cell surfaces of their host appears to be an essential requirement for colonization and in some cases for the subsequent invasion of the tissue. The molecular mechanisms underlying the intercellular recognition and cell–cell adhesion processes have been intensively investigated during the last few years. Some of the most interesting macromolecules postulated to be actively involved in many adhesion phenomena are carbohydrate-binding proteins with properties similar to the well-known plant lectins (Lis & Sharon 1980).

1981 Adhesion and microorganism pathogenicity. Pitman Medical, Tunbridge Wells (Ciba Foundation symposium 80) p 17-35

As we shall probably also hear from other contributors to this symposium, more and more evidence is becoming available that shows that a considerable number of bacterial strains contain carbohydrate-specific lectins on their cell surfaces which apparently mediate their adherence to receptors on epithelial cells of their host (Ofek et al 1977, Mirelman et al 1980). A lectin has also been shown to appear on the surface of slime moulds during the process of transformation from single amoebae to the cohesive state which forms the slug (Barondes & Rosen 1976). Moreover, during embryonic development, the emergence and presence of endogenous lectins seems to correlate with the adhesion process which leads to the formation of the tissues (Kobiler & Barondes 1977). Although these and other available examples do not yet constitute unequivocal proof for the role of lectins in adherence, they strongly suggest that lectins have a function in cell–cell interaction.

With this knowledge as a basis it was reasonable to assume that the adherence phenomenon observed for *Entamoeba histolytica,* a pathogen which causes amoebic dysentery in man, is mediated by a similar system.

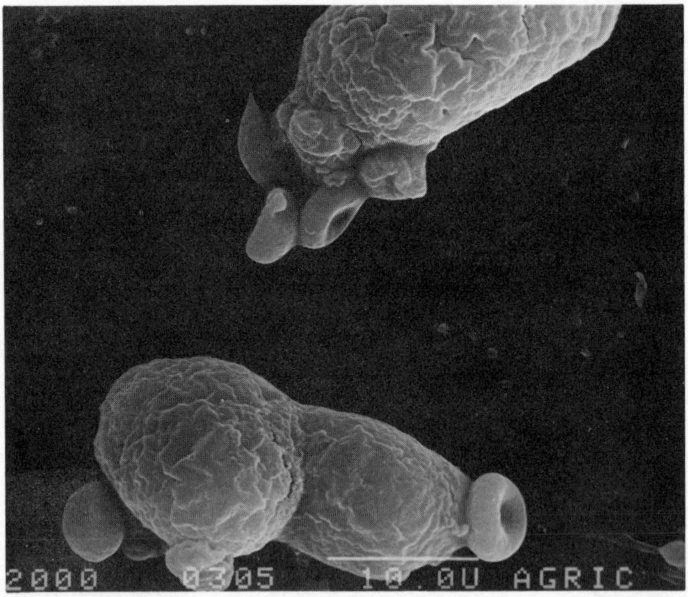

FIG. 1.(A) Adherence of erythrocytes to trophozoites of *E. histolytica* as seen by scanning electron microscopy (SEM). Human erythrocytes (type B) were incubated in TYI-S-33 medium for 2 min at 37 °C with trophozoites (strain HK-9) that had been grown on glass cover-slips. After incubation, the samples were fixed *in situ* by the addition of glutaraldehyde (2% final concentration), followed by dehydration through an ascending series of ethyl alcohol solutions to 100% alcohol. The glass cover-slips were mounted on SEM stubs and coated with gold. The material was examined with a Jeol JSM-35C SEM operated at an accelerating voltage of 25 kV.

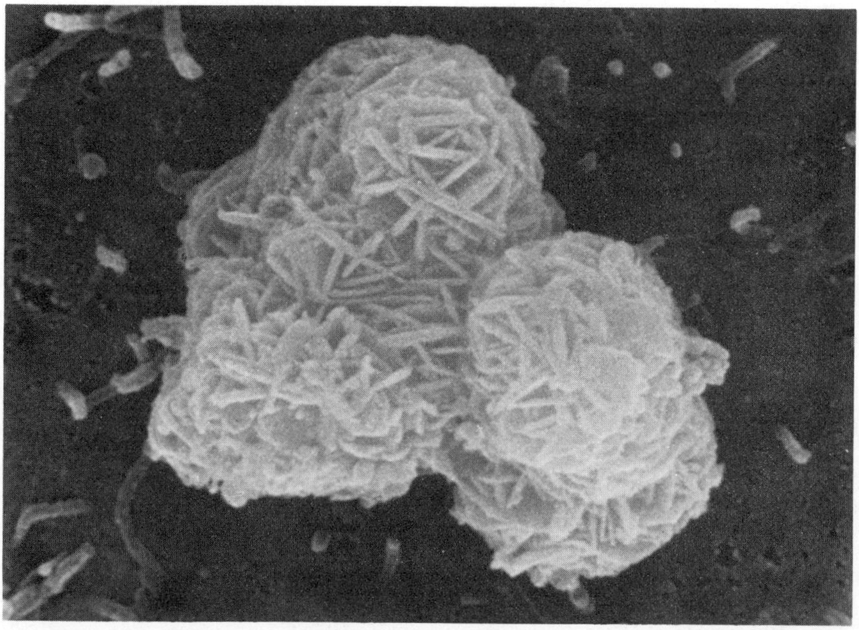

(B) Scanning electron micrograph which shows the remarkable adherence of *E. coli* cells to the surfaces of *E. histolytica* trophozoites. Trophozoites bound to glass covers were incubated with a suspension of *E. coli* cells (10^9/ml) for 15 min after which they were washed and fixed with glutaraldehyde (1%).

Entamoeba histolytica trophozoites have been shown to adhere to and phagocytize various types of cells, most notably erythrocytes and gut microflora (Hoare 1952, Trissl et al 1978) (Fig. 1). Furthermore, the initial step in amoebic pathogenesis is the attachment and proliferation of amoebae on mucosal surfaces of the caecum and colon (Chevez et al 1976). This is usually followed by invasion and penetration of amoebae through the intestinal wall, leading to secondary but more serious infections such as abscesses of the liver, brain or lung. The initial aim of our investigation was therefore to try and find an amoeba component which could be responsible for its attachment to other cells.

A lectin activity in E. histolytica

Since Louis Diamond (1968) first successfully managed to cultivate amoebae in the absence of other microorganisms (axenic cultures) with which amoebae usually associate, it has become possible to study specific properties as well as to isolate

TABLE 1 The agglutinability of different red blood cells by the amoeba lectin and wheat germ agglutinin

Red blood cells	*Activity*[a] *Amoeba lectin*[b]	*WGA*[c]
Human A	8	8
Human B	32	16
Human O	16	8
Rabbit	4	32
Guinea-pig	32	8
Rat	4	4
Sheep	8	32

[a]Titre of activity is defined as the reciprocal of the highest dilution of lectin solution that agglutinates an equivalent volume of 4% red blood cells.
[b]Crude extract (25 μl) of 10^6 amoeba/ml (strain HK-9) was used as the source of lectin activity.
[c]The concentration of wheat germ agglutinin (WGA) used was 0.25 mg/ml. The actual titre may vary slightly when tested with different batches of the same RBC.

TABLE 2 Properties of Entamoeba histolytica haemagglutinin

(1) Heat-labile; activity destroyed after heating (100 °C, 10 min)
(2) Autoproteolysis-sensitive; total loss of activity on incubation of extract (2 h, 37 °C). The activity is preserved by the addition of ϕ-methyl sulphonyl fluoride (PMSF, 5 μM).
(3) pH effect: optimum pH for agglutination is between 5.7 and 6.0. No activity observed at pH 7.2.
(4) Reducing agent requirement: β-mercaptoethanol (5 mM) prevents loss of activity on storage.

the responsible component from the isolated amoeba. The details of the axenic cultivation and maintenance of the various strains of amoeba have been well documented (see Diamond et al 1978). Most of our studies were carried out with the axenic strains HK-9, HM-1 and 200:NIH, obtained from Dr Diamond (NIH, Bethesda) (Kobiler & Mirelman 1980).

E. histolytica trophozoites, released from culture tubes by being chilled in an ice bath and washed a number of times with a solution of cold phosphate-buffered saline, were found by the classical titre plate technique to agglutinate a variety of glutaraldehyde-fixed red blood cells (Table 1). This activity was enhanced when amoebae were presonicated for a short period.

The optimal conditions for the haemagglutinating activity are listed in Table 2. The pH requirement of the sonicated preparation appears to be quite critical and its activity can be detected only between pH 5.7 and 6.0, whereas intact trophozoites also agglutinated red blood cells at pH 7.2. Autoincubation of the sonicated amoebae at 37 °C for 2 h completely abolished the agglutinating activity. In addition, overnight storage of the disrupted amoebae at 4 °C also led to considerable losses of

TABLE 3 Lectin activity in three Entamoeba histolytica strains

Strain	*Specific activity*[a]	*Activity per 10^5 amoeba*[a]
HK-9	73	3.2
200:NIH	32	1.8
HM-1	14	2.3

[a]Titre of activity is defined as the reciprocal of the highest dilution of extract that agglutinates an equivalent volume of 4% red blood cells. Specific activity is defined as titre divided by mg protein/ml extract.

activity (about 75%). These inactivations of haemagglutinating activity were prevented by the addition of ϕ-methyl sulphonyl fluoride (PMSF, 5 μM) and β-mercaptoethanol (BME, 5mM).

The three *E. histolytica* strains (HM-1, HK-9 and 200:NIH) tested all possessed haemagglutinating activity. No correlation was found between the amount or specific activity of the agglutinin and the relative virulence of the strain as known from studies of other investigators. The amoeba strain which had the highest agglutinin content was HK-9 (Table 3). Strains HM-1 and 200:NIH, however, are known to be more virulent, as determined by their enhanced ability to form abscesses in the livers of newly born hamsters (Mattern & Keister 1977) as well as by their cytotoxic effect on baby hamster kidney cells (Bos 1979). The possible connection between the availability of agglutinin in a given amoeba strain and its degree of pathogenicity will depend on a quantitative study of the subcellular distribution of the agglutinin and the relevant fraction where it is located, i.e. the amount of agglutinin that is present on the surface of each of the tested strains. For the agglutinin to be able to mediate adherence of the amoeba to other cells it has to be present on the cell surface or appendages of the amoeba.

Preliminary studies of the subcellular location of the amoeba agglutinin indicate that it is associated with the membrane fraction. When amoebae (strain HK-9) were disrupted by sonication, about two-thirds of the haemagglutinating activity was found in the soluble supernatant solution obtained after high speed centrifugation (100 000 *g*, 60 min) and only one-third of the activity remained associated with the particulate membrane pellet. However, when trophozoites were disrupted by being frozen and thawed, practically all the activity remained associated with the sedimentable membrane fraction (Table 4). The finding that the activity seems to be associated with the cell membrane is encouraging, but the location of the agglutinin on the cell surface must be more thoroughly investigated before firm conclusions can be drawn.

Lectins are commonly defined as carbohydrate-binding proteins which are able to agglutinate a variety of cells and whose agglutination is blocked by specific saccharides (Lis & Sharon 1980). To clarify the nature of the amoeba agglutinin,

TABLE 4 Separation of lectin and toxin-like activities by centrifugation of disrupted trophozoites from strains HM-1 and HK-9

	Lectin activity (Titre[a])		*Toxin activity*[b]	
Fraction	HM-1 (10^6/ml)	HK-9 (10^6/ml)	HM-1 (2.5×10^5/ml)	HK-9 (3.5×10^6/ml)
Freeze-thaw suspension	14	18	5+	4+
Supernatant	0	2	5+	5+
Pellet (100 000 *g* for 60 min)	28	36	0	0

Amoebae grown for 72 h in TYI-S-33 medium were washed with saline and resuspended in pH 5.7 buffer. The amoebae were disrupted by freezing the suspension in acetone–solid CO_2 for 5 min. After rapid thawing, the freeze-thaw suspension was sedimented for 1 h at 100 000 g, producing a supernatant and a pellet. The membrane pellet was resuspended in pH 5.7 buffer by sonication.

[a]Titre of activity is defined as the reciprocal of the highest dilution of extract that agglutinates an equivalent volume of 4% fixed red blood cells.
[b]5+ activity is defined as the extent of damage caused by a standard toxin solution obtained from 2.5×10^5 amoebae of strain HM-1 to a monolayer of epithelial cells. 0 is the value for a control without amoebic material. Damage was determined by the decrease in methylene blue binding to the residual tissue-cultured cells.

we tested the inhibitory potential of a large number of haptens on the haemagglutinating activity (Table 5). A partial acid hydrolysate of purified chitin at a final concentration of 100 μM (calculated on the basis of the *N*-acetylglucosamine (GlcNAc) monomer) inhibited 50% of the lectin activity. When the chitin hydrolysate was fractionated into the various oligosaccharide components, the lectin seemed most sensitive to the trimer and tetramer of *N*-acetylglucosamine.

The carbohydrate specificity of the amoeba lectin is thus similar to that found for potato tubers and *Datura stramonium* lectins (Lis & Sharon 1980). Wheat germ agglutinin has a sugar specificity similar to that of the amoeba lectin. In contrast to the latter it is also inhibited by monomeric *N*-acetylglucosamine as well as by sialic acid (Allen et al 1973, Adair & Kornfeld 1974). The lectins from potato tubers (Allen & Neuberger 1973) and *Datura stramonium* (Kilpatrick & Yeoman 1978) are also inhibited solely by oligomers of *N*-acetylglucosamine. The best inhibitor of the potato lectin is the β-(1→4)-linked pentasaccharide of *N*-acetylglucosamine. In eukaryotic cells, a lectin specific for *N*-acetylglucosamine residues was isolated from avian hepatic cells (Kawasaki & Ashwell 1977). In addition to the low molecular mass chitin oligosaccharides the amoeba agglutinin was inhibited by a variety of *N*-acetylglucosamine-containing compounds such as bacterial peptidoglycan, rabbit colonic mucus, bovine and human sera and an IgA fraction isolated

TABLE 5 (A) Inhibition of haemagglutinating activity of E. histolytica lectin by chitin oligosaccharides

Saccharide[a]	*Concentration that inhibits 50% of the activity*
Partial acid hydrolysate of chitin	100 μM
N-Acetylglucosamine	>50 mM
$(N\text{-Acetylglucosamine})_2$	50 μM
$(N\text{-Acetylglucosamine})_3$ + $(N\text{-Acetylglucosamine})_4$	25 μM
$(N\text{-Acetylglucosamine})_4$ + $(N\text{-Acetylglucosamine})_5$	50 μM
$(N\text{-Acetylglucosamine})_6$	100 μM

[a]Concentration of the fractions was calculated on the basis of *N*-acetylglucosamine monomer. Preparation of oligosaccharides was as described (Rupley 1964).

(B) Inhibition of lectin activity by N-acetylglucosamine-containing macromolecules

Compound	*Concentration that inhibits 50% of the activity (μg/ml)*
Chitin (fine powder)	1
Insoluble *Escherichia coli* peptidoglycan	2.5
IgA fraction from colostrum	50
Rabbit colonic mucus	50
Bovine serum	0.1%

(C) Saccharides with no effect (at 50 mM final concentration) on the haemagglutinating activity of E. histolytica

Monosaccharides:	Glu, Gal, Man, Fuc, Xyl
Disaccharides:	Suc, Lac, Mel, Mal
Amino sugars:	GlcNAc, GalNAc, ManNAc, NANA

from human colostrum (Table 5). Preliminary results also show that an IgG preparation from sera of amoebiasis patients inhibited the amoeba lectin, whereas no inhibition was observed with sera from human controls.

Partial purification of the soluble amoeba lectin was achieved by affinity chromatography. Based on its carbohydrate specificity, the lectin could be retained on either a chitin (Fig. 2) or a Sepharose–*N*-acetylglucosamine conjugate column similar to the one previously used for the purification of wheat germ agglutinin (Allen et al 1973). The elution of the lectin was based on the pH dependence of its activity. The crude soluble lectin was loaded on the column with a slightly acidic solution (pH 5.7) and was eluted from the column with phosphate-buffered saline

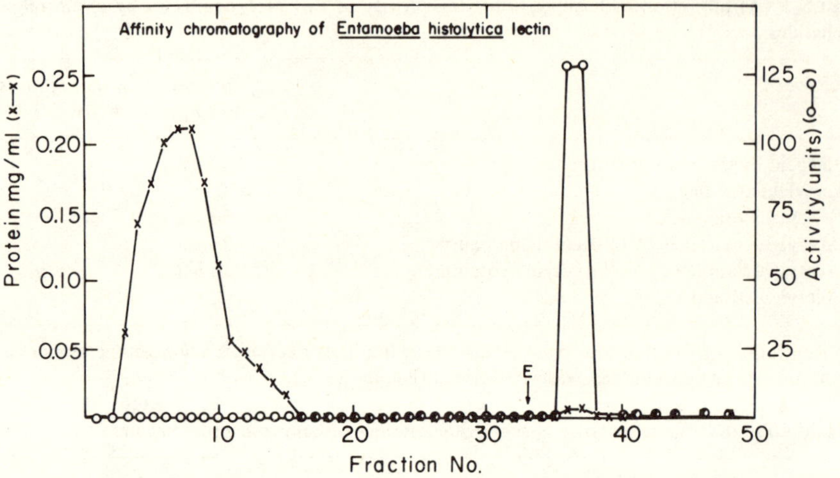

FIG. 2. Affinity chromatography of amoebic lectin on chitin column. A sample of amoebic extract (from 4 x 10^7 amoebae) was applied to the column in buffer pH 5.7 and the column was washed with the same buffer (100 ml). At the point indicated by the arrow, PBS pH 7.2 (containing PMSF and BME, as described in text) was added. Lectin activity was determined by haemagglutination (see text).

(pH 7.2). Based on protein determination the affinity chromatography step purified the lectin about 300 times and the recovery was 80%. The yield of lectin from trophozoites of strain HK-9 was about 10 μg protein/10^7 amoeba and its calculated specific activity was 20 000 titre units/mg protein (see Table 1).

Polyacrylamide gel electrophoresis of the lectin activity eluted from the affinity chromatography column showed that the lectin was not purified to homogeneity and a number of bands could be seen on the gel. Complementary purification steps including detergent solubilization and ion exchange chromatography are now under way and these we hope will yield purified lectin. We investigate the effect of a number of detergents on the solubilization of the amoeba lectin from the particulate membrane fraction as well as the preservation of its activity. Both Zwittergen TM312 (Calbiochem 0.5% w/v) and Sarcosyl NL-97 (0.1% w/v) are suitable for these purposes.

Adhesion of E. histolytica trophozoites to mammalian cells

The amoeba lectin seems to play an important role in the initial attachment of the amoeba to mammalian cells. The adhesion of red blood cells to the surfaces of intact amoeba trophozoites (Fig. 1A) can be prevented if chitin oligosaccharide haptens

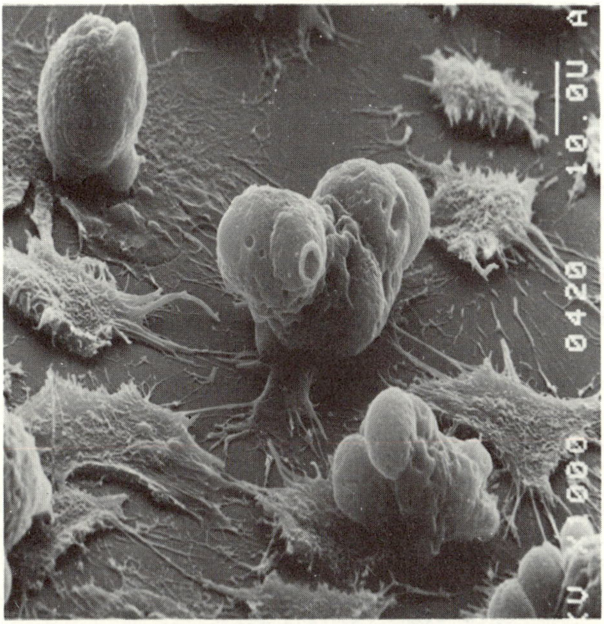

FIG. 3. Scanning electron micrograph showing the adherence of trophozoites of *E. histolytica* to a monolayer of human intestinal cells (Henle 407). Washed trophozoites were allowed to interact with the mammalian cells for 30 min at 37 °C. The micrograph was taken at a 60° angle.

(0.2 M), are present. In addition, *E. histolytica* trophozoites adhere very well to monolayers of a tissue-cultured line of human epithelial cells (Henle 407, CCL-6, ATCC) (Fig. 3).

A convenient method for quantitatively determining the adherence of amoebae to mammalian cells is by radioactive labelling. ^{35}S-labelled trophozoites were prepared by growing amoebae (3 x 10^6) for 4 h in a medium lacking cysteine but containing [^{35}S] cystine (0.2 μCi/ml, sp. act. 663 Ci/mMol), followed by repeated washings. The amount of labelled material incorporated was 100 c.p.m. per 1000-4000 amoebae, depending on the preparation.

The attachment of intact radioactively labelled amoeba trophozoites to glutaraldehyde-fixed monolayers of the human intestinal cell line was comparable to that of non-fixed monolayers. The attachment was dependent on the amoeba concentration (Fig. 4), time and temperature. Results obtained so far indicate that a number of the *N*-acetylglucosamine-containing macromolecules inhibit the adherence of intact amoeba trophozoites to the mammalian cells. Powdered chitin (0.1 mg/ml), insoluble *E. coli* peptidoglycan (1 mg/ml) and bovine serum (1%) caused about 50% inhibition of adherence.

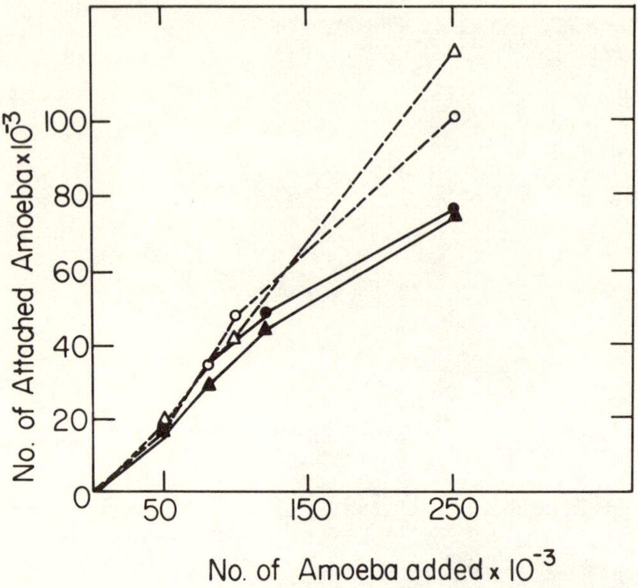

FIG. 4. Adherence of [^{35}S] cysteine labelled trophozoites of *E. histolytica* HK-9 (————) and HM-1 (--------) to a monolayer of human intestinal cell line (Henle 407). Adherence was determined to cells that were either prefixed in glutaraldehyde (0.25%) (△,▲) or non-fixed (○,●) in Costar plates (16 mm diameter). A suspension of trophozoites in culture medium (basal medium [Eagle] with Hanks' BSS medium [0.5 ml]) without serum was incubated with the confluent monolayer for 20 min at 37 °C, after which the remaining suspension was aspirated and the cells were washed twice with saline. The radiolabelled amoebae that remained bound to the cells were dissolved with sodium dodecyl sulphate (0.25%) and counted in a scintillation counter.

Moreover, the attachment of labelled intact trophozoites to sections (2 cm) of inverted (inside out) guinea-pig colon followed essentially the same pattern as we observed for the binding to tissue culture monolayers. In this case too, small amounts of serum, bacterial peptidoglycan, chitin or chitin oligosaccharides all inhibited adherence.

The inhibitory potency of GlcNAc-containing glycoconjugates implies that lectin activity is present on the surface of the amoeba. It also suggests that the mammalian cell receptor may be a surface component with a carbohydrate section containing GlcNAc residues in a glycosidic linkage or possible $(GlcNAc)_2$ moieties. Such glycoconjugate structures (Yamashita et al 1978, Mizuochi et al 1978) have been reported to be present in glycoproteins and these may be on the surfaces of different types of cells where they may serve as ligands for the attachment of the amoeba through its lectin. Another aspect of the affinity of the amoeba for GlcNAc-containing glycoconjugates is our novel finding that rabbit colonic mucus, which is continuously secreted by intestinal goblet cells, is a lectin inhibitor. Very little is known about the role of colonic mucus aside from its lubricating function. It is tempting to speculate that the intestinal mucus may serve also as a sort of rudimentary protective mechanism in that it will protect the underlying mucosa cells by effectively binding amoebae (or other pathogens) and removing them from the gastrointestinal tract by peristaltic motion. Our information about the factors that affect intestinal mucus secretion and its possible function in protecting the host is also limited. For obvious technical reasons it is difficult to study this problem but we are now doing some work with an animal model that will, we hope, help us to obtain more information on this question (see Katz et al, this volume).

Similarities and differences between the lectin and toxin-like activities of E. histolytica

Councilman & LaFleur (1891) first suggested that amoebae had a 'chemical solvent action' on infected tissues. Roos (1894) and later Shaudinn (1903) suggested that amoebae released a toxin which destroyed underlying epithelial cells and they therefore named it histolytica for its supposed cytolytic effect on tissues. Whole living trophozoites of *E. histolytica* have a cytotoxic effect when tested against a variety of mammalian cells (Jarumilinta & Kradolfer 1964). Lushbaugh et al (1979) and Bos (1979) have recently found that cell-free extracts prepared from sonicated preparations of axenically cultivated trophozoites contained cytopathic activity (they caused the rounding up and release of cells from tissue culture monolayers) as well as enterotoxic activity (they induced the secretion of fluid in ligated ileal loops of rabbits). Lectin-like properties have recently been attributed to the toxin, as its activity was prevented and reversed by serum, fetuin and *N*-acetylgalactosamine (Mattern et al 1980). In order to find out whether there is any connection between the lectin first found by us (Kobiler & Mirelman 1980) and the toxin, we compared the two types of activity.

A subcellular separation between the lectin and toxin-like activities was achieved when trophozoites (strains HM-1 and HK-9) disrupted by repeated freezing and thawing (Table 4) were fractionated by high speed centrifugation (100 000 *g*, 60 min). Practically all the lectin activity, as determined by haemagglutination, was associated with the sedimented membrane fraction. Portions of the resuspended membrane pellet did not have any toxic effect when incubated with monolayers of tissue-cultured fibroblasts or human epithelial cells. On the other hand the supernatant solution, which contained no lectin activity, damaged the tissue-cultured cells extensively, changing their morphology and detaching them from the plates. The toxin content of the virulent strain (HM-1) was at least 10 times more than in strain HK-9 (Table 4). In addition, when intact trophozoites were incubated in phosphate-buffered saline, toxin but no lectin activity was actively secreted into the medium. Both the lectin and toxin activities were heat-labile and were preserved by the presence of reducing agents and proteolytic enzyme inhibitors. In contrast to the toxin, the isolated lectin was inactive at pH 7.2 and active only at pH 5.7–6.0. Both the lectin and toxin were inhibited by a number of GlcNAc-containing compounds such as chitin, bacterial peptidoglycan, bovine or human serum, colonic mucus and an IgA fraction isolated from human colostrum.

Only the lectin activity, however, was inhibited by low molecular mass chitin oligosaccharides $(GlcNAc)_{n=2\text{-}6}$ or by peptidoglycan subunits obtained after lysozyme digestion. The toxin activity, on the other hand, was partially inhibited by fetuin and by a crude preparation of ox brain gangliosides.

Concluding remarks

The recent findings of the lectin and toxin activities in *E. histolytica* have given new impetus to attempts to understand the mechanism of pathogenicity of this parasite. Although our present results indicate that the lectin and toxin are two separate activities, they share many properties and this suggests that they are related. It is undoubtedly premature to claim that the amoeba toxin or the lectin, or both, are responsible for the remarkable virulence that this parasite displays towards host cells. Nevertheless, one cannot resist the temptation to make comparisons with the mode of action of bacterial enterotoxins such as those of *Vibrio cholera* or *Shigella dysenteria* that have been studied in more detail. It is well established today that cholera toxin contains two peptide subunits, one of which (B) specifically binds to GM_1 gangliosides with a lectin-like affinity and the other subunit (A) then penetrates the membrane and causes a biological response believed to be mediated by the activation of adenylate cyclase and the formation of cAMP (Gill 1978). Similar modes of action are thought to be the basis for the action of toxins from other pathogens of the small bowel such as the heat-labile toxin from enterotoxigenic *Escherichia coli* (Gill 1978). Shigella infections, on the other hand, are mainly found in the caecum and colon, and there they are known to produce a toxin which appears to have a different carbohydrate specificity; the binding of this toxin to a variety of cells is inhibited by oligosaccharides of GlcNAc (Keusch & Jacewicz 1977). This toxin was recently isolated by affinity chromatography on chitin columns (Olsnes & Eiklid 1980). It seems to us most interesting and intriguing that *Shigella* and *E. histolytica,* in addition to being colon pathogens and having similarities in the initial stages of the aetiology of dysentery, share the important property that their toxins recognize GlcNAc-containing molecules in the host.

It is a well known yet poorly understood phenomenon that each type of pathogen will usually cause its infection at well-defined sites along the intestinal tract of susceptible hosts. It would be interesting to know whether the availability of specific binding receptors on certain locations of the host mucosa is one of the prerequisites for subsequent colonization of the host by a particular pathogen.

As mentioned earlier, our preliminary studies have shown that GlcNAc-containing macromolecules are efficient inhibitors of the toxin-like activity of amoebae and will also prevent the trophozoite from adhering to the surface of a variety of host cells. It is hoped that once the lectin and the toxin have been isolated and purified they will serve, together with the antibodies that will be prepared against them, as potent tools for studies at the molecular level of the interaction of the amoeba trophozoite with receptors of the host cells. Such investigations may lead eventually to novel therapeutic approaches and to the development of new drugs.

Acknowledgements

We wish to thank Mrs Ada Wexler for excellent technical assistance. This research was supported by a grant from the Rockefeller Foundation, New York.

REFERENCES

Adair NL, Kornfeld S 1974 Isolation of the receptors for wheat germ agglutinin and the *Ricinus communis* lectins from human erythrocytes using chromatography. J Biol Chem 249:4696-4704

Allen AK, Neuberger A 1973 Purification and properties of the lectin from potato tubers, a hydroxyproline-containing glycoprotein. Biochem J 135:307-314

Allen AK, Neuberger A, Sharon N 1973 The purification, composition and specificity of wheat-germ agglutinin. Biochem J 131:155-162

Barondes SH, Rosen SD 1976 Cell surface carbohydrate binding proteins: role in cell recognition. In: Barondes SH (ed) Neuronal recognition. Plenum, New York, p 331-356

Bos HJ 1979 *Entamoeba histolytica:* cytopathogenicity of intact amebae and cell free extracts; isolation and characterization of an intracellular toxin. Exp Parasitol 47:369-377

Chevez A, Sepulveda B, Segura M, Corona D, Diaz J 1976 Initial phases of pathogenic activity of *E. histolytica* in colon and liver of hamsters. In: Sepulveda B, Diamond LS (eds) Proc Int Conf Amebiasis, 1976 Instituto Mexicano del Seguro Social, Mexico City, p 418-517

Councilman WT, LaFleur HA 1891 Amebic dysentery. Johns Hopkins Hosp Rep 2:393-512

Diamond LS 1968 Techniques of axenic cultivation of *Entamoeba histolytica* Schaudinn, 1903 and *E. histolytica*-like amebae. J Parasitol 54:1047-1056

Diamond LS, Harlow DR, Cunnick CC 1978 A new medium for the axenic cultivation of *Entamoeba histolytica* and other *Entamoeba.* Trans R Soc Trop Med Hyg 72:431-432

Gill DM 1978 Seven toxic peptides that cross cell membranes. In: Jeljaszewicz J, Wadström T (eds) Bacterial toxins and cell membranes. Academic Press, London, p 291-332

Hoare CA 1952 The food habits of *Entamoeba histolytica* in its commensal phase. Parasitology 42:43-47

Jarumilinta R, Kradolfer F 1964 The toxic effect of *Entamoeba histolytica* on leukocytes. Ann Trop Med Parasitol 58:375-381

Katz S, Izhar M, Mirelman D 1981 An *in vivo* model for studying adherence of intestinal pathogens. In this volume, p 94-97

Kawaskaki T, Ashwell G 1977 Isolation and characterization of an avian hepatic binding protein specific for N-acetylglucosamine-terminated glycoproteins. J Biol Chem 252:6536-6543

Keusch GT, Jacewicz M 1977 Pathogenesis of Shigella diarrhea. VII. Evidence for a cell membrane toxin receptor involving β1→4-linked N-acetyl-D-glucosamine oligomers. J Exp Med 146:535-546

Kilpatrick DC, Yeoman MM 1978 Purification of the lectin from *Datura stramonium.* Biochem J 175:1151-1153

Kobiler D, Barondes SH 1977 Lectin activity from embryonic chick brain, heart and liver: changes with development. Dev Biol 60:326-330

Kobiler D, Mirelman D 1980 A lectin activity in *Entamoeba histolytica* trophozoites. Infect Immun 29:221-225

Lis H, Sharon N 1980 Lectins in higher plants. In: Stumpf PK, Conn EE (eds) The biochemistry of plants: a comprehensive treatise. Academic Press, New York, vol 4, in press

Lushbaugh WB, Kairalla AB, Cantey JR, Hofbauer AF, Pittman FE 1979 Isolation of a cytotoxin/enterotoxin from *Entamoeba histolytica.* J Infect Dis 139:9-17

Mattern CFT, Keister DB 1977 Experimental amebiasis II. Hepatic amebiasis in the newborn hamster. Am J Trop Med Hyg 26:402-411

Mattern CFT, Keister DB, Caspar-Natovitz P 1980 *Entamoeba histolytica* "toxin": fetuin neutralizable and lectin like. Am J Trop Med Hyg 29:26-30

Mirelman D, Altmann G Eshdat Y 1980 Screening of bacterial isolates for mannose-specific lectin activity by agglutination of yeasts. J Clin Microbiol 11:328-331

Mizuochi T, Yonemasu K, Yamashita K, Kobata A 1978 The asparagine-linked sugar chains of subcomponent Cl_q of the first component of human complement. J Biol Chem 253:7404-7409

Ofek I, Mirelman D, Sharon N 1977 Adherence of *Escherichia coli* to human mucosal cells mediated by mannose receptors. Nature (Lond) 265:623-625

Olsnes S, Eiklid K 1980 Isolation and characterization of *Shigella shigae* cytotoxin. J Biol Chem 255:284-289

Roos E 1894 Zur Kenntniss der Amöbenenteritis. Arch Exp Pathol Pharmakol 33:389-392

Rupley JA 1964 The hydrolysis of chitin by concentrated hydrochloric acid and the preparation of low molecular weight substrates for lysozyme. Biochim Biophys Acta 83: 245-255

Schaudinn F 1903 Untersuchungen über die Fortpflanzung eininger Rhizopoden. Arb K Gesundheitsamte 19:547-576

Trissl D, Martinez-Palomo A, De la Torre M, De la Hoz R, De Suarez EP 1978 Surface properties of *Entamoeba:* Increased rates of human erythrocyte phagocytosis in pathogenic strains. J Exp Med 148:1137-1145

Yamashita K, Tachibana Y, Kobata A 1978 The structures of the galactose-containing sugar chains of ovalbumin. J Biol Chem 253:3862-3869

DISCUSSION

Taylor-Robinson: Do all the pathogenic and the less pathogenic strains of *Entamoeba histolytica* produce lectins in the same amounts?

Mirelman: The four strains we studied (HM-1, HK-9, 200:NIH and a non-axenic strain) have different quantities of lectin. The most pathogenic strain is not the one that contains the most lectin.

However, we tested that in sonicated preparations and this may be irrelevant. The difference in lectin content between the various strains should be assessed after subcellular fractionation of the amoebae. The lectin may be distributed among various organelles and the fraction that plays a role in adherence is most probably the one that is exposed on the surface. We hope to obtain antibodies to the lectin so that we can then determine how much of it is really relevant and accessible to the surface of the amoebae. The differences we saw in the quantity of lectin might be easier to understand if we knew how much was on the surface.

Taylor-Robinson: If you continue to passage the strains in medium, do they lose pathogenicity and also lectin activity?

Mirelman: The axenic strains we are using in our experiments have been passed many times in medium for at least four years. It is well known that axenic strains lose their virulence and regain it on passage through liver or growth with bacteria. The non-axenic strain has much more lectin than the other three. We are now trying to see whether, once the axenic strains are grown once more in the presence of a bacterial population, they will contain more lectin.

Newell: It would be useful to have a mutant that completely lacks lectin activity. With affinity chromatography it should be possible to pour a population of amoebae down the column and then pick up those that don't bind at the bottom. Lerner found with *Dictyostelium* that the mutants that lacked particular lectins were non-cohesive (Ray et al 1979). When revertants were obtained that regained their ability to cohere, the ability to form lectin was restored. Similar experiments in your system might be revealing.

Mirelman: Dr A. Martinez Palomo's group in Mexico City is looking for lectin-lacking amoebae in human carriers who appear to be non-symptomatic. They are also trying to select mutants that are devoid of lectin activity. A member of this group is going to join us this summer to test the hypothesis that the strains that don't cause disease lack lectin activity and to test new mutants by their inability to attach to chitin affinity chromatography columns in the way you mentioned.

Sharon: Do the amoebae bind to the column?

Mirelman: Yes.

Silverblatt: Is the semi-purified lectin itself an inhibitor of attachment or binding?

Mirelman: We have not looked at that, because we have very little lectin activity that we can spare to spread onto tissue culture. But wheat germ agglutinin that was prebound to tissue cultures specifically blocked the adherence of amoebae, whereas peanut agglutinin did not.

Feizi: Does your lectin retain its carbohydrate binding activity in the presence of detergents?

Mirelman: Yes. The two best detergents for solubilizing our material were Sarcosyl NL-97 (0.1% w/v) and Zwittergen TM312 (0.5% w/v). Triton at 0.1% can also be used but that doesn't keep the activity for long enough.

Elbein: Could any clinical correlation be made between susceptibility of individuals to this organism and their secretor state, i.e. the blood group activity or something like that?

Mirelman: I don't think anyone has tried to make that correlation but it is an interesting point. We suspect that people who are carriers may have an immunodeficiency, perhaps genetic, or there may be a nutritional deficiency that prevents them from producing enough of the mucus or protecting material in the intestine so that they become susceptible to the binding of the amoebae.

Elbein: Have you any idea what the *N*-acetylglucosamine-containing component in the mammalian cell is?

Mirelman: No. Mucus is a glycoprotein that contains considerable amounts of *N*-acetylglucosamine but we don't know yet what the component in the mucosa will be. It doesn't have to be a terminal *N*-acetylglucosamine.

Tramont: There aren't any cell susceptibility differences with species of *Amoeba*, as there are with *Giardia*.

Friend: You get inhibition by glycan and peptidoglycan but not by *N*-acetylglucosamine tetramers and trimers. Could the amoebae be hydrolysing the trisaccharides and tetrasaccharides but not the polysaccharides?

Mirelman: That is a possibility. The amoebae have very potent *N*-acetylglucosaminidase activity. But I suspect that many hapten inhibitors of low molecular mass will not really block the adherence of the intact amoeba which contains many lectin molecules with which it binds to many receptors. I don't know whether it is a problem of the binding affinity or binding constant of the lectin to the low molecular mass haptens or a matter of covering many receptors at once with a multivalent macromolecule containing many glucosamine residues.

Helenius: In *Dictyostelium* there is evidence that a glucose-binding lectin is involved in phagocytosis. Could your lectin have a dual function – attachment of the amoebae to cells and phagocytosis?

Mirelman: Even the low molecular mass chitin oligosaccharide blocks adherence of red blood cells to the amoeba. Of course it also blocks phagocytosis of red blood cells. It is possible that the binding of the red blood cells involves fewer receptors and thus the low molecular mass haptens are able to block this type of adherence. I don't know, however, whether one will be able to distinguish between the two functions, as you suggest.

Helenius: Could the pits you mentioned have something to do with phagocytosis?

Mirelman: Yes, I think there is a close connection. The moment an amoeba gets onto something it considers a good substrate it becomes energized and starts ingesting or phagocytizing anything that is near it. This shows up very clearly in our pictures.

Svanborg Edén: Is the blocking effect of IgA that you mentioned due to binding to the amoeba or is IgA covering the receptor site?

Mirelman: We don't know. The purified IgA was obtained from colostrum, not from a patient with amoebiasis. We are trying to see whether there is any specific secretory IgA in patients that can be shown to have an advantage over non-specific IgA in inhibiting or covering the receptor, as you mentioned.

Silverblatt: I thought you were making the point that the IgA contains *N*-acetylglucosamine.

Mirelman: That is our understanding at the moment. The IgA we tested was obviously a non-specific glycoprotein but perhaps one obtained from an amoebiasis patient would be an even better inhibitor.

Feizi: Have you tested other immunoglobulin inhibitors such as serum IgA or serum IgG?

Mirelman: We have tested only IgG from amoebiasis patients and IgG from normal individuals. The IgG fractions from normal humans do not inhibit lectin activity.

Feizi: How about IgA myeloma proteins?

Mirelman: They might be good compounds to test.

Feizi: It would be interesting to determine whether the inhibitory activity is confined to secretory IgA, or whether it is associated with serum IgA. There is considerable information on the carbohydrate chains of IgA.

Freter: You mentioned that mucus gel has binding affinity for amoebae. Could you elaborate on how you determine the binding activity?

Mirelman: We label mucus with an iodinated reagent and watch for the binding of the mucus to the amoeba. This gives us a semi-quantitative assay. Then in inhibition studies with red blood cells we can precoat the amoebae with the colonic mucus gel. This inhibits the agglutination of red blood cells very well. We have problems in determining the inhibition of adherence of intact amoebae to monolayers because of the viscosity of this material and its ability to stick by itself to the epithelial cells. We also take colon tissue from rabbits, invert it and observe the adherence of amoebae to an intact part of the intestine. There too these chitin and bacterial peptidoglycans inhibit adherence. The colonic mucus tends to associate very fast with the mucosal cells and we have difficulty in finding a good assay for the inhibitory effect of mucus on the adherence of amoebae.

Sharon: The possible dual role of adherence was mentioned. We (Bar-Shavit et al 1977) and Dr Silverblatt (Silverblatt et al 1979) have found that *E. coli* also adheres to phagocytes in a mannose-sensitive reaction. This raises the possibility, which I shall discuss in more detail in my presentation see p 132), of a dual role for the mannose-sensitive adherence (or mannose-specific lectin) of *E. coli:* it may help the organism to initiate infection, by binding to epithelial cells, but may also make it easier for the host to eliminate the invading bacteria, by their attachment to mannose residues on its phagocytes.

Howard: Inhibition of invasion of human red cells by *P. falciparum* with high concentrations of *N*-acetylglucosamine (above 25 mM) has recently been demonstrated (Weiss et al 1980). Several other sugars had no inhibitory effect. This may implicate recognition of carbohydrates in the parasite's invasion mechanism. However, additional studies will be required to demonstrate that *N*-acetylglucosamine does not non-specifically affect internal parasite processes unrelated to the specific invasion mechanism.

The concentrations of inhibitory sugars employed in your attachment studies, Dr Mirelman, are significantly lower and are more likely to affect only the membrane–membrane interaction. This general approach will obviously be important for studies of the interaction of protozoa with cells.

Sharon: A concentration of 25 mM *N*-acetylglucosamine is not very high.

Howard: Extracellular *N*-acetylglucosamine can affect the intracellular metabolism of carbohydrates at levels lower than 25 mM. In the assay of Weiss et al (1980) heavily infected cells were incubated with uninfected cells in a solution of the carbohydrates being tested. The number of parasites which entered uninfected cells after release from the infected cells was measured. In order to test the effect of added carbohydrates on invasion of uninfected cells, one must be quite sure that terminal events in parasite differentiation before release from infected cells are not affected by the extracellular carbohydrate, especially when the membranes of infected cells are known to be of increased permeability.

Wallach: Would it be possible to radiolabel the lectin in intact cells?

Mirelman: We have tried to radioiodinate intact amoebae. A lot of proteins seem to become labelled by the iodinating technique and we do not know yet which of them is the lectin. One of the problems is that amoebae constantly turn over their outer membranes. What you think of at one moment as an outer membrane component can be completely inside the amoeba after a very short time, and it regenerates constantly. Amoebae interacting with red blood cells or interacting with antibodies continually cap and shed materials that became associated with their outer surface.

Wallach: Is that temperature-sensitive?

Mirelman: Yes. For example if you do the experiment at 5 °C the amoebae at this temperature have a completely different shape than in the normal situation. They become rounded and then you can ask what is the consequence of this unnatural rounding-up on the labelling of surface proteins.

Wallach: One thing you would do is label the lectin if it stays at the surface.

Mirelman: The lectin is probably labelled but we don't know yet which of the proteins is the lectin. We can radioiodinate a whole amoeba and then get labelled proteins remaining on the affinity chromatography column so we are quite certain that we are labelling the lectin.

Howard: You could use photoaffinity labelling to identify the amoeba's surface lectin. As *N*-acetylglucosamine-containing oligosaccharides bind to the lectin on the amoeba's membrane, you could attach a photoactivatable reagent to an oligosaccharide and then use flash photolysis to covalently cross-link the bound oligosaccharide and lectin. This technique would allow very precise control of the labelling kinetics and may therefore overcome the problem of rapid membrane turnover.

Mirelman: We have someone working on that approach. I hope it can give us more information about the nature of the receptor.

Hughes: You don't have to wait for antibodies. You can get bovine serum albumin covalently linked to *N*-acetylglucosamine oligomers which is a very good cytochemical reagent (Schrevel et al 1979).

It is interesting that your lectin is preserved by β-mercaptoethanol, indicating that a free SH group is needed for activity. This seems to be a general property of animal lectins, which sharply distinguishes them from plant lectins (see for example Briles et al 1979).

Mirelman: The amoeba is apparently not the only parasite in which biologically active components are sensitive to oxidation. In work on *Trypanosoma cruzi* it was found that trypanosomes have a protease-like material, the activity of which disappears in the absence of β-mercaptoethanol and can be preserved with iodoacetate. I believe that the rapid loss of biological activity may be of importance for the ability of parasites to evade the host immune system. The host in many instances will be producing antibodies against the biological inactivated components of the parasite and these usually do not confer protection against the active species.

Taylor-Robinson: Is there anything relevant to say about the adhesive properties of other protozoa such as *Leishmania?*

Mirelman: I have been told that lectin-like material has been found in *Toxoplasma* but I don't know anything more about that.

Razin: You mentioned that extraction with chloroform–methanol removed the lipid but denatured the protein. You should consider using phospholipases as a means of hydrolysing phospholipids so that you can see whether the lectin is dependent on phospholipids for activity.

Silverblatt: Bacteria probably also have the ability to inactivate their surface lectin rapidly. It is not difficult to construct a hypothetical reason why this may be important for organisms which shed their ligands into the environment – were the ligand to remain active, the shed lectin would compete for binding sites with the bacteria.

REFERENCES

Bar-Shavit Z, Ofek I, Goldman R, Mirelman D, Sharon N 1977 Mannose residues on phagocytes as receptors for the attachment of *Escherichia coli* and *Salmonella typhi.* Biochem Biophys Res Commun 78:455-460

Briles EB, Gregory W, Fletcher P, Kornfeld S 1979 Vertebrate lectins. Comparison of properties of β-galactoside-binding lectins from calf and chicken tissues. J Cell Biol 81:528-537

Ray J, Shinnick T, Lerner R 1979 Mutation altering the function of a carbohydrate binding protein blocks cell-cell cohesion in developing *Dictyostelium discoideum.* Nature (Lond) 279:215-221

Schrevel J, Kieda C, Caigneaux E, Gros D, Delmotte F, Monsigny M 1979 Visualization of cell surface carbohydrates. Biol Cellulaire 36:259-266

Silverblatt FJ, Dryer JS, Schauer S 1979 Effect of pili on susceptibility of *Escherichia coli* to phagocytosis. Infect Immun 24:218-223

Weiss MM, Openheim JD, Vanderberg JP 1980 *Plasmodium falciparum: in vitro* assay for substances that inhibit penetration of erythrocytes by merozoites. Exp Parasitol, in press

Mechanisms of association of bacteria with mucosal surfaces

ROLF FRETER

Department of Microbiology and Immunology, The University of Michigan, Ann Arbor, Michigan 48109, USA

Abstract Bacterial association with host mucosal surfaces involves a large number of steps. Successful negotiation of each of these requires – or is at least facilitated by – the development of a distinct set of characteristics (virulence factors) by the bacterium. The major steps include: (a) chemotactic attraction of motile bacteria to the surface of the mucus gel, (b) penetration of and trapping within the mucus gel (which may be passive or can be promoted actively by bacterial motility and chemotaxis), (c) adhesion to receptors in the mucus gel or to mucosa-associated layers of the indigenous microflora, (d) adhesion to epithelial cell surfaces, and (e) multiplication of the mucosa-associated bacteria. Each reaction is further modified – or reversed entirely – by substances such as taxins, inhibitors of adhesion, and substrates for bacterial growth that are present in the mucosal microenvironment. Association with the mucosa is often important for bacterial colonization but can also lead to more effective elimination of the bacterium by the host. Bacteria lacking one or several of these virulence factors may still be successful colonizers if they show exceptionally high competence in relation to others. Examples are the strong adhesion to epithelial cells by *Escherichia coli* strains bearing the K88 antigen (such strains need not be motile in order to be pathogenic) or the active chemotactic association with mucus gel by cholera vibrios (some of which do not appear to adhere strongly to epithelial cells). Consequently a single *in vitro* assay for 'adhesion' can be expected to correlate with bacterial pathogenicity only when the assay is based on the same specific mechanism(s) which the bacterium under study actually uses for mucosal association *in vivo*.

It seems intuitively obvious that bacteria which inhabit or infect the body surfaces of a metazoan host must somehow hold onto these surfaces in order to avoid physical removal. It is therefore rather surprising that few relevant papers were published before the decade 1960–1970 when this concept was formulated on Theoretical grounds (Lankford 1960, Freter et al 1961, Smith & Halls 1967), when

---

1981 Adhesion and microorganism pathogenicity. Pitman Medical, Tunbridge Wells (Ciba Foundation symposium 80) p 36-55

colonization of body surfaces was demonstrated histologically (La Brec et al 1965, Dubos et al 1965, Hoffman & Frank 1966), and when the protective effect of local immunity was correlated with inhibition of adhesion (Freter 1969). Since the demonstration during the early 1970s of a close association between adhesive capacity and distribution *in vivo* of a variety of oral bacteria (reviewed by Gibbons & van Houte 1975) and between adhesion of *Escherichia coli* and its pathogenicity in animals (Jones & Rutter 1972) there has been a sudden surge of interest in the subject. The results of the early investigators had indicated that the adhesive reactions studied by them involved a direct interaction, presumably based on receptor–ligand binding, between the bacterial surface and that of mucosal cells or tooth enamel. Following this example, the numerous subsequent investigators who were interested in colonization similarly concentrated their studies on the role of surface–ligand interactions.

Since the initial exciting demonstration of a correlation between the presence of certain adhesins (K88, K99, CFA) and pathogenicity, it has become painfully difficult to correlate unequivocally bacterial adhesion, as measured *in vitro*, with the presumed pathogenic potential of bacteria that infect the mucosal surfaces of the human body. For example, Sugarman & Donta (1979) found no correlation between adherence to a variety of human and tissue culture cells and the presumed pathogenicity of enterobacteria. Likewise, Thorne et al (1979) described a variety of 'patterns' of mannose-resistant haemagglutination and adhesion to human buccal mucosal cells among toxigenic *E. coli* strains of human origin. While the results of Svanborg Edén & Jodal (1979) demonstrate a statistical correlation between adherence to human urinary sediment cells and the presumed pathogenic potential of urinary tract isolates, their data also show a large fraction of isolates from patients that did not adhere to cells at all. Such data suggest (or at least do not rule out the possibility) that surface–ligand interactions may indeed contribute in some way to bacterial colonization of many body surfaces, but that such adhesive reactions are complicated and modified *in vivo* by other intervening events that are not reproduced in the currently popular *in vitro* tests for bacterial adhesion. This paper therefore stresses the fact that the current emphasis on surface–ligand binding has somewhat obscured the recognition and study of other interactions which in many instances may be of major importance in the association of bacteria with body surfaces. Moreover, a number of such interactions are preliminary to the eventual ligand binding of bacteria at epithelial cell surfaces. Consequently, a bacterium may never advance *in vivo* to the point where ligand binding can take place unless it has the necessary equipment (virulence factors) to negotiate successfully the preliminary steps. In view of the fact that the primary ecological function of bacterial adherence to body surfaces, namely resistance to physical removal, may be accomplished by means other than surface–ligand interactions, the term 'association' with the mucosa will be used here when the nature of the relevant interactions is not known or not specified (Freter 1980). In contrast, the terms 'adhesion' or 'adherence' will be used synonymously to denote specifically surface–ligand binding between bacterial and mammalian cells which, by this definition, becomes a special case of mucosal association.

Mechanisms of mucosal association

The 'mucus' materials of mucosal surfaces are of two types. The first consists of glycoproteins and glycolipids synthesized by the epithelial cell, which form an integrated glycocalyx. The second type includes glycoproteins which differ in various characteristics from those of the glycocalyx. This material, hereafter referred to as 'mucus gel', is secreted by specialized cells of the mucosa and often forms a layer covering the epithelium. The degree to which mucus gel inhibits the access of bacteria and inert colloidal particles to the epithelial cells is a matter of controversy. For example, Edwards (1978) considered mucus to be a 'particle- and macromolecule-proof coating for cell surfaces' through which invading bacteria must 'bore a channel'. Actually, several reports indicate that macromolecules cross mucosal barriers with some regularity (reviewed by Warshaw et al 1977). Florey (1933) made direct stereomicroscopic observations on live intestinal mucosa and concluded that the mucus gel formed a discontinuous lacework of strands, which in certain areas was penetrable by colloidal carbon particles. The active motion of the villi caused particles to be 'rolled up' in mucus gel and thus to be expelled into the intestinal lumen (Florey 1933). The work of Schrank & Verwey (1976) suggested that mucus gel prevents the penetration of carbon particles to the epithelial surface. They showed further that efficient penetration required actively motile cholera vibrios.

Recent work from my laboratory has indicated that a small proportion of polystyrene spheres presented to the mucosa could penetrate rapidly (within 15 minutes or less) into the intervillous spaces and crypts of the mouse or rabbit small intestine (Freter et al 1978a, b, 1979a, b). Larger particles such as yeast cells penetrated less efficiently (*loc. cit.*). Non-motile mutants of *Vibrio cholerae* showed very slow penetration, at a rate resembling that of yeast cells. Penetration of normally motile but non-chemotactic vibrio mutants was only slightly more efficient than that of non-motile bacteria. In contrast, wild-type cholera vibrios of normal motility and normal chemotactic responsiveness showed rapid penetration into the intervillous spaces at a rate significantly faster than polystyrene spheres.

The evidence reviewed is consistent with the conclusions of some earlier workers that the mucus gel presents a barrier to the penetration of particles in the size range of bacteria. In contrast to some earlier assumptions, however, this barrier appears to be imperfect, because even inert particles may penetrate to the base of the villi. The rate of such passive penetration is relatively low, but can be increased significantly when bacteria are motile and are guided by chemotactic stimuli towards the bases of the villi. In this respect, bacterial motility without chemotactic guidance appears to be ineffective. This evidence was obtained with intestinal mucosa but it seems reasonable to assume that similar effects would occur in the mucus gel of other mucosae as well. It is also apparent that the seemingly simple process of bacterial adherence represents a potentially very complex sequence of host–bacterium interactions. The more obvious mechanisms of interaction will be listed briefly below. It should be emphasized that no single bacterial species can (or needs to) deploy every one of these mechanisms. On the other hand, each of these interactions has been demonstrated with at least some pathogenic or commensal bacteria. It is also important to realize that some of the 'intermediate' interactions to be discussed below, such as trapping in the mucus gel or adherence to layers of indigenous bacteria, may possibly represent the final stages of association for certain bacteria, in that these sites may confer sufficient protection from mechanical removal to permit bacterial colonization.

(1) Motile bacteria in the vicinity of a mucosal surface are attracted to it by means of a chemotactic gradient (Allweiss et al 1977). This interaction requires, of course, the presence of taxin gradients in the mucosa and, on the part of the bacteria, an ability to respond chemotactically to these taxins. Successful attraction of motile bacteria presupposes that a suitable environment must exist *in vivo* (or outside the body before infection) for the synthesis of flagella and of certain receptors for bacterial chemotaxis. The synthesis of flagella *in vitro* by *E. coli*, for example, depends critically on cultural conditions (Adler & Templeton 1967). Also, certain taxin receptors of *E. coli* are inducible – that is, are not present on the surface of bacteria grown in the absence of the specific taxin (Adler 1975). For this reason, bacteria classified by *in vitro* criteria as motile and chemotactic may not exhibit either of these traits in certain *in vivo* environments, and vice versa.

(2) Penetration of the mucus gel by bacteria located at the gel surface constitutes the next step of association. As discussed above, this may occur passively at a slow rate, but is significantly enhanced when bacterial motility is guided by chemotactic gradients. These interactions, like those discussed in (1), require the production of attractants (taxins) by the host as well as the corresponding responsive capabilities on the part of the bacteria. Recent results (Freter et al 1979a, b) indicate that the taxin gradient in mucus gel of small intestinal mucosa of mice and rabbits extends into the deep intervillous spaces and crypts and, for this reason, can indeed promote deep bacterial penetration.

(3) Although there are no relevant published data, the possibility of bacterial adhesins reacting with receptor substances in the mucus gel must certainly be considered as a potential additional mechanism which would promote mucosal association by retaining bacteria in the mucus gel. For example, microscopic observations in my laboratory (unpublished) of rabbit intestinal slices or of intestinal loops exposed to cholera vibrios, often show phase-contrast-dense material, embedded in the mucus gel, around which numerous vibrios are aggregated. These aggregated

bacteria have become immobilized and appear to adhere to the dense material.

(4) Bacterial adhesion to epithelial cell surfaces represents the final step in the process of association with the mucosa. As discussed above, it is also the mechanism that was identified earliest and, consequently, is being studied most widely. The adhesive interactions identified so far appear to involve specific receptor–ligand binding. This obviously requires *in vivo* synthesis of bacterial adhesins. In view of the stringent cultural conditions that are often required for adhesin production *in vitro* (Jones 1977) one cannot assume solely on the basis of *in vitro* data that a given bacterium must necessarily synthesize adhesin *in vivo*. A non-specific mechanism of adhesion – that is, one which depends on such forces as differences in surface charges or hydrophobic interactions – has been considered by several workers. The suggestion has been made that such non-specific surface interactions may be preliminary events which promote subsequent more specific mechanisms of adhesion (Brinton 1965, Isaacson 1977, Smyth et al 1978).

Mechanisms which modify bacterial association with mucosal surfaces

In the natural environment epithelial cell surfaces are bathed in glycoproteins from such sources as mucus gel or saliva. Many of these glycoproteins share receptor specificities with the glycocalyx of epithelial cells (Williams & Gibbons 1975, Etzler 1979). Forstner (1971) showed that glycoproteins of the glycocalyx may be released by mild proteolytic digestion, as may occur *in vivo*. In the *in vivo* environment substances from glycocalyx and mucus gel must therefore be expected to compete for bacterial adhesins with receptors on the epithelial cell surfaces. Moreover, the observations of Forstner et al (1973) and Etzler (1979) indicate that mucus gel may adhere to the epithelial cell surface with considerable tenacity, such that it can be removed only by vigorous washing. Consequently, bacteria may adhere *in vivo* to epithelial cells not only via receptors of the glycocalyx but also by reacting with receptors of mucus gel coating the cell surface. Bacterial adhesion to cell-bound mucus gel would also be weakened in the presence of competing free mucus gel. It therefore seems reasonable to assume that, if free mucus gel is present *in vivo*, bacterial adhesion to epithelial cells would often be relatively weak and highly reversible, unless the specific bond could be reinforced at the cell surface by a second adhesive mechanism such as hydrophobic interactions.

Bacterial and host enzymes may modify several host factors of importance in mucosal association. For example, the neuraminidase of cholera vibrios, which destroys red blood cell receptors for influenza virus or for *Mycoplasma pneumoniae*, can also *enhance* the adhesion of some bacteria to the altered cell surface (Sobeslavsky et al 1968). Schneider & Parker (1978) demonstrated that some protease- and neuraminidase-deficient mutants of *V. cholerae* show a 'dramatic' loss of virulence in experimental cholera. Parsons et al (1978) noted that neuraminidase enhanced the adhesion of *E. coli* to rabbit urinary bladders. Proteolytic enzymes also remove neuraminic acid (Parsons et al 1978) when this is a component of glycoproteins at cell surfaces (Forstner 1971). Peptic digestion of mucosal scrapings has been shown to release materials which inhibit cholera vibrios with respect to both chemotaxis and adhesion to isolated brush border membranes of epithelial cells (Freter & Jones 1976, Allweiss et al 1977). While the mechanisms involved in these phenomena obviously require further investigation, these examples illustrate that promotion as well as inhibition of bacterial association with body surfaces by microbial or host enzymes is not an unlikely possibility.

Most bacterial adhesins that have been studied in this respect appear to react with carbohydrate receptors on the mammalian cell surface (Jones 1977). Since most natural foodstuffs contain sugars, polysaccharides, glycoproteins or glycolipids, one must assume that many of these will antagonize the adhesion of bacteria in the mouth and gastrointestinal tract by competitive inhibition. For example, the unit membranes surrounding milk fat globules show serological cross-reactions with red blood cells. Reiter & Brown (1976) reported that agglutination of red blood cells by *E. coli* K88 and K99 antigens can be inhibited by milk fat globules or fat globule membranes. To the extent that the milk fat globule membrane carries other cell receptor moieties in addition to those reacting with K88 and K99 antigens, such inhibitory activity may extend to other adhesive pathogens as well.

While bacterial adhesins exhibit a remarkable degree of specificity for host species, for individual hosts and for cell types, there are also a surprising number of cross-reactions. For example, *V. cholerae* and human enteropathogenic *E. coli* strains adhere to the rabbit small intestine and agglutinate red blood cells from a variety of animal species (Sugarman & Donta 1979, Thorne et al 1979). It is likely that a number of such cross-reactions with cells of other species involve adhesin–receptor systems different in specificity from those operating in the natural host (Freter 1980, Jones 1977). For this reason, a meat diet may contain receptors for bacterial adhesion which can react in the host with indigenous or pathogenic bacteria, even though the dietary receptors may differ from those of the host. Obviously, bacteria bound to cells or cell debris of dietary meat products passing through the gut are likely to be diverted from association with mucosal epithelium, regardless of the specificities of the adhesive reactions involved. Moreover, deposition on a bacterial surface of molecules that are able to react with one set of specific adhesins will often inhibit the reactivity of other adhesins by mechanisms such as steric hindrance. Thus, receptor substances in the diet which do not match host cell receptors may still inhibit bacterial adhesion, as long as they are able to bind in some manner to the bacterial surface.

Carbohydrate-binding proteins (lectins) from plant and animal sources are constituents of the daily diet of the human species. Lectins may agglutinate cells, as well as inhibit adhesion. For example, concanavalin A inhibited adhesion of lactobacilli to epithelial cells and a variety of lectins inhibited the adhesion of *E. coli* to epithelial cell brush border membranes (reviewed by Jones 1977). Jones noted that the reaction was not specific, in that lectins with specificities for different carbohydrates inhibited adhesion of *E. coli* equally well, as long as the lectin was able to bind to the brush borders. For this reason he attributed the inhibition of adhesion to steric hindrance. The non-specific nature of inhibition by lectins marks this reaction as one of potentially broad significance in the inhibition of bacterial adhesion by dietary substances *in vivo.* Brady et al (1978) showed that wheat germ agglutinin, a lectin specific for *N*-acetylglucosamine, could be detected in the faeces of human subjects consuming a diet containing wheat germ, thus demonstrating the stability *in vivo* of this substance.

It seems obvious that practically all foodstuffs that are being digested as they pass through the mouth and gastrointestinal canal must elaborate taxins (e.g. sugars, amino acids) that can affect the chemotactic behavior of motile bacteria *in vivo*. Such taxins may function in several ways: (a) they may create a taxin gradient that has its highest concentration in the lumen, thereby drawing bacteria away from the mucus gel. (b) Alternatively, taxin present in relatively high contentrations may block the corresponding receptors on the bacterial surface, thereby abolishing the ability of the bacterium to respond chemotactically to that specific taxin (Adler 1975). If the blocking taxin or taxins are specific for the same bacterial receptors as the taxins that attract the bacteria into the mucus gel, active penetration of the mucus gel by the bacteria will be abolished. It is likely that this latter mechanism is responsible for the blocking effect of mucosal extracts on the penetration of cholera vibrios into intestinal mucosa (Freter et al 1979b).

Special emphasis must be given to the effect of the indigenous microflora. Components of this flora have been shown to degrade glycoproteins. Hoskins & Boulding (1976) have presented evidence suggesting that the human intestine selects for indigenous bacteria that are capable of utilizing blood-group-specific glycoproteins of the host. Systemic infections with *Bacteroides* species are known to bring about phenotypic changes in a patient's blood group (T or Tk polyagglutination) by cleaving terminal sugar residues on the red blood cells (Inglis et al 1975). There can be no doubt, then, that the nature of host cell receptors in areas normally inhabited by an indigenous microflora must be determined to a considerable extent by the activities of that flora. Moreover, since the indigenous flora forms thick, many-layered populations on some mucosal surfaces, the initial adhesion of other microorganisms may be to the indigenous bacteria themselves, rather than to the epithelial surface. This may conceivably account for certain unexpected experimental findings, such as the failure to detect an adhesive capacity of pneumococci for human pharyngeal cells (Selinger & Reed 1979). A similar mechanism has been postulated for the adhesion of free *Actinomyces* to pellicle-bound streptococci in the mouth (Cisar et al 1979). It goes without saying that colonization of mucosae requires that association with the mucosa is followed by bacterial multiplication. The nutritional substrates available for multiplication depend to a large extent on host- or diet-derived substances that are altered quantitatively and qualitatively by the indigenous flora of the microenvironment. As discussed elsewhere in this symposium, virus infection can also alter the adhesive properties of mammalian cells.

More than a decade ago, local antibodies were shown to prevent bacterial association with intestinal mucosa (Freter 1969). Various workers have subsequently extended this finding to other mucosae, to different adhesive bacteria and to macromolecules. Local antibodies may bring about inhibition of association by a variety of mechanisms, such as bacterial clumping, inhibition of bacterial motility and direct inhibition. Walker and co-workers showed that soluble immune complexes can stimulate the release of mucus gel from intestinal goblet cells, and that soluble antigen alone will have this effect in immunized rats (Lake et al 1979). If bacteria entering the mucus gel of the immune host can also trigger mucus flow, one would have to consider it a potentially effective immune mechanism that counteracts mucosal association with microorganisms by means of the flushing action of increased mucus flow.

One must conclude, in summary, that the seemingly simple process of bacterial association with the mucosa can be surprisingly complex. This complexity results from the presence of a mucus gel on the epithelial surface as well as from the influence of a large variety of intervening factors. A bacterium must therefore proceed sequentially through a large number of different reactions, some of which promote or retard its progress in approaching the epithelial cells, whereas others promote or antagonize its eventual adhesion to the epithelial cell surface. Bacterial association with the mucosa is therefore determined by the final equilibrium that is established as a consequence of various synergistic and antagonistic reactions. An understanding of such a complex, interdependent system of reactions cannot be gained solely by studying each of its component parts in isolation. It therefore appears necessary to pursue studies in animal models, where the sequential functions and, most importantly, the interdependence of individual host and virulence factors can be reproduced.

Relation of mucosal association and colonization

Rigorous proof that mucosal association of bacteria is a prerequisite for colonization and subsequent disease demands that association occurs regularly at least in the early stages of infection, that inhibition of association brought about by a variety of means always prevents colonization and disease and, most importantly, that mutants which lack the ability to associate are avirulent. Strong evidence for a role in pathogenesis, based on these various types of data, is available for a number of adhesins including K88, K99, the CFA antigens and the fimbriae of *E. coli* strain 987 (reviewed by Jones 1977 and Duguid & Old 1980). All these adhesins appear to mediate surface–ligand binding between bacteria and epithelial cells. One must ask, therefore, whether an alternative type of mucosal association described in the present article, namely association with the mucus gel, can also be shown to promote bacterial colonization.

As reported earlier from this laboratory (Freter et al 1978b, 1979a), non-chemotactic mutants of *Vibrio cholerae* that do not readily associate with intestinal mucus gel *in vivo* or *in vitro* were also impaired in their ability to colonize the gut of germ-free mice or to multiply in intestinal loops of adult rabbits. In contrast, the parent strain or chemotactic revertants readily colonized these *in vivo* models and, in mixed infections, outgrew and displaced the non-chemotactic mutants (Freter et al 1978b, 1979a). In all instances, light microscopic observation of frozen sections of infected intestinal mucosa revealed that the vibrios were located in the mucus gel, with no indication of adhesion to the mucosa (R. Freter, unpublished). The parent strain used in these studies had been selected for optimal chemotactic responsiveness by repeated reisolation from the spreading edge of colonies in semi-solid agar. The resulting isolate had lost its original ability (described by Jones & Freter 1976) to adhere to rabbit intestinal brush border membranes. It also did not adhere to brush border membranes of human or mouse small intestine. Interestingly, a similar effect of this type of selection on adhesiveness had been reported for *E. coli* by Swaney et al (1977). The parent vibrio strain was the cause of an accidental laboratory infection of cholera. When reisolated from the patient, it still did not produce adhesins for intestinal brush border membranes of rabbit, mouse or human origin. The results summarized in this paragraph may thus be interpreted to show that the superior colonization potential of the parent strain in the animal models (and most likely in the patient as well) was due to its ability to associate with intestinal mucus gel and was not a consequence of its adherence to epithelial cells. The mutant strains which lack chemotactic reactivity – the principal factor promoting invasion of the mucus gel by motile bacteria – were for this reason inferior to the parent strain in colonizing the animal models.

Association with mucosal surfaces is generally regarded as a mechanism that promotes bacterial colonization and many of the results from my laboratory are consistent with this view. We have, however, recently obtained evidence in the infant mouse that non-chemotactic mutants of *V. cholerae* are more virulent than the chemotactic parent or revertant strains, and that the mutants actually outgrow parent or revertant strains in mixed infections of these animals (Freter et al 1978b). We had originally speculated that the chemotactic bacteria might be drawn into the intestinal lumen by taxins emanating from the coagulated milk which characteristically fills the small intestine of infant mice (Freter et al 1978b). Such a reaction would, of course, result in the preferential elimination of chemotactic bacteria from the gut. Subsequent (unpublished) studies have disproved this hypothesis: histological examination of infected baby mice revealed that, as in all other animal models, chemotactic vibrios were considerably more active in associating with small intestinal mucus gel than the non-chemotactic mutant. Furthermore, extensive serial cultures of infected animals revealed that the relatively superior net rate of multiplication of the mutant strain in baby mice was due to an antibacterial effect directed against the parent, and did not result from stimulation of the growth rate of the mutant. Since the parent strain associated preferentially with the mucosa, one may hypothesize that this habitat had a detrimental (i.e. inhibitory or bactericidal) effect on cholera vibrios. It should be noted that antibacterial mechanisms at mucosal surfaces have been reported earlier, such as for immune intestinal mucosa (Freter 1970) and the normal bladder (see Norden et al 1968 and earlier references quoted therein). In such instances a superior tendency to associate with the mucosal surface must be expected to correlate with a decrease in colonization potential of the bacterium. This, then, is another important factor which must be determined before the significance of results obtained with *in vitro* or *in vivo* adhesion tests can be evaluated properly.

Conclusions

The evidence discussed in this article indicates that simple *in vitro* tests for bacterial adhesion have been of value in the past but that their usefulness is rapidly diminishing. The reasons for this are threefold: (1) there is more than one basic mechanism by which bacteria may associate with the mucosa; (2) numerous intervening reactions in the mucosal microenvironment modify the various steps leading to association; and (3) mucosal association may sometimes be detrimental to a bacterium. A better understanding of the relationships between mucosal association and the colonization potential or virulence of microorganisms will require a more thorough study in animal models of the various mechanisms that promote or modify mucosal association of bacteria. Only when the nature and, most importantly, the interdependence of such mechanisms have been delineated in some detail will it become possible to design laboratory procedures which can test for the possible role of such mechanisms in human infections.

REFERENCES

Adler J 1975 Chemotaxis in bacteria. Annu Rev Biochem 44:341-356

Adler J, Templeton B 1967 The effect of environmental conditions on the motility of *Escherichia coli.* J Gen Microbiol 46:175-184

Allweiss B, Dostal J, Carey KE, Edwards TF, Freter R 1977 The role of chemotaxis in the ecology of bacterial pathogens of mucosal surfaces. Nature (Lond) 266:488-450

Brady PG, Vannier AM, Banwell JG 1978 Identification of the dietary lectin, wheat germ agglutinin, in human intestinal contents. Gastroenterology 75:236-239

Brinton CC Jr 1965 The structure, function, synthesis and genetic control of bacterial pili and a molecular model for DNA and RNA transport in gram negative bacteria. Trans NY Acad Sci 27:1003-1054

Cisar JO, Kolenbrander PE, McIntire FC 1979 Specificity of coaggregation reactions between human oral streptococci and strains of *Actinomyces viscosus* or *Actinomyces naeslundii.* Infect Immun 24:742-752

Dubos RJ, Schaedler RW, Costello R, Hoet P 1965 Indigenous, normal and autochthonous flora of the gastrointestinal tract. J Exp Med 122:67-76
Duguid JP, Old DC 1980 Adhesive properties of Enterobacteriaceae. In: Beachey EH (ed) Bacterial adherence. Chapman & Hall, London, p185-217
Edwards PAW 1978 Is mucus a selective barrier to macromolecules? Br Med Bull 34:55-56
Etzler ME 1979 Lectins as probes in studies of intestinal glycoproteins and glycolipids. J Clin Nutr 32:133-138
Florey HW 1933 Observations on the functions of mucus and the early stages of bacterial invasion of the intestinal mucosa. J Pathol Bacteriol 37:283-289
Forstner GG 1971 Release of intestinal surface membrane glycoproteins associated with enzyme activity by brief digestion with papain. Biochem J 121:781-789
Forstner J, Taichman N, Kalnins V, Forstner G 1973 Intestinal goblet cell mucus: Isolation and identification by immunofluorescence of a goblet cell glycoprotein. J Cell Sci 12:585-602
Freter R 1969 Studies of the mechanism of action of intestinal antibody in experimental cholera. Texas Rep Biol Med 27 (Supp 1):299-316
Freter R 1970 Mechanism of action of intestinal antibody in experimental cholera II. Antibody mediated antibacterial reaction at the mucosal surface. Infect Immun 2:556-562
Freter R 1980 Association of pathogenic bacteria with the mucosa of the small intestine – mechanisms and pathogenic implications. In: Ouchterlony O, Holmgren J (eds) Cholera and related diarrheas. Karger, Basel (Proc Nobel Symp 43) p155-170
Freter R, Smith HL Jr, Sweeney FJ 1961 An evaluation of intestinal fluids in the pathogenesis of cholera. J Infect Dis 109:35-42
Freter R, Jones GW 1976 Adhesive properties of *Vibrio cholerae:* nature of the interaction with intact mucosal surfaces. Infect Immun 14:246-256
Freter R, Allweiss B, O'Brien PCM, Halstead SA 1978a The role of chemotaxis in the virulence of cholera vibrios. In: Proceedings of the thirteenth joint US–Japan conference on cholera. Government Printing Office, Washington DC (DHEW publ. No. NIH 78-1590) p152-181
Freter R, O'Brien PCM, Halstead SA 1978b Adhesion and chemotaxis as determinants of bacterial association with mucosal surfaces. Adv Exp Med 107:429-437
Freter R, O'Brien PCM, Macsai MS 1979a Effect of chemotaxis in the interaction of cholera vibrios with intestinal mucosa. J Clin Nutr 32:128-132
Freter R, O'Brien PCM, Macsai M 1979b Correlation between in vitro chemotactic attraction of vibrios into the deep intervillous spaces of the intestinal mucosa. In: Takeya K, Zinnaka Y (eds) Symposium on cholera: proceedings of the fourteenth joint US–Japan conference on cholera. (Japanese Cholera Panel), US–Japan Cooperative Medical Science Program, p 94-101
Gibbons RJ, van Houte J 1975 Bacterial adherence in oral microbial ecology. Annu Rev Microbiol 19:19-44
Hoffman H, Frank ME 1966 Microbial burden of mucosal squamous epithelial cells. Acta Cytol 10:272-285
Hoskins LC, Boulding ET 1976 Degradation of blood group antigens in human colon ecosystems. J Clin Invest 57:74-82
Inglis G, Bird GWG, Mitchell AAB, Milne GR, Wingham J 1975 Effect of *Bacteriodes fragilis* on the human erythrocyte membrane: pathogenesis of Tk polyagglutination. J Clin Pathol 28:964-968
Isaacson RE 1977 K99 surface antigen of *Escherichia coli:* purification and partial characterization. Infect Immun 15:272-279
Jones GW 1977 The attachment of bacteria to the surfaces of animal cells. In: Reissig JL (ed) Microbial interactions. Chapman & Hall, London, p 139-176
Jones GW, Freter R 1976 Adhesive properties of *Vibrio cholerae.* Infect Immun 14:240-245
Jones GW, Rutter JM 1972 Role of the K88 antigen in the pathogenesis of neonatal diarrhea caused by *Escherichia coli* in piglets. Infect Immun 6:918-927
La Brec EH, Sprinz H, Schneider H, Formal SB 1965 Localization of vibrios in experimental cholera: a fluorescent antibody study in guinea pigs. In: Bushnell OA, Brookhyser CS (eds) Proceedings of the cholera symposium, Honolulu, Hawaii. Government Printing Office, Washington DC, p 272-276

Lake AM, Bloch KJ, Neutra MR, Walker WA 1979 Intestinal goblet cell mucus release. II. In vivo stimulation by antigen in the immunized rat. J Immunol 122:834-837
Lankford CE 1960 Factors of virulence of *Vibrio cholerae.* Ann NY Acad Sci 88(5):1203-1212
Norden CW, Green GM, Kass EH 1968 Antibacterial mechanisms of the urinary bladder. J Clin Invest 47:2689-2700
Parsons CL, Shrom SH, Hanno PM, Mulholland SG 1978 Bladder surface mucin. Examination of possible mechanisms for its antibacterial effect. Invest Urol 16:196-200
Reiter B, Brown T 1976 Inhibition of haemagglutination of red blood cells by K88 and K99 adhesin using milk fat and fat globule membrane. Proc Soc Gen Microbiol 3:109
Schneider DR, Parker CD 1978 Isolation and characterization of protease-deficient mutants of *Vibrio cholerae.* J Infect Dis 138:143-151
Schrank GD, Verwey WF 1976 Distribution of cholera organisms in experimental *Vibrio cholerae* infections: proposed mechanisms of pathogenesis and antibacterial immunity. Infect Immun 13:195-203
Selinger DS, Reed WP 1979 Pneumococcal adherence to human epithelial cells. Infect Immun 23:545-548
Smith HW, Halls S 1967 Observations by the ligated intestinal segment and oral inoculation methods on *Escherichia coli* infections in pigs, calves, lambs and rabbits. J Pathol Bacteriol 93:499-529
Smyth CJ, Jonsson P, Olsson E, Soderlind O, Rosengren J, Hjerten S, Wadstrom T 1978 Differences in hydrophobic surface characteristics of porcine enteropathogenic *Escherichia coli.* Infect Immun 22:462-472
Sobeslavsky O, Prescott B, Chanock RM 1968 Adsorption of *Mycoplasma pneumoniae* to neuraminic acid receptors of various cells and possible role in virulence. J Bacteriol 96:695-705
Sugarman B, Donta ST 1979 Specificity of attachment of certain Enterobacteriaceae to mammalian cells. J Gen Microbiol 115:509-512
Svanborg Edén C, Jodal U 1979 Attachment of *Escherichia coli* to urinary sediment epithelial cells from urinary tract infection-prone and healthy children. Infect Immun 26:837-840
Swaney LM, Liu Y, To C, To C, Ippen-Ihler K, Brinton CC Jr 1977 Isolation and characterization of *Escherichia coli* phase variants and mutants deficient in type 1 pilus production. J Bacteriol 130:495-505
Thorne GM, Deneke CF, Gorbach SL 1979 Hemagglutination and adhesiveness of toxigenic *Escherichia coli* isolated from humans. Infect Immun 23:690-699
Warshaw AL, Bellini CA, Walker WA 1977 The intestinal mucosal barrier to intact antigenic protein: difference between colon and small intestine. Am J Surg 133:55-58
Williams RC, Gibbons RJ 1975 Inhibition of streptococcal attachment to receptors on human buccal epithelial cells by antigenically similar salivary glycoproteins. Infect Immun 11:711-718

DISCUSSION

Taylor-Robinson: One of your main points, Dr Freter, is that we must be very careful about directly associating events which occur *in vivo* with things we see in *in vitro* tests.

Svanborg Edén: Professor Freter's results do not necessarily contradict ours. Binding to mucus or association with mucus may be necessary for colonization of any surface in a fluid system, and may be true for any organism colonizing a mucus surface. For bacteria that induce infection through a toxin the binding to mucus may allow them to be present in large numbers and secrete sufficient toxin to produce disease. The attachment to cells of the epithelial surface, on the other hand, is probably only necessary for organisms that invade underlying tissues. There are several populations of pili or binders present on *E. coli* associated with urinary tract infections. There are indications that one population mediates binding to mucus and that this is found on most strains independently of the type of disease produced. The other type of pili is associated with disease and with binding to epithelial cells.

Howard: It is conceivable that there is some specificity in the binding of bacteria to either the epithelium or mucus, or both. What criteria would you suggest, Professor Freter, for the demonstration of specificity? One method I can suggest which would show that a particular bacterial surface antigen was important for binding would be to attempt blocking of binding with Fab fragments of monoclonal antibody directed against various bacterial surface components. It may even be possible to transfer monoclonal antibody passively to the intestinal lumen *in vivo* and demonstrate specific inhibition of adherence. I should point out that passive transfer of monoclonal antibodies (directed against surface determinants on malaria parasites) to mice infected with *P. yoelii* malaria has been shown to decrease parasite growth, probably through blocking parasite attachment and invasion of erythrocytes (Freeman et al 1980).

Freter: I do not believe that inhibition of adhesion by specific or even monoclonal antibody necessarily identifies adhesins or receptors. It is clear from our work with cholera vibrios that one can get inhibition of adhesion with antibodies specific for a variety of surface antigens, not necessarily only with antibody against an antigen that is active in adhesion. In such instances antibody may block adhesion by steric hindrance or by other non-specific mechanisms. Competition experiments with purified adhesin would help but these do not necessarily provide conclusive proof either. Purified materials extracted from bacteria may, for example, block adhesion by mechanisms (e.g. effects on surface tension or surface charge) that are unrelated to ligand-binding. The most conclusive evidence I can think of might come from non-adhesive mutants which could be studied for differences from the parent strain.

Bredt: If the organism is associated only with the mucus and not with the cell membrane it ought to be washed away. If the organism stimulates fluid secretion the mucus would be washed away even faster and the organisms also would be removed. How can they stay in the intestine? Is the mucus so tenacious a thing that it is not removed by the disease process?

Freter: Pathogenicity is not necessarily beneficial to the bacterium, as the organisms are eventually eliminated from (or with) the host. The various mechanisms of association appear to work differently for different organisms. For example, an *E. coli* with K88 antigen can be highly pathogenic, yet the organism need not be motile. A few cells of this type of bacterium may pass through the mucus passively and, by adhering strongly to the epithelium, may multiply into large populations. All cholera vibrios we have tested show, in contrast, a very strong active penetration of the mucus gel and some (but probably not all) adhere only weakly to mucosal cells. Such weakly adherent pathogens must be very active in 'swimming upstream' continuously through the mucus, because even normally mucus gel flows towards the lumen. When the mucus flow becomes excessive (e.g. at the height of the diarrhoea) there is probably no longer any need for the microorganisms to maintain large populations on the mucosa. Indeed, they don't remain very long – just for a few days. The residual action of the toxin alone can perpetuate the clinical disease.

Sharon: Freedman (1967) found that diuretic mice are 10^6 times more susceptible to infection by *E. coli* than normal mice. This is contrary to expectation, since the flushing of the urinary tract should remove the bacteria.

In your experiment on inhibition with fucose did you also achieve displacement of preattached bacteria with fucose?

Freter: We could not do that. The bacteria were spontaneously released rather quickly (within 15 minutes) from the brush borders. The bacteria lost their adhesive activity within an hour of incubation in buffer (or sometimes sooner), even in the absence of substrates for adhesion.

Sharon: To what extent is your *V. cholerae* mannose-sensitive? In India Bhattacharjee & Srivastava (1978) observed the mannose inhibits *V. cholerae* El Tor.

Freter: Garth Jones (Jones & Freter 1976) also showed in his system that mannose was inhibitory, though not to the same degree as fucose. Among *E. coli* there are many adhesive systems reacting with different mammalian cells, for example. The question we are debating here is which of these numerous adhesins that can be found in most, and probably in all, microorganisms are relevant *in vivo*, and which ones represent *in vitro* phenomena without function in disease.

Sharon: You have no idea whether the same type of ligand or lectin is present in *V. cholerae*? Incidentally, for the sake of consistency we should decide what we call the substance on the bacteria and what we call the substance on the animal cell. The one on the cell is usually called the receptor and the one on bacteria is called the ligand or lectin or adhesin.

Taylor-Robinson: I like the recommendation that Shmuel Razin has made about mycoplasmas, referring to the receptor site on the epithelial cell and the binding site on the organism (Razin 1978).

Sharon: Do you think the binding sites for fucose and mannose are distinct structures on distinct molecules, or are they part of the same molecule?

Freter: We haven't followed that up. We concentrated our efforts on studying association with mucus when it became apparent that this phenomenon might be more important with our vibrios. Inhibition by mannose was not additive; that is, when adhesion was blocked by fucose it could not be reduced further by adding mannose. This would be consistent with the assumption that mannose may be a part of the fucose-containing receptor.

Taylor-Robinson: Can you say any more about binding sites on *V. cholerae*?

Sharon: Are they distinct sites, distinct molecules?

Freter: We have no information on that.

Taylor-Robinson: Presumably they are carbohydrate in nature, as the attachment of other vibrios, such as *V. parahaemolyticus,* can be inhibited by treatment with periodate (Carruthers 1977).

Richmond: On the question of a correlation between association and pathogenicity: one thing we mustn't forget is that the host or the source of the cells is very important, as we and others have shown. If you use buccal cells from one individual you can get a relatively coherent picture. If you then switch to buccal cells from other individuals you find totally different binding patterns amongst a range of coliform bacteria. So one source of the lack of correlation is almost certainly that when you look at the characters of pathogenicity you are looking at a range of different phenomena. For example there is the effect of human histocompatibility types. This must be investigated.

It is also helpful to think about a situation of this type in evolutionary terms, particularly thinking about the selection pressures. In the gut the production of mucus may be convenient for certain microorganisms because it could be a source

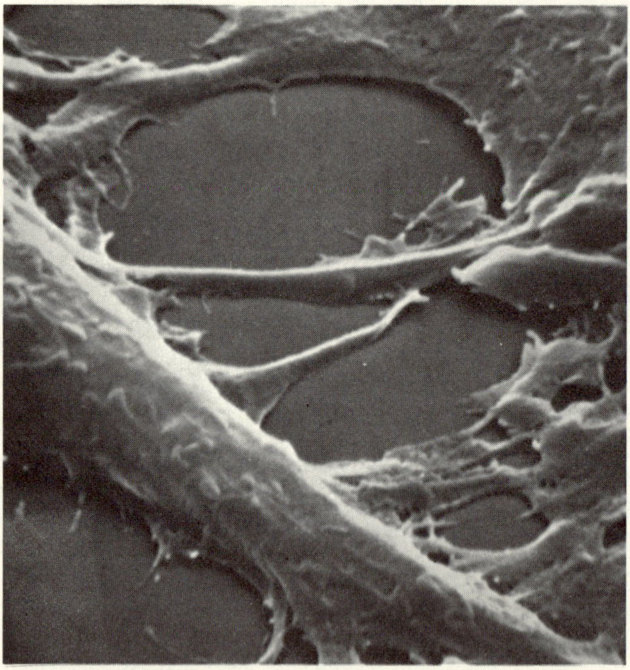

(a)

FIG. 1 (Richmond). The production of processes associated with the adherence of prokaryotic to eukaryotic cells. (Photographs provided by Dr G. Chabanon and Mrs P. Stirling.)
(a) Electron scanning photograph of normal eukaryotic cells.
(b) Electron scanning photograph of eukaryotic cells with adhering prokaryotes.
(c) A section showing the trapping of bacterial cells by the processes arising from the eukaryotic cells.

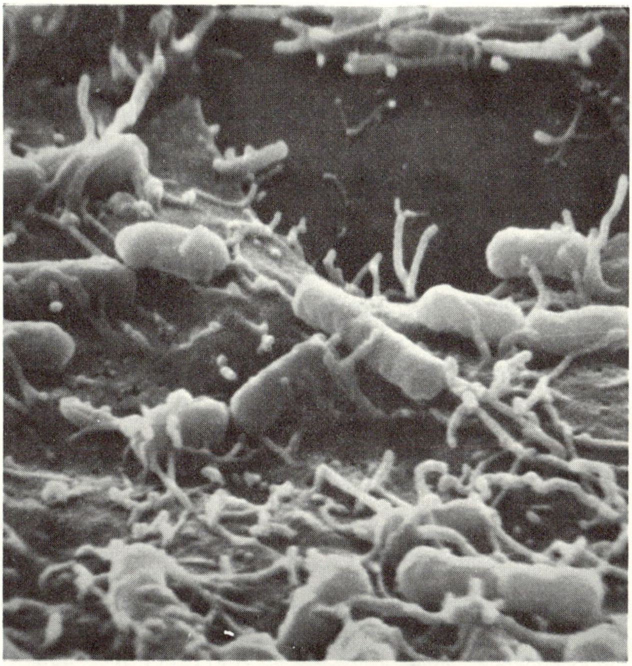

(b)

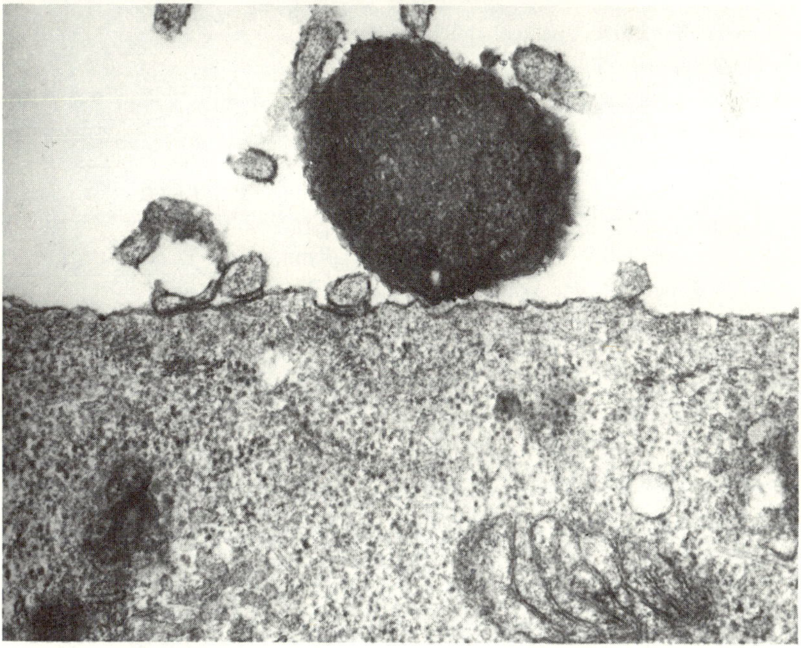

(c)

of carbon and nitrogen for growth. The gut might even have made an advantage of this. I can imagine that to have noxious bacteria approaching too close to the endothelium could be undesirable. Clearly in certain circumstances it is undesirable. There might then be more beneficial associations between certain 'benign' bacterial lines and the surface of eukaryotic cells. There may even be a strictly symbiotic association. So the second source of any lack of correlation may be that some bindings are beneficial and, as it were, inhibitory to pathogenic effects.

Thirdly, there is some evidence that ligand binding is not the only thing that occurs. Dr Chabanon in our laboratory has been looking at certain *E. coli* strains. He finds that certain mannose-resistant *E. coli* with no pili seem to provoke certain types of cell to produce finger-like processes which envelop the bacteria (see Fig. 1). If we stain with ruthenium red, we find evidence of a glycocalyx lying between the bacteria and the processes (Fig. 2). So ligand adhesion is certainly one mechanism but there are probably a large number of other types of biochemical interactions. And on top of this I think there are various phenomena which stimulate the eukaryotic cells to alter the liquidity and the conformation of their membranes. We mustn't be too narrow-minded in thinking about the basis of adhesion.

Taylor-Robinson: What you say reminds me of some of the electron micrographs we took of Fallopian tube organ cultures experimentally infected with gonococci. There was undoubtedly a disturbance of the epithelial cell membrane which appeared to form little cups to partially surround the organisms (Taylor-Robinson et al 1974).

Richmond: What makes me feel that the phenomenon is significant is that only certain bacterial lines do it.

Svanborg Edén: A specific antibody that interferes with adherence is in the long run most interesting from the point of view of making a vaccine, whereas in the natural situation of defence against disease, as Professor Freter mentioned, any antibody directed to the surface with the capacity of agglutinating the organism or blocking transport to the host tissue is likely to keep it from the surface. For cholera a synergistically protective effect of antibodies to the lipopolysaccharide and the toxin has been shown (Svennerholm & Holmgren 1976).

Mirelman: We must differentiate between mucosa and mucus in the baby and in the older animal. There is a very high incidence of neonatal diarrhoea and a baby can be much more easily infected than an adult animal.

Together with Dr Thaler and the late Dr Hirschberger I did some studies (Hirschberger et al 1977) on whether there was any difference between the affinity of binding of a mannose-positive *E. coli* to the intestinal mucosa of newborn and older animals. These experiments were done in inverted loops as well as with scraped mucosal cells from different parts of the intestine. A mannose-sensitive strain of *E. coli* bound with very high affinity (by a mannose-sensitive mechanism) to the intestine of baby rats and rabbits. In the older animals the same *E. coli* cells had a lower binding capacity and the binding was not mannose-sensitive. Obviously

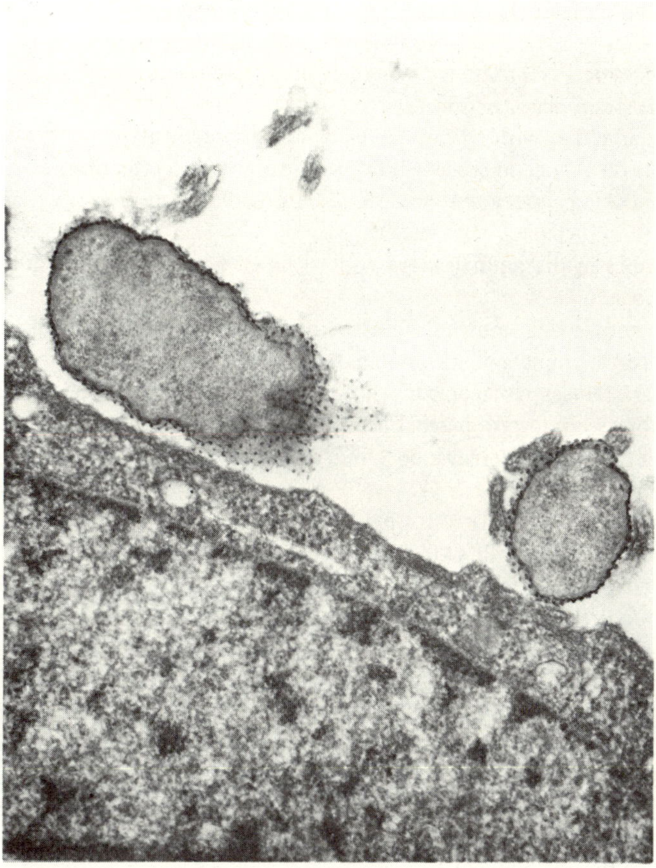

FIG. 2 (Richmond). Section stained with ruthenium red, showing the glycocalyx surrounding adhering bacteria. (Photograph provided by Dr G. Chabanon and Mrs P. Stirling.)

we are observing a change in the adherence mechanism that differs with age of the mammal. Is any difference known between the quantity, structure and consistency of intestinal mucus in animals of different ages? This might explain the differences Professor Freter mentioned here.

Hughes: The most striking difference is the increase in disulphide cross-linking as respiratory mucus matures in older animals.

Levine: A paper by Runnels et al (1980) shows that with mannose-resistant K99 pili adhesion to intestinal cells of mice, calves and pigs is related to age of the donor. Pigs and calves are the normal hosts of this pathogen. With mucosal cells from the guts of very young animals, K99-piliated bacteria would stick, whereas they wouldn't stick to cells from older individuals. As a control they used *E. coli* strain 123 which has type 1 pili. This strain showed little adhesion to cells of young or old animals. This suggests that with type 1 pili there is no age-related difference.

Mirelman: Was that strain mannose-sensitive?

Levine: I would assume that with a type 1 pilus it is mannose-sensitive.

Sharon: Was there a difference in susceptibility of the animals to infection?

Levine: There is a striking age-related susceptibility to K99, much more so than to K88.

Mirelman: Our studies in the animals were with strain O111, which is known to cause intestinal diarrhoea.

Feizi: There are of course well known developmental changes in the carbohydrate moieties of mucins, for example colonic mucins. It has been shown by Szulman (1964) that in human fetuses of 'secretor' type, mucins produced by epithelial cells of the colon express blood group ABH activities. At about the time of birth there occurs a change in distal colon: blood group activities are no longer detectable in the mucins they secrete.

Hughes: That is on the very early time-scale, isn't it?

Feizi: There is much susceptibility to enteritis in the neonate. It would be interesting to investigate whether the bacteria involved show preferential reactivity with carbohydrate structures in neonatal intestine.

Levine: In relation to the differences in host cell susceptibility to *V. cholerae*, there are epidemiological data that show fairly convincingly that in the Philippines (Barua & Paguio 1977) and Bengal (Chaudhuri & De 1977) individuals of blood group O more commonly get admitted to hospital with cholera than those of other groups. There are also data supporting a relationship between blood group O and an increased severity of cholera, showing that this is not related to any known HLA-A, B or C allotype (Levine et al 1979). Blood group antigens are secreted in the gastrointestinal tract in adults in small quantities.

This increased severity with blood group O raises several possibilities. One is that there is an enhanced susceptibility to the effect of toxin, which in theory seems unlikely. Then there is the possibility of these factors being there or not being there in the mucus gel, or of them being on the surface of the mucosa.

Sharon: In O blood group determinants fucose may be more accessible.

Freter: Our stereomicroscopic observations of the gut of baby animals and adults suggest that the intestinal flora has a considerable effect. The mucus gel in baby animals is similar to that in germ-free animals and is much thicker than in conventional adult animals. Conventional animals have thicker layers of mucus gel in the upper small intestine (where there is little indigenous flora) than in the large intestine. Certainly, the indigenous flora of the large intestine is very beneficial: it is the most important defence mechanism against infection that we have. The anaerobes in this flora are largely responsible for the breakdown of mucopolysaccharides, which may explain the relative scarcity of mucus in regions of the gut that harbour an indigenous microflora. The indigenous bacteria form thick layers on the gut wall and, obviously, only the first layer can adhere to the epithelial cells. Therefore, initial adhesion of pathogenic invaders in the lower ileum and in the large intestine may be predominantly to the indigenous bacteria rather than to epithelial cells. These various effects of the indigenous microflora are important reasons why the susceptibility to intestinal pathogens of baby animals which do not yet have an adult type of indigenous microflora is different from that of adults.

REFERENCES

Barua D, Paguio AS 1977 ABO blood groups and cholera. Ann Hum Biol 4:489-492

Bhattacharjee JW, Srivastava BS 1978 Mannose-sensitive haemagglutinins in adherence of *Vibrio cholerae* eltor to intestine. J Gen Microbiol 107:407-410

Carruthers MM 1977 In vitro adherence of Kanagawa-positive *Vibrio parahaemolyticus* to epithelial cells. J Infect Dis 136:588-592

Chaudhuri A, De S 1977 Cholera and blood groups. Lancet 2:404

Freedman LR 1967 Experimental polynephritis. XIII. On the ability of water diuresis to induce susceptibility to *E. coli* bacteriuria in the normal rat. Yale J Med 39:255-266

Freeman RR, Trejdosiewicz AJ, Cross GAM 1980 Protective monoclonal antibodies recognizing stage-specific merozoite antigens of a rodent malatia parasite. Nature (Lond) 284:366-368

Hirschberger M, Thalee MM, Mirelman D 1977 Mechanisms of attachment by a pathogenic strain of *E. coli* (O111/B4) to intestinal mucosa in pre and postweanling rats. Pediatr Res 11:500

Jones, GW, Freter R 1976 Adhesive properties of *Vibrio cholerae.* Infect Immun 14:240-245

Levine MM, Nalin DR, Rennels MB, Hornick RB, Sotman SS, Van Blerk G, Hughes TP, O'Donnell S 1979 Genetic susceptibility to cholera. Ann Hum Biol 6:369-374

Razin S 1978 The mycoplasmas. Microbiol Rev 42:44-470

Runnels PL, Moon HW, Schneider RW 1980 Development of resistance with host age to adhesion of K99$^+$ *Escherichia coli* to isolated intestinal epithelial cells. Infect Immun 28:298-300

Svennerholm A-M, Holmgren J 1976 Synergistic protective effect in rabbits of immunization with *Vibrio cholerae* lipopolysaccharide and toxin/toxoid. Infect Immun 13:735-740

Szulman AE 1964 The histological distribution of the blood group substances in man as disclosed by immunofluorescence. J Exp Med 119:503-523

Taylor-Robinson D, Whytock S, Green CJ, Carney FE, Jr 1974 Effect of *Neisseria gonorrhoeae* on human and rabbit oviducts. Br J Vener Dis 50:279-288

The mechanism of entry of viruses into plant protoplasts

J. W. WATTS, J. R. O. DAWSON and JANET M. KING

John Innes Institute, Colney Lane, Norwich NR4 7UH, UK

Abstract Plant protoplasts may be efficiently infected with viruses. Attachment to the plasmalemma is a critical step in infection; positively charged viruses readily adhere and infect but negatively charged viruses require the presence of a polycation (e.g. poly-L-ornithine) before infection can occur. The role of the polycation appears to be twofold: to modify the charge of the virus and to damage the surface of the protoplast. Virus probably enters the protoplast during repair of damaged regions of the plasmalemma. Pinocytosis does not appear to play a significant part in inoculation.

The plant cell consists of the living protoplast surrounded by a rather rigid wall. The wall is typically composed of cellulose and related polymers and in plant tissues the cells are cemented together with pectin and related substances. This is of course a gross oversimplification but it is adequate for our purposes. The living contents, the protoplast, may be visualized as the equivalent of an animal cell although there are considerable structural differences and the composition of the plasmalemma differs from that of an animal cell. Further, since most plant cells are highly vacuolate, the protoplast is almost invariably spherical if freed from the restraint of the wall and is stable only in a medium of relatively high osmolarity (equivalent to 0.4–0.5M-sucrose). Protoplasts can be handled by some of the techniques commonly used in animal tissue culture.

Cocking & Pojnar (1969) showed that it was possible to infect tomato protoplasts with tobacco mosaic virus (TMV). Their techniques allowed them to make electron microscopic studies but were unsuitable for biochemical studies. Soon afterwards Takebe and his collaborators, who had already developed methods for producing large quantities of protoplasts from tobacco leaves, showed that they could be efficiently and synchronously infected with TMV (Takebe & Otsuki 1969).

1981 Adhesion and microorganism pathogenicity. Pitman Medical, Tunbridge Wells (Ciba Foundation symposium 80) p 56-71

It was then possible, therefore, to study the process of infection with some precision and with the advantage that, compared with animal viruses, the yields of virus were high, e.g. 10^6–10^8 particles per cell.

Since the work discussed here has been done almost entirely with protoplasts it may be useful if we outline the methods of preparation (Takebe & Otsuki 1969). The following method applies to tobacco protoplasts but is of general use. The lower epidermis is removed from a tobacco leaf with forceps. The leaf is then cut into pieces about 2 cm square and shaken with a solution of crude cellulose in 0.7 M-mannitol. Individual cells are released over a period of about 45 min, collected by centrifugation, washed and then shaken with a solution of crude pectinase in 0.7 M-mannitol. The wall dissolves and releases the protoplasts, which are collected and washed as before. Protoplasts are inoculated in 0.7 M-mannitol containing 10 mM-citrate buffer (pH 5.2), virus (1–5 μg/ml of inoculum) and poly-L-ornithine (1 μg/ml) for 10 min.

When Takebe & Otsuki (1969) published their original studies on infection of tobacco protoplasts they observed that TMV would not infect protoplasts unless a polycation of high molecular weight was present. The most effective of those they examined was poly-L-ornithine. The effect of poly-L-ornithine was clear-cut; it allowed almost 100% infection when used at a concentration of 1–2 μg/ml in the inoculum. Since poly-L-ornithine was essential for infection there was virtually no secondary infection by any virus appearing in the culture medium after the polycation had been removed, and the course of infection was synchronous. The stimulatory effect of poly-L-ornithine varied with molecular mass and the best results were obtained with material of relative molecular mass about 10^5. Poly-L-ornithine stimulates pinocytosis in animal cells (Ryser 1967) and although there is little evidence for this process in plant cells (Cram 1980) it was concluded that virus was entering the protoplasts by pinocytosis (Takebe & Otsuki 1969).

Most of our own studies with viruses in protoplasts have used brome mosaic virus (BMV) and cowpea chlorotic mottle virus (CCMV) in tobacco protoplasts (Motoyoshi et al 1973a, 1974a, b). These viruses are closely related but have rather different host ranges (Lane 1974). They have a divided genome distributed among three almost identical particles which can be distinguished by density in caesium chloride gradients. There are four different species of RNA of which only the three heaviest are essential for infection; the lightest RNA appears to be the messenger for coat protein but the gene for coat protein is on one of the heavier RNA molecules (RNA 3). Initial attempts to infect protoplasts of legumes (cowpea, pea, broadbean) with CCMV were unsuccessful but the virus grew well in tobacco protoplasts. A polycation was essential for infection with CCMV and it was observed that there was interaction between the virus and polycation so that at concentrations around 100 μg/ml they co-precipitated (Motoyoshi et al 1973a). It is commonly observed that for maximum stimulation to be produced, virus and polycation must be preincubated for about 10 min before inoculation (Table 1); this indicates that a

TABLE 1 The effect of preincubation of poly-L-ornithine and cowpea chlorotic mottle virus on the subsequent inoculation of tobacco protoplasts

Duration of preincubation (min)	*% infected protoplasts*
2	2.2
5	27
10	28

Poly-L-ornithine and virus were used in concentrations of 1 and 0.5 μg/ml respectively.

complex interaction plays a part in the process. CCMV is negatively charged in the conditions used in inoculation (pH 5.2); the polycation has a positive charge so that some degree of interaction may be anticipated with the production of aggregates with modified surface charges. It seemed reasonable to assume that charge modification was connected with stimulation. The plasmalemma of the protoplasts has a negative charge which will repel the negatively charged CCMV particles; if the charge on the virus is reduced or its sign is changed, inoculation should occur more readily. If stimulation were due simply to pinocytosis it should be possible to add the polycation to the protoplasts and then add viruses, or vice versa; no combination of manipulations can in fact substitute for preincubation followed by inoculation. Fig. 1 is our picture of the mechanism of attachment.

Preincubation is characterized by another unusual feature that favours the idea that aggregation plays a vital role in infection. There is an optimal ratio of virus to polycation; with a fixed concentration of poly-L-ornithine, say 1 μg/ml, the percentage infection increases as the concentration of CCMV is increased up to about 10 μg/ml, when it begins to fall (Fig. 2).

If charge is a major consideration in inoculation – that is, in the attachment of the virus to the protoplast and its subsequent entry – viruses with positive charge might be expected to behave rather differently from TMV and CCMV. This proves to be the case; the positively charged BMV and pea enation mosaic virus (PEMV) both infect protoplasts in the absence of a polycation and it is a simple matter to show that they bind spontaneously and strongly to the protoplast surface (Motoyoshi et al 1974a, Motoyoshi & Hull 1974). Indeed, it is possible to use BMV as an agent for causing protoplasts to clump. BMV and CMV are closely related, the major physical differences being the protein composition of the coat and the consequent difference in net charge of the virus particles. It seems plausible therefore to conclude that charge is the factor responsible for the difference in response to added cation.

Although BMV and PEMV can infect in the absence of polycation, addition of poly-L-ornithine increases the percentage infection. The effect does not therefore seem to be solely due to charge effects.

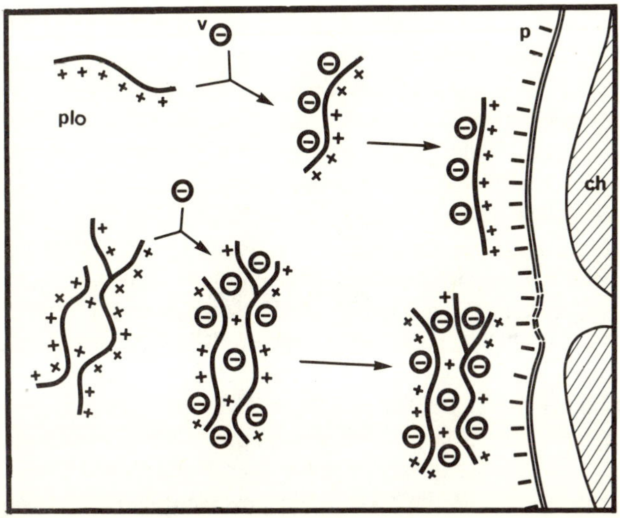

FIG. 1. Suggested scheme for interaction of negatively charged viruses and polycations permitting attachment to the plasmalemma. V: virus particles; plo: polycation; ch: chlorplast; p: plasmalemma.

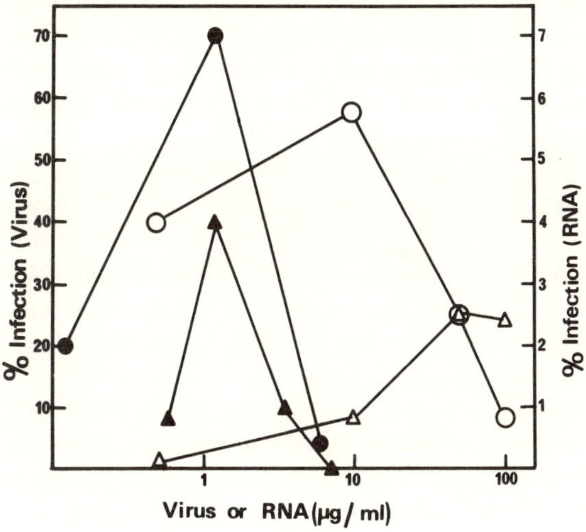

FIG. 2. Effect of different ratios of virus and RNA to poly-L-ornithine on infection of tobacco protoplasts. The inocula contained 1 μg poly-L-ornithine and the indicated concentrations of virus or RNA, ○: CCMV; △: BMV; ●: CCMV RNA; ▲: BMV RNA.

Protoplasts can be relatively easily inoculated with free viral RNA. The preferred method of inoculation is now by means of polyethylene glycol in the presence of calcium ions (Dawson et al 1978), but it is possible to obtain high levels of infection of protoplasts by using polycations (Beier & Bruning 1976, Dawson et al 1978). The requirement for polycation is similar for all types of viral RNA, whether derived from negatively or positively charged virus. What is of particular interest with RNA is the way in which the optimal ratio of polycation to RNA differs from that observed with the virus (Fig. 2). Of the bromoviruses both CCMV and BMV RNA show an optimal ratio of about 1.5 μg RNA to 1 μg poly-L-ornithine, compared to about 12:1 for CCMV and no stoichiometry for BMV (Motoyshi et al 1973a, 1974a). The amount of RNA is thus about 10% that of CCMV for maximum infectivity. The most obvious reason for this difference is charge: RNA is much more strongly negatively charged than CCMV, so more poly-L-ornithine is needed to modify the charge before RNA can attach to the protoplasts.

The use of protoplasts allows precise questions to be asked about the number of particles associated with an infection. With CCMV it was possible to inoculate with ^{14}C-labelled virus and poly-L-ornithine, exhaustively wash the protoplasts and then determine the number of virus particles per protoplast and the percentage infection. Since at least three particles, one of each type, are required for infection with CCMV, a minimum of about 10 particles per protoplast might be expected for infection to occur. In practice a minimum of about 10^3 particles was attached to the infected protoplasts (Motoyoshi et al 1973b). This reflects in part the low specific infectivity of CCMV but there is also the possibility that infection occurs by the uptake of large aggregates of virus and poly-L-ornithine. With BMV the picture was very different; large amounts of inoculum became attached to the protoplasts and exhaustive washing failed to displace the BMV (Motoyoshi et al 1974a). The published data show that the amount of virus attached initially was of the same order as that recovered from the infected protoplasts after 48 h in culture. There can be little doubt, therefore, that charge is the principal factor controlling attachment of the virus or its RNA to the surface of the protoplast.

The initial observations on the interaction of polycations and viruses have been extended and amplified by several workers (Takebe 1977). One of the more interesting observations concerns the role of the buffer during inoculation. The bromoviruses and TMV were inoculated in a medium buffered with sodium citrate but it was later shown that for some viruses, notably tobacco rattle virus and TMV, the specific infectivity of the virus was much higher if the medium was buffered with phosphate (Takebe 1977). Other workers have shown that tomato protoplasts are best infected by TMV in a medium buffered with Tris (Motoyoshi & Oshima 1976). Evidence has been presented to support the view that the buffer itself may become involved in the aggregation that occurs during preincubation of the inoculum (Mayo et al 1979). The phenomenon is not universal, however, and the use of other buffers has given no significant improvements with the bromoviruses.

Although the requirement for polycation has been presented so far as being almost entirely related to charge, some exceptions may occur (Takebe 1977). Cucumber mosaic virus and cowpea mosaic virus both infect cowpea protoplasts in the absence of poly-L-ornithine, although polycation is normally included during inoculation to improve reproducibility of infection. Both viruses should be negatively charged at the pH used in inoculation but their isoionic points are close enough to that of the inoculum buffer to suggest that the net charge is not sufficient to prevent these viruses approaching the cowpea protoplasts. More work is needed on these exceptions.

As already mentioned, poly-L-ornithine improves the efficiency of inoculation with positively charged viruses, although it is not essential. It presumably assists entrance of the virus into the protoplast, but in what way? It has been suggested that pinocytosis is responsible for the transfer of virus into the protoplast (Takebe 1977). Several studies have shown that particles of widely differing sizes, including viruses, bacteria and chloroplasts, can be taken up by protoplasts but it is not clear how this occurs. There is little or no evidence to suggest that pinocytosis is an important process in intact plant cells and a recent review has argued that it would be impossible for pinocytosis, or rather endocytosis, to play any significant role (Cram 1980). The wall is too rigid a barrier. Once the cell wall is removed, however, that constraint disappears and pinocytosis might be thermodynamically possible. Furthermore, only a small amount of pinocytosis would be necessary to account for infection by viruses and several workers have published studies with the electron microscope to support this idea (Cocking & Pojnar 1969, Honda et al 1974). Burgess et al (1973a,b) consider, however, that the evidence obtained by electron microscopy offers no real support for pinocytosis, certainly not the type observed in animal cells (Silverstein et al 1977). Burgess et al (1973b) suggested that the presence of poly-L-ornithine produces characteristic lesions that are concerned in the uptake of virus and it is during either the formation or the repair of these that virus passes into the protoplast. Beyond this the problem cannot be resolved by electron microscopy.

The argument in favour of pinocytosis would be more attractive if the response to poly-L-ornithine were less absolute. The positively charged viruses infect readily without a polycation so the mechanism for entry (the putative pinocytosis) occurs spontaneously – unless of course we assume that positively charged viruses stimulate pinocytosis. Why then is there absolutely no infection with TMV and CCMV in the absence of a polycation? If entrance is through pinocytic vesicles we should expect some virus to enter whatever the charge, since the medium typically contains of the order of 10^{11} particles/ml during inoculation. There is fortunately some evidence of a more biochemical nature which will now be considered.

No discussion of the mechanisms of infection of protoplasts would be complete without some reference to the way in which inoculation is influenced by the physiological condition of the plant from which the protoplasts were prepared.

Stable protoplasts – that is, protoplasts that do not disintegrate during or shortly after isolation – can be prepared only from plants grown under appropriate conditions, typically for tobacco a 15-h day, 10^4 lux illumination, temperature 22–25 °C. In addition to these requirements, however, there are other, poorly defined, factors which determine whether the protoplasts are susceptible to infection. In one series of experiments (Motoyoshi et al 1974b) in which leaves were taken at different stages of growth it was found that susceptible (infectible) protoplasts could be prepared from only one leaf, used over a restricted time period. This problem of reproducibly obtaining and infecting protoplasts is so acute that some workers have abandoned their use in virology.

Once the protoplasts have been prepared, however, it is possible to handle them so that susceptibility to infection is increased. Certain treatments that may be expected to damage the plasmalemma increase susceptibility to infection. Repeated centrifuging illustrates this effect. Takebe (1975) discusses the way cultures of protoplasts rapidly accumulate substances that inhibit inoculation. The recommended procedure is that protoplasts should be centrifuged and resuspended in fresh medium immediately before inoculation. Even a delay of a few minutes may be sufficient to allow the release of enough inhibitor to reduce the percentage infection considerably. Takebe (1975) also reports that the medium out of which the protoplasts have been centrifuged is able to inhibit inoculation. The interfering substances may be compounds such as polyphenols that are secreted by damaged cells, and the amount of damage will be related to the physiological condition of the cells in the original plant. It is a common experience that stable protoplasts can be subjected to remarkable treatments without visible damage whereas less robust protoplasts show browning and disruption even during isolation. The inhibitory effect of these substances appears to operate by absorption on the protoplast surface. If instead of a simple inoculation the protoplasts are inoculated, immediately spun down and then resuspended in the supernatant fluid, the level of infection can be significantly increased; for example, in one experiment the infection increased from 14 to 22% (Motoyoshi et al 1974b). This type of result suggests that the process of repeated centrifugation and resuspension somehow activates the protoplasts. In another experiment an inoculum containing suboptimal concentrations of virus and polycation was used and the protoplasts were treated with successive portions of fresh inoculum. Infection increased almost linearly and the end-result was twice the infection produced by a similar amount of virus and polycation used in a single inoculation (Motoyoshi et al 1974b). One explanation of these results is that centrifuging and resuspension disturb an inhibitory surface layer that obscures sites of entry. If the ideas of Burgess et al (1973a, b) on the role of damage in infection are correct, however, the process may also result in fresh damage, so providing more sites for entry.

Pinocytosis in animal cells is energy-dependent (Silverstein et al 1977) and is depressed by low temperatures. However, when protoplasts are inoculated either

TABLE 2 The effect of temperature during inoculation on the infection of tobacco protoplasts with cowpea chlorotic mottle virus

Expt no.	*Temperature during inoculation (°C)*	*Period at 0 °C after inoculation (h)*	*% protoplasts infected*
1	25	0	23
	0	3	17
2	25	0	52
	0	2	43
	0	6	40

with virus or viral RNA at 0 °C there is virtually the same percentage infection as at 25 °C (Table 2). With RNA, which is subject to rapid degradation by nucleases at room temperature, inoculation at low temperatures has some advantages and has become a normal condition (Aoki & Takebe 1969, Watts et al 1975, Motoyoshi & Oshima 1979). The limiting step in inoculation therefore seems to be energy-independent. Pinocytosis is inhibited at low temperatures but particles adsorbed to the surface of animal cells at low temperature are engulfed pinocytically when the temperature is raised (Silverstein et al 1977). It might be argued that this is also the case with protoplasts and viruses, but the behaviour of viral RNA suggests otherwise. If RNA were only adsorbed to the surface and not taken into the protoplast, it would still be exposed to nuclease activity; there would then be no advantage in inoculating at low temperature. It seems probable, however, that even at low temperatures viral RNA passes readily into the cytoplasm. Further, unless the pinocytic vesicles were very short-lived they would be unlikely to be the vehicle of entry of viral RNA because again degradation would continue in the vesicle.

Sodium azide is an efficient inhibitor of the metabolism of protoplasts; for example, 10^{-4} M-azide rapidly and reversibly inhibits protein synthesis by over 99%. Azide prevents the development of infection but does not prevent successful inoculation of protoplasts when applied before, during or after inoculation (Table 3). Protoplasts that have been inoculated and stored for many hours in medium containing azide proceed to develop normal infections when azide is removed (Motoyoshi et al 1974b). Again the inference is that the limiting stage in inoculation is energy-independent, and that means that pinocytosis is unlikely to play an important role in the mechanism of infection.

The effect of poly-L-ornithine in stimulating infection has been an important part of the evidence in favour of a pinocytic mechanism of infection. Poly-L-ornithine is known to stimulate pinocytosis in animal cells and several other polycations have been found to have a similar effect (Ryser 1967). It is characteristic of this phenomenon that it shows biological specificity and the responses to D and

TABLE 3 The effect of sodium azide on infection of tobacco protoplasts with cowpea chlorotic mottle virus (data from Motoyoshi et al 1974b)

Azide treatment (0.1 mmol/l)	*% infected protoplasts*
None (control)	59
Azide during inoculation	59
Azide for 10 min before inoculation	66
Azide for 10 min before and 10 min during inoculation	52
Azide after inoculation for (h)	
0 (control)	79
6.5	59
14.5	44
21.5	37

TABLE 4 Effect of different polycations on infection of tobacco protoplasts with cowpea chlorotic mottle virus

Polycation	*M_r ($\times 10^{-3}$)*	*% infected protoplasts*	*Increase in pinocytosis in animal cells (control = 1)*
Poly-L-ornithine	120	80	15
Poly-D-lysine	70	75	5
Poly-L-lysine	50	56	2
Poly-L-arginine	55	45	—
DEAE dextran	500	3	10

Polycations were used at 1 μg/ml during inoculation. The data for animal cells are calculated from figures in Ryser (1967); other data from Motoyoshi et al (1974b).

L forms of the same polymer and to different polymers are very different. This does not appear to be the case with virus infection of protoplasts. A range of polymers, including poly-D- and poly-L-lysine, gave results that did not differ significantly from those obtained with poly-L-ornithine (Motoyoshi et al 1974b). If pinocytosis occurs it must be a very different phenomenon from that observed in animal cells (Table 4).

We may summarize the evidence as follows. Three factors can be distinguished as critical to successful inoculation:

(1) The physiological condition of the protoplast; this in turn depends on the condition of the original plant from which the protoplasts were prepared.

(2) The approach and attachment of virus to the protoplast, which depends on the charge on the virus.

(3) The movement of the virus from outside to inside the plasmalemma, which seems to occur during repair of damaged areas of the surface of the protoplast.

REFERENCES

Aoki S, Takebe I 1969 Infection of tobacco mesophyll protoplasts by tobacco mosaic virus ribonucleic acid. Virology 39:439-448

Beier H, Bruning G 1976 Factors influencing the infection of cowpea protoplasts by cowpea mosaic virus. Virology 72:363-369

Burgess J, Motoyoshi F, Fleming EN 1973a Effect of poly-L-ornithine on isolated tobacco mesophyl protoplasts: evidence against stimulated pinocytosis. Planta (Berl)111:199-208

Burgess J, Motoyoshi F, Fleming EN 1973b The mechanism of infection of plant protoplasts by viruses. Planta (Berl) 112:323-332

Cocking EC, Pojnar E 1969 An electron microscopic study of the infection of isolated tomato fruit protoplasts by tobacco mosaic virus. J Gen Virol 4:305-312

Cram WJ 1980 Pinocytosis in plants. New Phytol 84:1-17

Dawson JRO, Dickerson PE, King JM, Sakai F, Trim ARH, Watts JW 1978 Improved methods for infection of plant protoplasts with viral ribonucleic acid. Z Naturforsch Sect C Biosci 33:548-551

Honda Y, Matsui C, Otsuki Y, Takebe I 1974 Ultrastructure of tobacco mesophyll protoplasts inoculated with cucumber mosaic virus. Phytopathology 64:30-34

Lane L 1974 The bromoviruses. Adv Virus Res 19:151-220

Mayo MA, Roberts IM 1979 Some effects of buffers on the infectivity and appearance of virus inocula used for tobacco protoplasts. J Gen Virol 44:691-698

Motoyoshi F, Hull R 1974 The infection of tobacco protoplasts with pea enation mosaic virus. J Gen Virol 24:89-99

Motoyoshi F, Oshima N 1976 The use of tris-HCl buffer for inoculation of tomato protoplasts with tobacco mosaic virus. J Gen Virol 32:311-314

Motoyoshi F, Oshima N 1979 Standardization in inoculation procedure and effect of a resistance gene on infection of tomato protoplasts with tobacco mosaic virus RNA. J Gen Virol 44: 801-806

Motoyoshi F, Bancroft JB, Watts JW, Burgess J 1973a The infection of tobacco protoplasts with cowpea chlorotic mottle virus and its RNA. J Gen Virol 20:177-193

Motoyoshi F, Bancroft JB, Watts JW 1973b A direct estimate of the number of cowpea chlorotic mottle virus particles absorbed by tobacco protoplasts that become infected. J Gen Virol 21:159-161

Motoyoshi F, Bancroft JB, Watts JW 1974a The infection of tobacco protoplasts with a variant of brome mosaic virus. J Gen Virol 25:31-36

Motoyoshi F, Watts JW, Bancroft JB 1974b Factors influencing the infection of tobacco protoplasts by cowpea chlorotic mottle virus. J Gen Virol 25:245-256

Ryser H J-P 1967 A membrane effect of basic polymers dependent on molecular size. Nature (Lond) 215:934-936

Silverstein SC, Steinman RM, Cohn ZA 1977 Endocytosis. Annu Rev Biochem 46:669-722

Takebe I 1975 The use of protoplasts in plant virology. Annu Rev Phytopathol 13:105-125

Takebe I 1977 Protoplasts in the study of plant virus replication. In: Fraenkel-Conrat H, Wagner RR (eds) Comprehensive virology. Plenum Press, New York, vol 11:237-283

Takebe I, Otsuki Y 1969 Infection of tobacco mesophyll protoplasts by tobacco mosaic virus. Proc Natl Acad Sci USA 64:843-848

Watts JW, Cooper D, King JM 1975 Plant protoplasts in transformation studies; some practical considerations. In: Markham R et al (eds) Modification of the information content of plant cells. North-Holland, Amsterdam (Proc 2nd John Innes Symp) p 119-131

DISCUSSION

Pearce: Do you centrifuge protoplasts first and then add virus?

Watts: Yes. We have found, however, that repeated centrifuging in the presence of virus and polycation will further increase infection.

Pearce: For chlamydiae the effects of centrifugation on cell cultures are fully reversible. At the end of centrifugation cells appear to return to normal (Allan & Pearce 1979). The requirement for repeated centrifuging with protoplasts supports the idea that damage is necessary for susceptibility.

Taylor-Robinson: Does polycation damage help chlamydial entry?

Pearce: Polycation such as DEAE–dextran, when used to pretreat cell monolayers, enhance infection by poorly infective *Chlamydia trachomatis* strains by increasing attachment (Kuo & Grayston 1976). They do not enhance infection by the more highly infective lymphogranuloma venereum *C. trachomatis* strains. Yet all of these strains and the cell surface have an overall net negative charge (Kraaipoel & Van Duin 1979). So possibly enhancement by polycations results from neutralization of local repulsive charge in the vicinity of receptors.

Howard: What is the mechanism of viral attachment and entry in the natural situation when there is a cell wall, Dr Watts? Is there perhaps a polycation equivalent in the cell wall and you are substituting for that? Is there specificity for binding to the cell walls of some plants and not others?

Watts: About 90% of viruses enter plants by mechanical inoculation, for example, injection by a feeding aphid. The usual method of inoculation in the laboratory is by abrasion of the leaf with a mixture of carborundum dust and virus solution. Entry of the virus is very fast and it is probably during the healing process as the abrasions seal that virus is somehow incorporated. If poly-L-ornithine is also present during inoculation infection may be increased (Shaw 1972). In contrast, work with polyanions shows that these seem to protect against infection (Stein & Loebenstein 1972), so maybe there are some polycations naturally in the damaged area.

Silverblatt: Have you looked at polymers that have secondary or other kinds of amines? I am thinking of aminoglycoside antibiotics, which are used by farmers to combat infections in plants due to *Erwinia* and other bacterial pathogens. In animal cells the polybasic antibiotics bind to pinocytic receptors to stimulate pinocytosis (Silverblatt & Kuehn 1979).

Watts: There doesn't seem to be any real evidence for pinocytosis in plant cells, certainly in terms of endocytosis. There is no reason why one should expect pinocytosis to occur. Cram (1980) recently argued that thermodynamically it is not possible. Pinocytosis may be a factor in infection of protoplasts but we have no evidence from electron microscopy of anything that looks like a pinocytic vesicle. We have no evidence even that when we see virus taken in, that corresponds to an infection. We really know very little, largely I suspect because having got the virus in we were more interested in what happened subsequently.

Choppin: The main difference between the initiation of infection by animal viruses and plant viruses is perhaps that the plant virus is usually introduced by insects. The plant virus therefore doesn't need a mechanism of its own to negotiate the cell wall, whereas animal viruses have to be equipped with a specific absorption and penetration mechanism. However, the parallel is there; that is there are instances in which one can demonstrate increased uptake of certain animal viruses by polycations. Most animal viruses appear to be negatively charged. In the haemagglutination reaction, which has been well studied, cations must be present for haemagglutination to occur. The virus is negatively charged, the cell is negatively charged, and if there are no cations in the system the virus can't get close enough for the specific reaction between the reactive sites on the viral protein and on the cell to occur. Therefore cations must be present but in myxovirus haemagglutination the cation can be sodium; divalent cations are not required.

On the other hand the RNA tumour virus appears to resemble the system you are dealing with, Dr Watts. There are some mutants or variants of this virus which lack or are deficient in the surface glycoprotein, which is normally involved in the absorption and penetration of the virus into the cell. One can dramatically increase the infectivity of that type of RNA tumour virus by adding DEAE–dextran, whereas the wild-type RNA virus which has a normal complement of glycoprotein is not significantly enhanced by the addition of DEAE–dextran. Adding a polycation virus appears to promote the attachment of the mutant to the cell surface, whereas the wild-type virus has a reactive site on the glycoprotein to accomplish that.

Watts: This is what we seem to lack. DEAE–dextran is the least effective of all the polycations. If we add calcium, for example, to the inoculum we get no infection, presumably because the calcium ions compete with the polycation in the aggregates and do not allow the virus to approach the membrane. This is highly speculative but we have looked at a lot of these ionic effects and all fit this model.

Taylor-Robinson: What has been said about some viruses being negatively charged and being repelled by a negatively charged cell is also true for some bacteria. Heckels et al (1976) considered that the pili on gonococci were required to overcome the barrier created by the negative charge on the cell surface and the negative charge on the bacteria. When polycations were used in the system the pili weren't needed for the organisms to attach to the cells.

Friend: Does the infectivity of viruses for protoplasts alter according to the age of the protoplasts? When protoplasts are regenerating new cell walls are there differences in charge on the outside of the protoplast?

Watts: The susceptibility of protoplasts to infection decreases with time but this does not necessarily correlate with the production of cell wall; for example, susceptibility to BMV falls faster than that to CCMV.

Razin: Can calcium or magnesium ions fulfil the role of the polycations?

Watts: No, they are inhibitory. It is not a covalent association. At millimolar concentrations these ions compete with the polycation and I think they dissociate the virus–polycation aggregate.

Razin: Does calcium or magnesium alone, without a polycation, allow infection?

Watts: No. Sarkar et al (1974) reported that high pH allows infection in the presence of Mg^{2+} but these conditions create charge effects so that even strongly polyanionic molecules such as RNA can approach the plasmalemma. Another method uses polyethylene glycol and Ca^{2+} which cause aggregation of the virus and disturbance of the plasmalemma so that virus or RNA readily infect (Dawson et al 1978).

Candy: A requirement for divalent cations for firm adhesion has also been proposed for *Pseudomonas aeruginosa* and for *Klebsiella pneumoniae* (Johanson et al 1979). We don't have divalent cations in our *in vitro* assay system. We have found that inclusion of EDTA up to a concentration of 12 mM in the incubation medium does not markedly affect adhesion of a strain of *E. coli* to buccal epithelial cells. Has anybody else studying bacterial adhesion used divalent cations in the medium?

Tramont: Divalent cations improve the attachment of gonococci. I think what is going to develop is a unifying theme because you are dealing with the same system. There is a negative charge on the cell and as far as I know most bacteria have a negative charge.

Candy: Would monovalent cations have the same effect?

Tramont: They don't in gonococci and I don't think they do in other bacteria either.

Svanborg Edén: EDTA does not affect binding in our system. We have to add monovalent cations or anions up to a very high concentration before we see any effect on binding.

Razin: We see no significant effect of EDTA on *M. gallisepticum* attachment to glass (Kahane et al 1979).

Beachey: The same is true for streptococci.

Vosbeck: In our system, where *E. coli* sticks to tissue culture cell lines, adhesion increases greatly if we add divalent cations (calcium). But when we determine the kinetics we find that the binding is no longer linear with time. It goes up right away and then stays constant. Under the microscope we can see that this is due to clumping of the bacteria. So one has to be careful whether an apparent increase of binding is a true increase in adhesion or simply bacterial clumping.

Helenius: I would like to bring up a general point which applies to both plant and animal viruses. In infected cells where virus is being produced there is frequently a high concentration of virus particles or nucleocapsids. These particles are not being uncoated. On the other hand the virus particles that enter a new host cell from outside are efficiently uncoated. The newly synthesized virus or virus nucleocapsid must be different from the incoming virus. Could polycations have an effect on the uncoating of the virus particles?

Watts: Free viral RNA, which can infect, is already uncoated and the polycation has substantially the same effect on it as on intact negatively charged viruses (see Fig. 2, p 59). I don't think the virus is different but the intracellular environment has changed. With some of the plant viruses it is very obvious that something like specific organelles are set up – the chloroplasts may be modified – to produce an environment in which RNA and protein for the virus are brought into close association. The viruses that I have described reconstitute themselves *in vitro* without any difficulty. They are self-assembling.

We are now looking at mixed infections with BMV and CCMV. Inside the cell the virus occupies certain sites; once infection is established all these are occupied. Viruses that enter subsequently and which require the same specific sites are thus unable to infect.

Pearce: Perhaps one shouldn't press parallels between mammalian cells and protoplasts too far. However, Stephen and colleagues (Wolstenholme et al 1977) in testing virus cytopathic factors on cell cultures observed a very nice correlation between the cell vesiculation response to magnesium sulphate and cell susceptibility to cytopathic factors added in the presence of magnesium sulphate. It appeared likely that the cell lines most sensitive to cytopathic factors were those showing most marked vesiculation. Does treatment of protoplasts with poly-L-ornithine lead to a vesiculation response?

Watts: There is a controversy here. There are those who say it is endocytosis (pinocytosis) that is causing the effect. Burgess et al (1973a, b) conclude that the electron microscopic evidence is against pinocytosis, at least of the type shown by animal cells, and our studies of the energy requirements and the effects of polycations support this view. Polycations are very toxic: when poly-L-ornithine is put in, the plasmalemma literally blows bubbles and the protoplast begins to fall to pieces. Burgess thinks that these bubbles are the sites of infection. Personally I am more agnostic and take the view that the blebs correlate with infection but may not be where the virus enters.

Choppin: The aggregation phenomenon applies not only to CCMV but also to TMV. I can see why you would have to aggregate something with a multipartite genome, but it is not clear to me why there is an advantage in the aggregation of TMV.

Watts: I don't think there is. In fact the specific infectivity of TMV in this sort of set-up is two orders of magnitude better than that of CCMV. It doesn't need more than one or two particles, on average, for infection, but it does need charge modification. I think that is the critical thing with TMV.

Choppin: If I understand correctly, you are not postulating that aggregation of TMV is needed, just charge modification. In most animal virus systems infectivity goes down, not up, when there is aggregation. I was wondering whether there was some specific need in TMV for aggregation, and the answer is apparently no.

Watts: I am saying that, but all those experiments were in a citrate-buffered mannitol. If we use a phosphate-buffered mannitol with TMV in infected tobacco protoplasts some sort of aggregation occurs that enhances the infectivity of TMV enormously. This is observed even when we inoculate leaves. I don't know the cause of that. When we work with tomato protoplasts a completely different buffer (Tris) is used to get a similar sort of effect with TMV. I don't know what is going on.

Choppin: Maybe when you have to go through a wound there is some mechanical advantage in having a clump.

Taylor-Robinson: You mentioned that about 1000 CCMV particles attached but what was the number of BMV?

Watts: For CCMV the number of attached particles was 1000 per protoplast. That is two orders of magnitude higher than one might expect but then CCMV has a low specific infectivity anyway. BMV just confuses the issue, literally. The number of particles attaching is uncountable. We can have more virus (BMV) attached to the protoplast at the start than we have at the end, in terms of infectivity.

Taylor-Robinson: You mean it might be 10^6?

Watts: Yes, or even more than that.

Elbein: Did you try different sizes of polylysine molecules?

Watts: Yes, there seems to be evidence that a relative molecular mass (M_r) of about 100 000 gives the best results. I always specify a molecular mass of 120 000.

Elbein: When you mix the two viruses, one being positive and the other negative, do you put them in together?

Watts: Yes, BMV can infect as well with CCMV as with poly-L-ornithine present. If we put the two viruses together they associate. These are preliminary results but the controls give 50% infection with poly-L-ornithine and the viruses on their own; when we mix the two viruses the positively charged virus (BMV) gives maybe 70% infection and the negatively charged virus (CCMV) fails to infect.

Howard: Is it true to comment that the chemical nature of the charge–charge interaction in virus and animal cell interaction is different to that in virus and plant cell interaction because of the lack of sialic acid in plants?

Watts: Yes, there is a difference.

Choppin: There are many animal viruses that don't need sialic acid for adsorption. Myxoviruses and paramyxoviruses have a specific requirement for sialic acid, but most other viruses do not. We shouldn't leave this discussion with the idea that animals are monolithic in that requirement.

Watts: And we shouldn't leave plant viruses thinking they are all the same as the ones I discussed. We have worked on only a few, and others may have very different mechanisms of attachment and entry.

REFERENCES

Allan I, Pearce JH 1979 Modulation by centrifugation of cell susceptibility to chlamydial infection. J Gen Microbiol 111:87-92

Burgess J, Motoyoshi F, Fleming EN 1973a Effect of poly-L-ornithine on isolated tobacco mesophyll protoplasts: evidence against stimulated pinocytosis. Planta (Berl) 111:199-208

Burgess J, Motoyoshi F, Fleming EN 1973b The mechanism of infection of plant protoplasts by viruses. Planta (Berl) 112:323-332

Cram WJ 1980 Pinocytosis in plants. New Phytol 84:1-17

Dawson JRO, Dickerson PE, King JM, Sakai F, Trim ARH, Watts JW 1978 Improved methods for infection of plant protoplasts with viral ribonucleic acid. Z Naturforsch Sect C Biosci 33:548-551

Heckels JE, Blackett B, Everson JS, Ward ME 1976 The influence of surface change on the attachment of *Neisseria gonorrhoeae* to human cells. J Gen Microbiol 96:359-364
Johanson WG Jr, Woods DE, Chaudhuri T 1979 Association of respiratory tract colonization with adherence of gram-negative bacilli to epithelial cells. J Infect Dis 139:667-673
Kahane I, Gat O, Banai M, Bredt W, Razin S 1979 Adherence of *Mycoplasma gallisepticum* to glass. J Gen Microbiol 111:217-222
Kraaipoel RJ, Van Duin AM 1979 Isoelectric focusing of *Chlamydia trachomatis*. Infect Immun 26:775-778
Kuo CC, Grayston JT 1976 Interaction of *Chlamydia trachomatis* organisms and Hela 229 cells. Infect Immun 13:1103-1109
Sarkar S, Upadhya MD, Melchers G 1974 A highly efficient method of inoculation of tobacco mesophyll protoplasts with ribonucleic acid of tobacco mosaic virus. Mol Gen Genet 135:1-9
Shaw JG 1972 Effect of poly-L-ornithine on the attachment of tobacco mosaic virus to tobacco leaves and on the uncoating of viral RNA. Virology 48:380-385
Silverblatt FJ, Kuehn C 1979 Autoradiography of gentamicin uptake by the rat proximal tubule cell. Kidney Int 15:335-345
Stein A, Loebenstein G 1972 Induced interference by synthetic polyanions with the infection of tobacco mosaic virus. Phytopathology 62:1461-1466
Wolstenholme J, Burgoyne RD, Stephen J 1977 Studies on the $MgSO_4$-induced cytoplasmic uptake of proteins by cells in culture. Exp Cell Res 104:377-388

Models for studying the adhesion of enterobacteria to the mucosa of the human intestinal tract

D. C. A. CANDY*, T. S. M. LEUNG, A. D. PHILLIPS†, J. T. HARRIES and W. C. MARSHALL

Institute of Child Health, Guilford Street, London WC1N 1EH and †Queen Elizabeth Hospital for Children, Hackney Road, London E2 8PS

Abstract Intestinal adhesion of enterobacteria in humans is being studied in several models. *E. coli* previously shown to adhere to human fetal intestine (O26:K60:H11) and not to adhere (O1:K1:H7) were tested for adhesion to buccal epithelial cells (BEC) from eight healthy adults. The adhesive *E. coli* (O26:K60:H11) adhered to 72-100% of BEC, and the non-adhesive *E. coli* (O1:K1:H7) to 0-6% of BEC. Adhesion was confirmed by electron microscopy.

The BEC adhesion assay was then applied to study an outbreak of acute diarrhoea in children. BEC adhesive strains were all subsequently shown to produce a haemagglutinin resembling colonization factor II (CFA/II). CFA/II strains were obtained from six patients, and five of these were of the O9 serogroup. The non-adhesive *E. coli*, which did not produce CFA/II, were from a variety of serogroups. Three of the patients excreting CFA/II *E. coli* had no diarrhoea but were significantly older than those with diarrhoea.

The non-adhesive *E. coli* was tested with BEC from healthy adults and infants, and infants with acute or protracted diarrhoea. This strain of *E. coli* adhered to more than 25% of BEC from 0 of 10 healthy adults, 2 of 13 healthy infants, and 0 of 10 infants with acute diarrhoea. However, with BEC from infants with protracted diarrhoea, *E. coli* adhered to more than 25% of BEC from seven of eight infants (range 18-96%) and was thus significantly more adhesive to BEC from these infants ($P<0.001$; χ^2 test).

This finding suggests that mucosal cells from infants with protracted diarrhoea are intrinsically more receptive to bacterial adhesion.

In order to determine whether adhesion to BEC reflects adhesion to small intestine, other models, utilizing human jejunal mucosal biopsies and isolated enterocytes from human fetal jejunum, are being investigated.

Studies in experimental animals and humans have provided convincing evidence that adhesion to small intestinal mucosa is an essential virulence factor in *E. coli* that cause diarrhoea. Wider application of these models should increase knowledge of the mechanisms mediating bacterial adhesion to the human intestine and lead to the development of measures that inhibit adhesion of pathogenic bacteria.

---

* *Present address:* Institute of Child Health, The Nuffield Building, Birmingham Children's Hospital, Ladywood, Birmingham B16 8ET

1981 Adhesion and microorganism pathogenicity. Pitman Medical, Tunbridge Wells (Ciba Foundation symposium 80) p 72-93

During the last decade important advances have been made in our understanding of the pathogenesis of bacterial diarrhoea, and studies of experimental animals have provided a basis for the investigation of human diarrhoeal disease.

In 1971 Smith & Linggood first showed that the plasmid-encoded fimbrial antigen designated K88 was an essential virulence factor for the majority of strains of *Escherichia coli* that cause diarrhoea in piglets, and they suggested that K88 mediated the adhesion of these *E. coli* to the mucosal surface of the small intestine of piglets. This suggestion was confirmed by Jones & Rutter (1972), who subsequently demonstrated that in piglets from certain litters K88-producing ($K88^+$) *E. coli* were unable to adhere to the intestinal brush borders (Sellwood et al 1975). The 'non-adhesive' pig phenotype was inherited in an autosomal recessive fashion and conferred relative resistance to *E. coli* diarrhoea (Rutter et al 1975). These classical experiments illustrated the importance of mucosal adhesion as a prerequisite for bacterial proliferation and development of diarrhoea. Moreover, they drew attention to the importance of considering host–pathogen adhesive interactions (i.e. 'cohesion').

At the same time Ørskov et al (1975) described K99, a fimbrial surface antigen analogous to K88, which mediated mucosal adhesion, and thus diarrhoea caused by *E. coli*, in calves and lambs.

These animal studies provided the incentive to develop models for investigating bacterial adhesion in human diarrhoeal disease, as described below.

Human models

The relevance of mucosal adhesion to human diarrhoeal disease was suggested by McNeish et al (1975), who showed that enterotoxigenic strains of *E. coli* from humans adhered in greater numbers to human fetal intestine than did control *E. coli* that had not been implicated as causative agents in acute diarrhoea. Adhesion was species-specific, and the adhesive property of one strain (O26:K60:H11) was transferable via a plasmid (Williams et al 1978).

Evans et al (1975) examined another human enterotoxigenic strain of *E. coli* (O78:K80:H11) which could adhere to and colonize the infant rabbit intestine. The plasmid-encoded fimbrial antigen involved in intestinal adhesion in this *E. coli* was designated colonization factor antigen I (CFA/I) (Fig. 1) and, like K88 and K99, CFA/I has been shown to be an essential virulence factor in human challenge studies (Satterwhite et al 1978). CFA/I also agglutinates human erythrocytes, and the screening of enterotoxigenic *E. coli* pathogenic for humans by haemagglutination led to the identification of a second plasmid-encoded fimbrial antigen, colonization factor antigen II(CFA/II) (Evans & Evans 1978), which mediates agglutination of bovine erythrocytes and also adhesion to rabbit small intestine. Species-specific haemagglutination, which is not inhibited by the presence of D-mannose (i.e. it is

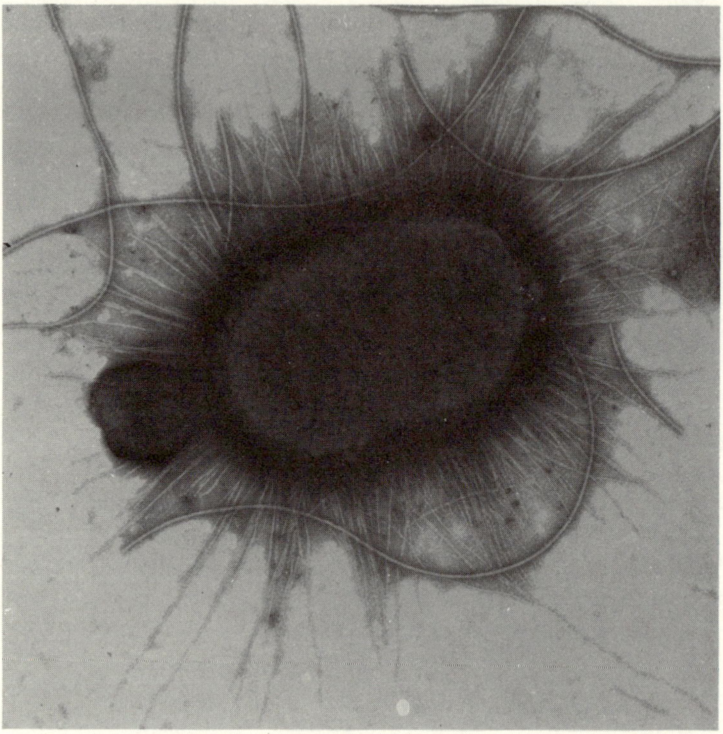

FIG. 1. Electron micrograph of negatively stained preparation showing morphology of colonization factor I fimbriae. × 60 000. (*Reduced to 67%*)

mannose-resistant), is also a property of K88 and K99 (see Table 1); here we have yet another example of the analogies between acute diarrhoea in humans and animals (Jones & Rutter 1972, Burrows et al 1976).

Type 1, or common, fimbriae are distinguished from specific colonization factor type fimbriae by the fact that agglutination of guinea-pig erythrocytes by these fimbriae is inhibited by D-mannose (i.e. it is mannose-sensitive). Although type 1 fimbriae allow *E. coli* to adhere to enterocytes from the small intestine (Isaacson et al 1978) their role in the pathogenesis of disease is not clear at present.

Buccal epithelial cells

In 1977 Ofek et al proposed that buccal epithelial cells (BEC) could be useful model for investigating the adhesion of *E. coli* in humans. These mucosal cells

TABLE 1 Surface antigens of E. coli involved in adhesion to the small intestine

Antigen	*Host*	*Mannose-resistant agglutination of RBC from*
K88	Pig	Guinea-pig
K99	Calf, lamb	Sheep
CFA/I	Human	Human
CFA/II	Human	Bovine
Type 1 (common) fimbriae	Not specific	Mannose-sensitive agglutination of guinea-pig RBC

have the advantage of accessibility, and repeated sampling is feasible for longitudinal studies in all age groups. Host susceptibility to adhesion and thus colonization can be investigated by comparing results with BEC obtained from different groups of individuals, and organisms can be studied with cells from the individual from whom the bacteria were obtained ('host specificity'). Although there are obvious morphological and functional differences between BEC and enterocytes that line the small intestine, the mouth is nevertheless embryologically related to the small intestine. Moreover both mucosal surfaces are bathed in enzyme-rich exocrine secretions, and certain intestinal diseases also involve the mouth (Simpson 1977). There is some evidence that both mucosal surfaces absorb glucose and galactose by a sodium-dependent, carrier-mediated transport system (Manning & Evered 1976). We were further encouraged to use BEC to study bacterial adhesion by reports that the mouth becomes colonized in *Vibrio cholerae* infections (Gorbach et al 1970) and that the same strain of *E. coli* can be recovered from the mouth, jejunum and colon in infants with protracted diarrhoea (Challacombe et al 1974).

For these reasons we have used a modification of the method of Ofek et al (1977) to study bacterial adhesion to BEC obtained from healthy adults and infants and from infants with acute and protracted diarrhoea (Candy et al 1978). Protracted diarrhoea of infancy of undetermined cause has been defined as the passage of four or more loose stools per day for longer than two weeks during which the patient either fails to gain weight or loses weight (Larcher et al 1977).

Scanning and transmission electron microscopy confirmed that the adhesive *E. coli* (O26:K60:H11) reported by McNeish et al (1975) adhered to BEC, without invasion (Figs. 2, 3). In all BEC adhesion assays we used this organism as the adhesive control strain and an *E. coli* (O1:K1:H7) which did not adhere to human fetal intestine as the non-adhesive control (these *E. coli* were provided by Dr N. Evans). Adhesion was quantified by examining 100 BEC and recording the percentage of BEC with adherent *E. coli*. The adhesive control *E. coli* adhered to 72–100% BEC

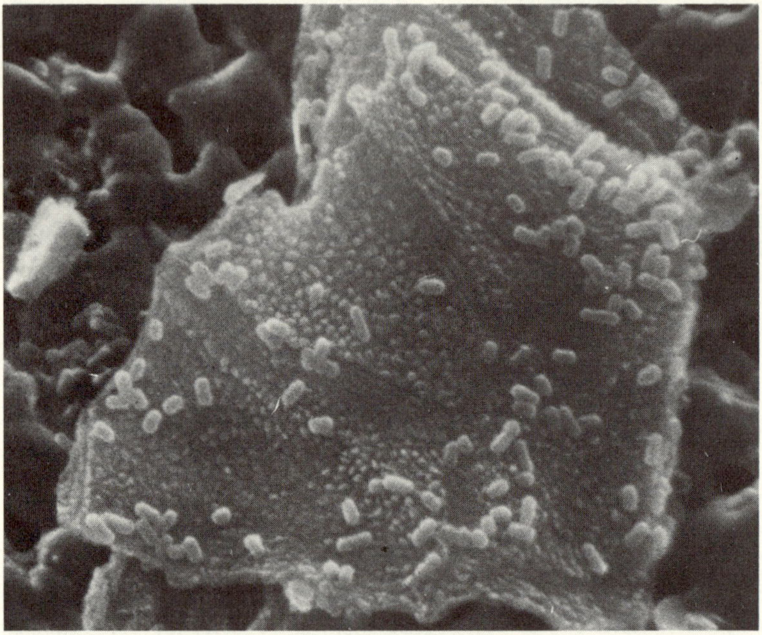

FIG. 2. Scanning electron micrograph of *E. coli* (O26:K60:H11) adhering to the surface of a human buccal epithelial cell. × 3700. (*Reduced to 90%.*)

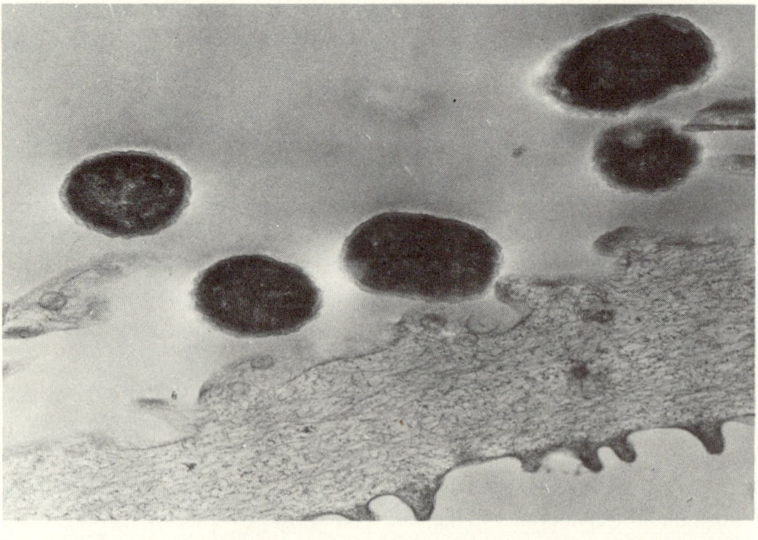

FIG. 3. Transmission electron micrograph of adhesive control *E. coli* (O26:K60:H11) adhering to the surface of a human buccal epithelial cell. × 54 000. (*Reduced to 67%.*)

TABLE 2 Adhesion of O26:K60:H11 (adhesive control) and O1:K1:H7 (non-adhesive control) E. coli to buccal epithelial cells (BEC) from healthy adults

Donor		*% BEC with adherent E. coli*	
		O26:K60:H11	*O1:K1:H7*
1		96	2
2		97	0
3		76	3
4		95	6
5		100	0
6		72	5
7		97	2
8		73	1
	Mean	88	2
	SEM	±4	±1
	Range	72–100	0–6

from eight healthy adults, compared with 0–6% for the non-adhesive control (Table 2). The light microscopy appearance of BEC incubated with the adhesive and non-adhesive control *E. coli* is shown in Fig. 4a, b.

Study of an outbreak of diarrhoea

An outbreak of mild diarrhoea occurred in a paediatric ward in which 13 patients aged between 1 and 21 months developed diarrhoea which lasted for between 3 and 26 days. The remaining six patients in the ward who did not develop diarrhoea were considerably older (4–12 years) and the significance of this age difference will be discussed later. *E. coli* isolated from the stools of both groups of patients agglutinated with a commercial O78 antiserum and were forwarded to us for studies of BEC adhesion.

E. coli from six patients with diarrhoea and four without diarrhoea were tested with BEC from three to six patients with diarrhoea and three to four healthy adults. The results were expressed as mean percentage of BEC with adherent *E. coli* (Table 3). There was good correlation between the percentage adhesion scores with BEC from patients and adults. Adhesion was mannose-resistant, suggesting that a colonization factor antigen was involved (Table 1). There was, however, no correlation between adhesion and diarrhoea, since three of six *E. coli* from patients with diarrhoea adhered to a mean of 75% or more of BEC from adults, and three of four *E. coli* from patients without diarrhoea also adhered to 75% or more of BEC.

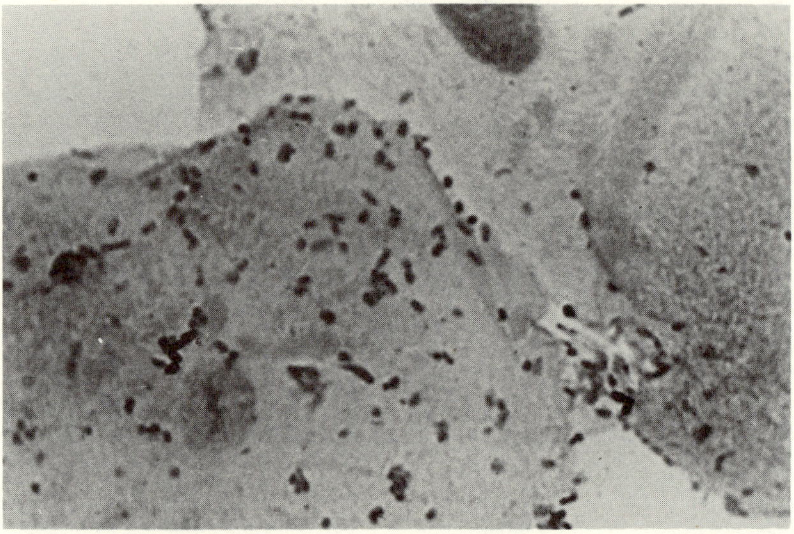
(a)

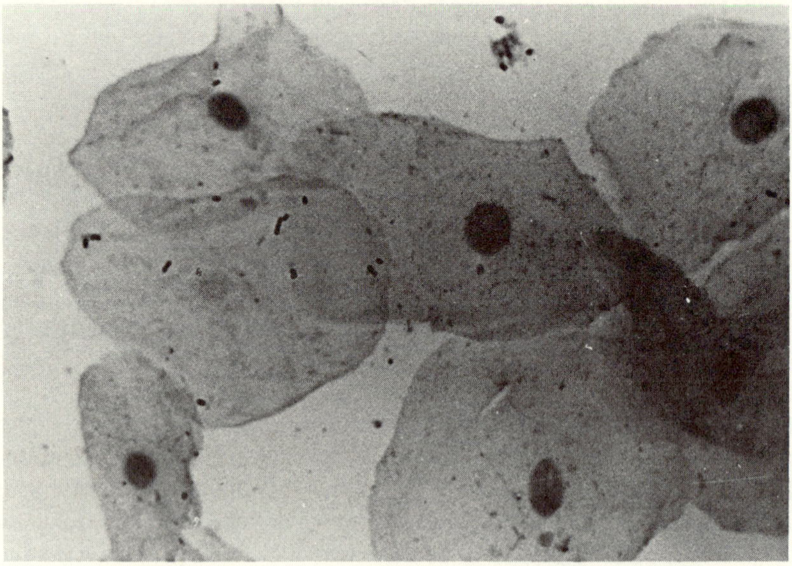
(b)

FIG. 4. Appearance of buccal cells incubated with (a) the adhesive control (O26:K60:H11) and (b) the non-adhesive control (O1:K1:H7) *E. coli*. Note numerous bacteria adhering to cells incubated with the adhesive control *E. coli*.

TABLE 3 Adhesion of E. coli isolated from an outbreak of diarrhoea: mean percentage adhesion of E. coli to buccal epithelial cells from patients with diarrhoea and from healthy adults

Patients	*Patient adhesion (%)*	*Adult adhesion (%)*
1[a]	99	88
2[a]	100	99
3[a]	99	75
4[a]	9	22
5[a]	NT	12
6[a]	NT	22
7	99	88
8	NT	93
9	NT	100
10	12	7

[a]: patients with diarrhoea
NT: not tested

TABLE 4 Adhesion of E. coli isolated from an outbreak of acute diarrhoea: relationship between symptoms, adhesion to buccal epithelial cells (BEC) from healthy adults, production of colonization factor antigen and O group

Patients	*BEC adhesion*	*CFA/I*	*CFA/II*	*O serogroup*
1[a]	Yes	–	+	O9
2[a]	Yes	–	+	O9
3[a]	Yes	–	+	O?
4[a]	No	–	–	O1, O111
5[a]	No	–	–	O?
6[a]	No	–	–	O25
7	Yes	–	+	O9
8	Yes	–	+	O9
9	Yes	–	+	O9
10	No	–	–	O92

[a]: patients with diarrhoea
Yes: adhesion to ⩾ 75% of BEC
No: adhesion to < 25% of BEC

After we had done these BEC experiments, the production of colonization factor antigens (see Table 1) as judged by D-mannose-resistant agglutination of appropriate erythrocytes was determined by Dr M. M. Levine. O serotypes of the *E. coli* were reassessed by the Enteric Reference Laboratories, Colindale, UK. The results of

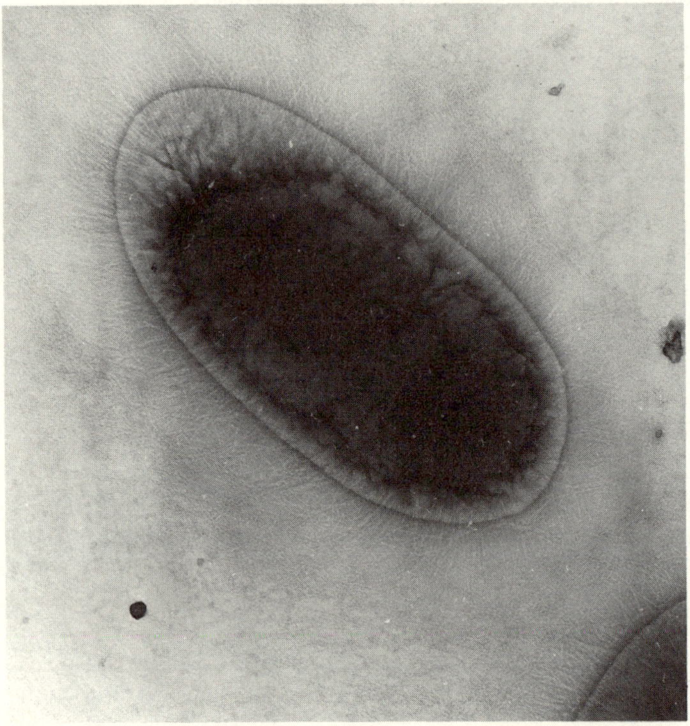

FIG. 5. Electron micrograph of negatively stained preparation showing morphology of fimbriae produced by CFA/II *E. coli* isolated during an outbreak of diarrhoea. × 90 000. (*Reduced to 50%.*)

these studies are shown in Table 4. *E. coli* which adhered to a mean of 75% or more of BEC from healthy adults are designated as adhesive whereas those which adhered to less than a mean of 25% of BEC are considered non-adhesive. All the adhesive organisms produced CFA/II, whereas none of the non-adhesive ones produced CFA/II. Slide agglutination with O78 antiserum was non-specific since, with the Reference Laboratory sera, five out of the six adhesive *E. coli* were shown to be serogroup O9, the remaining adhesive organism being untypable. The non-adhesive organisms displayed a variety of serogroups. None of the *E. coli* agglutinated with O78 antisera at the Reference Laboratory. The morphological appearance of the CFA/II produced by the adhesive *E. coli* is shown in Fig. 5. The CFA/II-like fimbrial haemagglutinin described in the study of an outbreak of *E. coli* diarrhoea has since been shown to be antigenically distinct from CFA/II (M. M. Levine, personal communication) and it thus represents a new CFA.

All *E. coli* were tested for enterotoxin production; none could be shown to produce heat-labile enterotoxin or Vero toxin with the Vero cell assay (Speirs et al 1977, Konowalchuk et al 1977), or heat-stable enterotoxin with the infant mouse assay (Dean et al 1972). Thus adhesiveness was the only virulence factor which could be identified in this outbreak of acute diarrhoea.

One can only speculate on the possible role of this CFA/II-producing strain of *E. coli* in the genesis of diarrhoea in these patients. There are, however, certain similarities between our findings and those of Smith & Linggood (1971) in young piglets. Smith & Linggood (1971) found that young, but not older, animals developed mild diarrhoea when challenged with a non-enterotoxigenic K88 strain of *E. coli*. It is therefore of interest that CFA/II *E. coli* were associated with diarrhoea in only the younger patients in our study.

Studies with the non-adhesive E. coli (O1:K1:H7)

The results of testing the non-adhesive control *E. coli* with BEC obtained from healthy adults and infants, infants with acute diarrhoea, and infants with protracted diarrhoea are shown in Fig. 6. The proportion of donors in the four groups with more than 25% of BEC with adherent *E. coli* was 0 out of 10 for healthy adults, 2 out of 13 for healthy infants, 0 out of 10 for infants with acute diarrhoea and 7 out of 8 for infants with protracted diarrhoea. Thus this 'non-adhesive' strain of *E. coli* adhered to BEC from a highly significantly greater proportion of infants with protracted diarrhoea compared with the other three groups ($P<0.001$; χ^2 test). These findings are reminiscent of those of Kallenius & Winbert (1978), who demonstrated increased 'cohesion' between a known pyelonephritic strain of *E. coli* and uroepithelial cells from girls with recurrent and unexplained urinary tract infections, suggesting that the mucosal surfaces of certain individuals are intrinsically more receptive to bacterial adhesion. It is of interest that bacterial overgrowth of the small intestine is a common finding in infants with protracted diarrhoea (Challacombe et al 1974, Gracey et al 1969, Gracey & Stone 1972, Heyworth & Brown 1975).

Correlation between adhesion to BEC and fetal enterocytes

Jejuna from human fetuses were obtained from the Tissue Bank of the Royal Marsden Hospital, London, UK. Fetal tissues were collected from surrounding hospitals and were therefore likely to be of varying viability; and McNeish et al (1975) have suggested that poor viability of intact fetal small intestine results in

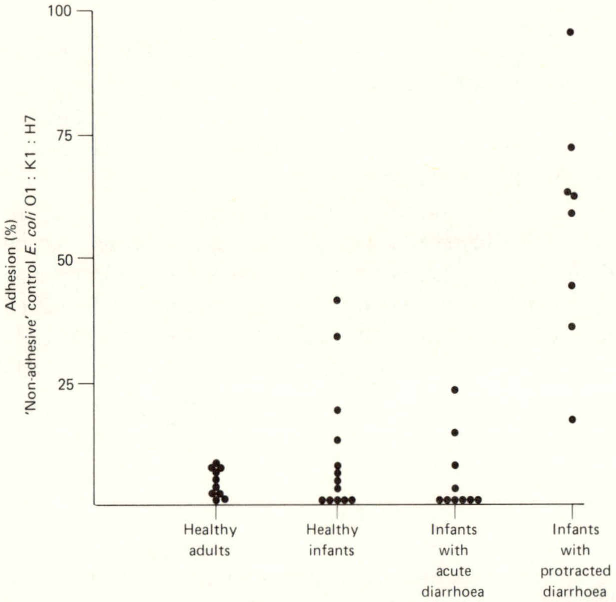

FIG. 6. Adhesion of the 'non-adhesive' control *E. coli* (O1:K1:H7) to buccal epithelial cells from adults, infants with acute diarrhoea and infants with protracted diarrhoea.

non-specific bacterial adhesion. In contrast, isolated enterocytes from porcine small intestine that were stored in buffer for up to eight weeks retained their adhesive properties with $K88^+$ *E. coli* (Isaacson et al 1978). We therefore prepared suspensions of human fetal jejunal enterocytes for adhesion assays using a similar method to that used in the BEC studies.

Fetal jejuna were transported at 4 °C from the Tissue Bank in RPMI Tissue Culture Medium (Gibco Bio-cult, Paisley, UK) omitting penicillin and streptomycin. Phosphate-buffered saline (PBS) was gently injected into the jejunal lumen to remove any intraluminal contents. Closed loops about 5 cm long were constructed by ligatures and distended with the buffer containing 96 mM-NaCl, 15 mM-KCl, 6 mM-Na_2HPO_4, 8 mM-KH_2PO_4 and 27 mM-Na citrate, pH 7.5, osmolality 295 mosm/kg. The loops were incubated at 37 °C for 15 min, immersed in PBS in a Petri dish, and opened; the enterocytes were then expelled by flushing with PBS and gentle manual expression.

The resulting preparations, when examined by light microscopy, contained erythrocytes, occasional squamous cells, and columnar and cuboidal cells. The percentage adhesion score was obtained as for the BEC assay, that is by examining 100 columnar cells and recording the percentage of cells with adherent *E. coli*.

TABLE 5 Adhesion of E. coli O26:K60:H11 (adhesive control) and O1:K1:H7 (non-adhesive control) to isolated enterocytes from human fetal jejuna

	Fetal gestation (weeks)	*% Enterocytes with adhering E coli* O26:K60:H11	O1:K1:H7
	15	100	2
	18	90	0
	20	68	5
	21	81	0
	22	75	5
	23	94	5
Mean	20	85	3
SEM		±5	±1
Range	15–23	68–100	0–5

Whereas up to 400 *E. coli* could adhere to a single BEC it was unusual for more than 30 *E. coli* to adhere to the smaller isolated enterocytes.

Adhesion of adhesive (O26:K60:H11) and non-adhesive (O1:K1:H7) E. coli to fetal enterocytes

The mean adhesion scores for adhesive and non-adhesive control *E. coli* were 85% and 3% when these were tested for adhesion to isolated fetal enterocytes (Table 5). The corresponding mean scores with BEC from eight healthy adults were 88% and 2% for the adhesive and non-adhesive control *E. coli* (Table 2).

Adhesion of enterobacteria to BEC and fetal enterocytes

Enterobacteria were cultured from the small intestine of four infants with protracted diarrhoea, one of whom (patient 4) had diarrhoea after resection of multiple jejunal atresias (see Table 6). Strains of *E. coli* were also cultured from the stools of an infant with acute diarrhoea, and from normal stools from three infants, one of whom (patient 6) had recovered from an episode of protracted diarrhoea. These enterobacteria were studied for adhesion to BEC from the patients from whom the organism had been obtained, to BEC from a healthy adult, and to fetal enterocytes (Table 6). Enterobacteria from patients 1, 2, 5 and 6 were tested with enterocytes from a 21-week gestation fetus, and those from patients 3, 4, 7 and 8 with enterocytes from a fetus of 23 weeks' gestation.

Enterobacteria from patients 1, 2 and 6 adhered to more than 50% of BEC from

TABLE 6 Adhesion of enterobacteria to buccal epithelial cells (BEC) from patients, a healthy adult, and isolated human fetal jejunal enterocytes

Patient	*Gastrointestinal symptoms*	*Species*	*Source*	*Adhesion*[a] *BEC from patient*	*BEC from adult*	*Fetal enterocytes*
1	Protracted diarrhoea	*E. coli* O?	Small bowel	92	52	98
2	Protracted diarrhoea	*P. mirabilis*	Small bowel	100	83	84
3	Protracted diarrhoea	*E. cloacae*	Small bowel	90	5	92
4	Protracted diarrhoea	*E. coli* O?	Small bowel	24	35	90
5	Acute diarrhoea	*E. coli* O127:K63	Stool	0	0	20
6	Nil	*E. coli* O?	Stool	100	70	66
7	Nil	*E. coli* O?	Stool	96	11	0
8	Nil	*E. coli* O26:K60	Stool	7	11	88

[a]Adhesion = % of cells with adhering enterobacteria

the patients and an adult, as well as to fetal enterocytes. A strain of *E. coli* from patient 4 adhered to 24% of the infant's own BEC, to 35% of adult BEC, and to 90% of fetal enterocytes. Two strains of *E. coli* from patients 5 and 8 failed to adhere to BEC from the patients or from an adult; one of these strains (from patient 5) adhered to 20% of enterocytes, the other to 88%. Enterobacteria from patients 3 and 7 adhered to BEC from the patients but not to BEC from an adult; again one strain adhered to fetal enterocytes (patient 3) and the other did not. Thus, all four isolates from the small bowel (from patients 1–4) adhered strongly to fetal enterocytes, and the three enterobacterial strains which adhered to BEC from both the patients and the adults also adhered to fetal enterocytes.

Effect of D*-mannose on adhesion of the adhesive control E. coli to BEC and fetal enterocytes*

D-Mannose (1%; 56 mM) reduced the mean adhesion score of the adhesive control *E. coli* from 84% to 31% when this strain was incubated with BEC from nine adult donors. In contrast, enterocyte adhesion was only reduced from 78% to 66% when enterocytes from the jejuna of three fetuses were used. These results suggest that type 1 fimbriae (see Table 1) may be more important in adhesion to BEC than in adhesion to enterocytes, and that adhesion to BEC and fetal enterocytes may be mediated by different systems.

Adhesion of enterobacteria to biopsy material from the small intestine

Another potential source of mucosal tissue from the human intestine is peroral jejunal biopsies. In preliminary experiments we have incubated portions of jejunal biopsies in PBS with suspensions of *E. coli* from humans with acute diarrhoeal disease and examined the villi by scanning electron microscopy. Rod-shaped bacteria can be seen adhering to the mucosal surface (Fig. 7). Adhesion was quantified by examining 20 randomly chosen fields at a constant magnification (X 5000). We have compared the adhesive properties of a CFA/I *E. coli* (O78:K80:H11) to a mutant of this strain which has lost the plasmid coding for CFA/I production (Evans et al 1975). The organisms were incubated with biopsy material obtained from two children who were undergoing investigation for suspected malabsorption but were subsequently proved to be normal; the histological appearance of the jejunal mucosa was normal under light microscopy. In one patient 586 coliforms were seen adhering in 20 high-power fields after the biopsy material had been incubated with the CFA/I *E. coli,* compared with no adherent coliforms after incubation with the mutant strain. Corresponding values in the second patient were

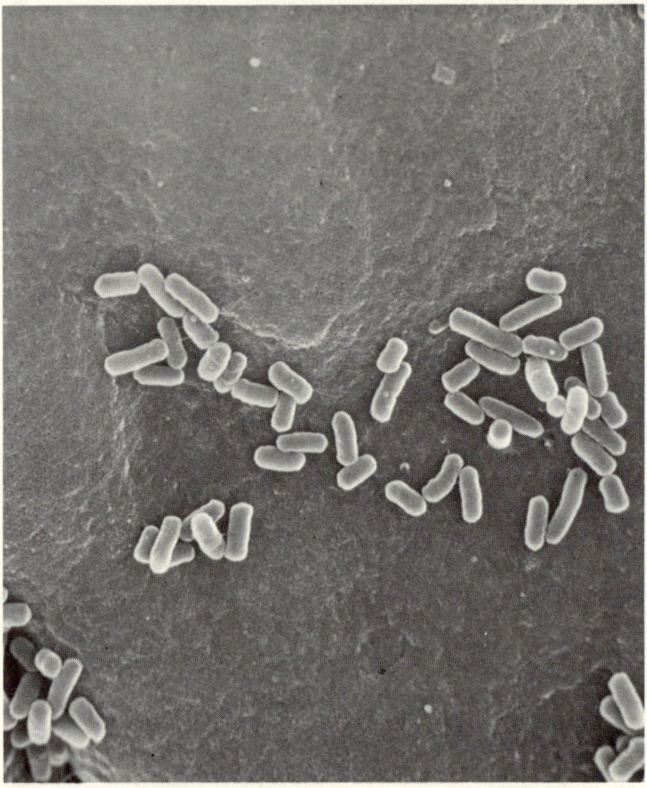

FIG. 7. Scanning electron micrograph of *E. coli* (O78:K80:H11) adhering to the mucosal surface of a human small intestinal biopsy *in vitro*. × 5000.

57 and 3. These results emphasize the importance of CFA/I in adhesion to the small intestine.

Conclusions

Ingested enteropathogenic bacteria possess specific properties which facilitate their adhesion to the mucosal surface of the intestinal tract, allowing them to proliferate and colonize the intestine and to produce overt disease.

Studies in the experimental animal have provided convincing evidence that adhesion is an essential virulence factor, and that adhesion is mediated by interactions between the organism and the mucosa of the host (i.e. 'cohesion'). Such cohesion counteracts attempts by the host to expel the enteropathogen from the intestinal tract (Dixon 1960) by peristalsis. Plasmid-encoded surface antigens which mediate adhesion have been identified in certain enterotoxigenic strains of *E. coli*, designated K88 in piglets, and K99 in calves and lambs. Two analogous fimbrial antigens had previously been described in human strains of *E. coli:* colonization factor antigen I and colonization factor antigen II.

These findings have motivated studies directed towards the development of human models for the investigation of 'cohesive mechanisms' which participate in bacterial adhesion to the human intestinal tract, and for identifying whether suspected enteropathogenic bacteria recovered from the small intestinal lumen or faeces of patients possess adhesive properties.

Buccal epithelial cells, enterocytes from intact fetal intestine or isolated enterocytes, and jejunal biopsies from healthy subjects and patients or enterocytes isolated from such material are all 'candidate models' which are currently being appraised. Encouraging results are emerging from studies on such human models, but a number of controversial issues remain to be resolved.

Acknowledgements

The authors are grateful to D. Jackson for technical assistance, to J. Beasley, of the Wellcome Research Laboratories, Beckenham, UK, for the electron micrograph shown in Fig.5, and to the Rayne Foundation for financial support for D. C. A. Candy.

REFERENCES

Burrows MR, Sellwood R, Gibbons RA 1976 Haemagglutinating and adhesive properties associated with the K99 antigen of bovine strains of *Escherichia coli.* J Gen Microbiol 96: 269-275

Candy DCA, Chadwick J, Leung T, Phillips A, Harries JT, Marshall WC 1978 Adhesion of Enterobacteriaceae to buccal epithelial cells. Lancet 2: 1157-1158

Challacombe DN, Richardson JH, Rowe B, Anderson CM 1974 Bacterial microflora of the upper gastrointestinal tract in infants with protracted diarrhoea. Arch Dis Child 49:270-277

Dean AG, Yi-Chuan C, Williams RG, Harden LB 1972 Test for *Escherichia coli* enterotoxin using infant mice: application in a study of diarrhea in children in Honolulu. J Infect Dis 125:407-411

Dixon JMS 1960 The fate of bacteria in the small intestine. J Pathol Bacteriol 79:131-140

Evans DG, Evans DJ Jr 1978 New surface-associated heat-labile colonization factor antigen (CFA/II) produced by enterotoxigenic *Escherichia coli* of serogroups O6 and O8. Infect Immun 21:638-647

Evans DG, Silver RP, Evans DJ Jr, Chase DG, Gorbach SL 1975 Plasmid-controlled colonization factor associated with virulence in *Escherichia coli* enterotoxigenic for humans. Infect Immun 12:656-667

Gorbach SL, Banwell JG, Jacobs B, Chatterjee BD, Mitra R, Brigham KL, Neogg KN 1970 Intestinal microflora in asiatic cholera. II. The small bowel. J Infect Dis 121:38-45

Gracey M, Stone DE 1972 Small intestinal microflora in Australian aboriginal children with chronic diarrhoea. Aust NZ J Med 2:215-219

Gracey M, Burke V, Anderson CM 1969 Association of monosaccharide malabsorption with abnormal small intestinal flora. Lancet 2:384-385
Heyworth B, Brown J 1975 Jejunal microflora in malnourished Gambian children. Arch Dis Child 50:27-33
Isaacson RE, Fusco PC, Brinton CC, Moon HW 1978 In vitro adhesion of *Escherichia coli* to porcine small intestinal cells: pili as adhesive factors. Infect Immun 21:392-397
Jones GW, Rutter JM 1972 Role of K88 antigen in the pathogenesis of neonatal diarrhoea caused by *Escherichia coli* in piglets. Infect Immun 6:918-927
Källenius G, Winberg J 1978 Bacterial adherence to periurethral epithelial cells in girls prone to urinary tract infections. Lancet 2:540-543
Konowalchuk J, Speirs JI, Stavric S 1977 Vero response to a cytotoxin of *Escherichia coli.* Infect Immun 18: 775-779
Larcher VF, Shepherd R, Francis DEM, Harries JT 1977 Protracted diarrhoea in infancy. Analysis of 82 cases with particular reference to diagnosis and management. Arch Dis Child 52:597-605
Manning AS, Evered DF 1976 The absorption of sugars from the human buccal cavity. Clin Sci Mol Med 51:127-132
McNeish AS, Turner P, Fleming J, Evans N 1975 Mucosal adherence of human enteropathogenic *Escherichia coli* Lancet 2:946-948
Ofek I, Mirelman D, Sharon N 1977 Adherence of *Escherichia coli* to human mucosal cells mediated by mannose receptors. Nature (Lond) 265:623-625
Ørskov I, Ørskov F, Smith HW, Sojka WJ 1975 The establishment of K99, a thermolabile, transmissible *Escherichia coli* K antigen previously called 'Kco' possessed by calf and lamb enteropathogenic strains. Acta Pathol Microbiol Sect B Microbiol Immunol 88:31-36
Rutter JM, Burrows MR, Sellwood R, Gibbons RA 1975 A genetic basis for resistance to enteric disease caused by *E. coli.* Nature (Lond) 257:135-136
Satterwhite TK, Evans DG, Dupont HL, Evans DJ Jr 1978 Role of *Escherichia coli* colonisation factor antigen in acute diarrhoea. Lancet 2:181-184
Sellwood R, Gibbons RA, Jones GW, Rutter JM 1975 Adhesion of enteropathogenic *Escherichia coli* to pig intestinal brush borders: the existence of two pig phenotypes. J Med Microbiol 8:405-411
Simpson HE 1977 Oral manifestations of Crohn's disease: studies in the pathogenesis. Proc R Soc Med 70:55 (letter)
Smith HW, Linggood MA 1971 Observations on the pathogenic properties of K88, Hly and ENT plasmids of *Escherichia coli* with particular reference to porcine diarrhoea. J Med Microbiol 4:467-485
Speirs JI, Stavric S, Konowalchuk J 1977 Assay of *Escherichia coli* heat-labile enterotoxin with Vero cells. Infect Immun 16:617-622
Williams PH, Sedgewick MI, Evans N, Turner PJ, George RH, McNeish AS 1978 Adherence of an enteropathogenic strain of *Escherichia coli* to human intestinal mucosa is mediated by a colicinogenic conjugative plasmid. Infect Immun 22:393-402

DISCUSSION

Taylor-Robinson: My first reaction is that your data do not show a correlation between the colonization factor (CFA/II) and diarrhoea.

Candy: CFA/II strains were excreted by infants with and without diarrhoea but the infants who didn't develop diarrhoea and excreted this strain were significantly older.

Tramont: You can look at that the other way round. There is a correlation of developing diarrhoea with colonization factor but not everyone who has that strain develops diarrhoea. That is a common occurrence with other infectious agents as well.

Candy: Yes, and there are other factors to consider. For example the *E. coli* that we studied happened to agglutinate weakly with the 078 antisera and this subsequently proved to be a non-specific finding. This meant that we could study only one *E. coli* colony from each patient instead of our usual five colonies. Furthermore we were unable to study patients who developed diarrhoea at the onset of the outbreak. Finding any common strain was surprising enough to us. Again, in epidemiological work with *E. coli* one is lucky to find the same pathogenic organism in two-thirds of the affected patients.

Richmond: In our experience most buccal cells are dead when we stain them. This means that if you look carefully for the cells that are alive you find that the binding characteristics of the live cells are not necessarily the same.

Candy: I agree that if you stain buccal cells with trypan blue most of them are dead. Isolated enterocytes kept for eight weeks in phosphate buffer at +4 °C are also presumably dead, but they still discriminate between $K88^+$ and $K88^-$ *E. coli* (Isaacson et al 1978). A cell doesn't necessarily have to be alive to be used as a marker for adhesion.

Richmond: You talked about the CFA^- mutants. Is CFA the only thing missing?

Candy: To my knowledge the only difference between the two strains was the absence of CFA/I. The CFA/I^- strain (H10407P) was still toxigenic but adhesive (Evans et al 1975).

Levine: Strains H10407P and H10407 are quite different. The plasmid in H10407 that encodes for CFA/I also encodes for heat-stable enterotoxin (ST); so H10407 lacks ST as well as CFA. We have some data to suggest that H10407P is a defective organism.

Richmond: A big plasmid will code for at least 60 products. If you lose the plasmid you are losing 59 you don't know about.

Tramont: It depends how you define disease relative to an epidemiological marker. If you define CFA/I^- by non-adhesion and you make the observation that that correlates with disease, that is a single valid observation. It does not address the question of which antigens are involved with attachment.

Richmond: In your experiments you have a plasmid-carrying cell which adheres and you have a plasmid-less cell which does not. In losing the plasmid you have lost 60 characters. You can't conclude that it is the CFA character that is responsible just because it is lost when the plasmid is lost. Adhesion could be the consequence of one of the other plasmid-mediated proteins.

Tramont: But it doesn't bind to a cell, by definition.

Candy: It is suggestive evidence but I agree it is not conclusive.

Tramont: I agree with your argument. My point is that it all depends on how you define what you are looking at. If you say that CFA/I is lacking and you define that by adhesion, then by definition it is not adherence. The exact attachment factors involved are irrelevant at that point. You can define the strains by O types or any other parameter. There is always a problem of correlating specific antigens or immunological phenomena with epidemiological observations. Correlating any kind of marker with the epidemiological system, or any other system, all depends on how you define the markers *vis à vis* the disease.

Richmond: So you are saying that from the results of this experiment CFA must be a complex of a large number of characters?

Candy: The incubation of CFA/I and CFA/I$^-$ strains with pieces of human jejunal mucosal biopsies was done with coded bacterial cultures. When the code was broken it was clear that the CFA/I strain was adhering more than the CFA/I$^-$ strain. Whether the loss of adhesiveness and loss of CFA/I were definitely related is not certain.

Tramont: That is basically what I am saying. In working with mutants one doesn't know whether there is one, a few or many missing factors. But the observation that a change took place that affected virulence is valid.

Richmond: If you are dealing with plasmids and you lose the plasmid you are inevitably dealing with the loss of a lot of genetic information.

Tramont: That may be irrelevant in terms of the observation whether disease occurred or not.

Richmond: But that is your bad luck. This is the trouble with using plasmid-deficient mutants. People usually greatly underestimate the number of characters specified by plasmids. They assume that when the plasmid is purged the only changes are the characters of particular concern to them.

Taylor-Robinson: Would you also agree that it makes no difference whether dead or live buccal cells are used, Dr Tramont?

Tramont: Again it depends on what system you are talking about and what correlation you are trying to make. Asking whether the general phenomenon of adhesion correlates with disease is different from asking whether adhesion correlates with a certain attachment factor. The question of live and dead cells and what receptors are being expressed is critical in the latter. For the question Dr Candy was asking, it makes absolutely no difference. The presence of or lack of adhesion of the organism correlated with what he saw epidemiologically, which is what he was addressing. The actual antigen involved was not addressed.

Feizi: What happens to the adhesive properties of buccal cells when your patients recover from diarrhoea, Dr Candy?

Candy: It took us a while to realize that buccal cells from patients with protracted diarrhoea were behaving differently from cells from other groups of individuals. Also some of this group recovered, or died before being studied again, and so there are very few longitudinal studies. One child we were able to study remained ill and continued to show increased adhesion with the so-called non-adhesive strain (O1:K1:H7). Another child we studied serially during her recovery showed decreased adhesion with both the adhesive (O26:K60:H11) and the non-adhesive strains. The problem was that during the course of her illness her buccal cells became colonized by an *E. coli*. It was therefore difficult to interpret what we saw. Once you have an *E. coli* on your control cells you can't say whether the Gram-negative bacteria observed on buccal cells are the test strain. It may well be that the *E. coli* that colonized her mouth *in vivo* competed with the control strains for adhesion sites *in vitro*.

Tramont: The question is: is the ability to support attachment in a state of flux? The answer to that is yes.

Candy: The answer to that is that more work is required.

Feizi: In dehydrated patients with a lot of dead buccal cells what changes do you observe in the adhesive properties?

Candy: We studied children with acute diarrhoea but they weren't necessarily dehydrated. Certainly exogenous factors can modify buccal cell adhesion. For example brief treatment of buccal cells with trypsin will enhance the number of *Pseudomonas aeruginosa* and *Klebsiella pneumoniae* that can subsequently adhere to the buccal cells, so exogenous factors can be important (Johanson et al 1979). Buccal cells can also vary endogenously. For example streptococcal adhesion to buccal cells from newborn infants increases during the first 72 hours of life, suggesting that receptors for streptococcal binding on buccal cells develop, or become unmasked, in early postnatal life (Ofek et al 1977).

Taylor-Robinson: How do you know which side of the buccal cell you are looking at?

Candy: The two sides of the cell may be different morphologically when seen in the electron microscope. By light microscopy the cells are so thin that bacteria can be seen adhering to either side of the cell.

Levine: Was the *E. coli* O9 strain grown at 18 °C and then looked at by electron microphotography for pili?

Candy: They are all cultured at 37 °C overnight on nutrient or CFA agar slopes (McNeish et al 1975, Evans et al 1977). I don't know whether low temperature inhibits the formation of this colonization factor. (Subsequent experiments have shown that culture of the O9 strain at 18 °C suppresses fimbriation, haemagglutination and BEC adhesion.)

Freter: Sugarman & Donta (1979) and Thorne et al (1979), among others, have published studies with buccal and red blood cells which showed that presumably saprophytic and enteropathogenic *E. coli* strains exhibit a great diversity of 'patterns' of adherence. This suggests that there are a great number of different adhesive systems and certainly not all of these can be related to pathogenicity. Do you think that further pursuit of this type of study would be useful in identifying other suspected *E. coli* strains for which we must ask the question: do they have adhesive systems and, if so, which of those relate to pathogenicity? I can't see how one can get out of this vicious circle of working with an unknown adhesive mechanism reacting with an unknown mammalian cell receptor for bacteria of unknown pathogenicity.

Candy: The question of what we are looking at with buccal cell adhesion is certainly important. If biopsies are needed for clinical reasons we can isolate enterocytes from the biopsy material, take buccal cells from the same child, and simultaneously compare the adhesion of the child's own bacterial flora to buccal cells and enterocytes. We have started to do this but both the children whom we studied had buccal cells colonized with *E. coli*. It is too early to say how adhesion to buccal cells compares with adhesion to enterocytes but the work with fetal enterocytes, for example, suggests that there may be important differences between buccal cell and gut adhesion.

Richmond: There must be a lot of different adherence mechanisms at work. It will be difficult to investigate these together with their pathogenic implications. I think one has to examine one or two defined systems at the molecular level and forget about the pathogenic implications for the moment. At some stage we might give some thought to what systems to tackle. One that comes to mind is *E. coli* K12 with the CFA plasmid and some eukaryotic cell line. One could clone bits of the plasmid and try to find out which bits control what.

Levine: I am not familiar with any publication so far that documents transfer of CFA/I to K12 or any other *E. coli*, in contrast to K88 and K99. One of the frustrations of working with the human strains is the mad rush to clone the genes without having the tools that people have who work with K88 or K99.

Candy: You can transfer the adhesive factor plasmid from O26:K60:H11 to K12 (Williams et al 1978).

Levine: But CFA/I and II have not been transferred, as far as I know.

Richmond: Presumably you can clone bits of the plasmid?

Levine: Yes. People are working at cloning CFA/I but K12 with CFA/I doesn't exist.

Richmond: So one would presumably attempt to clone bits of the CFA/I plasmid in some plasmid which will survive in *E. coli* K12.

Sussman: The discussion about dead or live buccal cells implies that there may be important differences in their properties. Can anyone tell us about defined differences between dead and live buccal cells with regard to the adherence of any given organism?

Svanborg Edén: The attachment to dead urinary tract epithelial cells (i.e. cells taking up trypan blue) is lower than to unstained cells. This does not seem to be true for pneumococci, which bind to the same extent to stained or unstained pharyngeal cells (B. Andersson et al, unpublished work).

REFERENCES

Evans DG, Silver RP, Evans DJ Jr, Chase DG, Gorbach SL 1975 Plasmid-controlled colonization factor associated with virulence in *Escherichia coli* enterotoxigenic for humans. Infect Immun 12:656-667

Evans DG, Evans DJ Jr, Tjoa W 1977 Hemagglutination of human group A erythrocytes by enterotoxigenic *Escherichia coli* isolated from adults with diarrhea: correlation with colonization factor. Infect Immun 18:330-337

Isaacson RE, Fusco PC, Brinton CC, Moon HW 1978 In vitro adhesion of *Escherichia coli* to porcine small intestinal cells: pili as adhesive factors. Infect Immun 21:397-399

Johanson WG Jr, Woods DE, Chaudhuri T 1979 Association of respiratory tract colonization with adherence of gram-negative bacilli to epithelial cells. J Infect Dis 139:667-673

McNeish AS, Turner P, Fleming J, Evans N 1975 Mucosal adherence of human enteropathogenic *Escherichia coli.* Lancet 2:946-948

Ofek I, Beachey EH, Eyal F, Morrison JC 1977 Postnatal development of binding of streptococci and lipoteichoic acid by oral mucosal cells of humans. J Infect Dis 135:267-274

Sugarman B, Donta ST 1979 Specificity of attachment of certain Enterobacteriaceae to mammalian cells. J Gen Microbiol 115:509-512

Thorne GM, Deneke GF, Gorbach SL 1979 Hemagglutination and adhesiveness of toxigenic *Escherichia coli* isolated from humans. Infect Immun 23:690-699

Williams PH, Sedgewick MI, Evans N, Turner PJ, George RH, McNeish AS 1978 Adherence of an enteropathogenic strain of *Escherichia coli* to human intestinal mucosa is mediated by a colicinogenic conjugative plasmid. Infect Immun 22:393-402

Short communication

An *in vivo* model for studying adherence of intestinal pathogens

SHMUEL KATZ, MORDEHAI IZHAR* AND DAVID MIRELMAN*

*Department of Surgery, Hadassah Medical Center, Jerusalem, and *Department of Biophysics, Unit for Molecular Biology of Parasitic Diseases, Weizmann Institute of Science, Rehovoth, Israel*

Abstract A new method for preparing an isolated colonic loop in a living rabbit is described. The loop with its intact neurovascular supply can be used as a 'living test tube' to study the adherence of microorganisms to intestinal mucosa. Moreover, the clear colonic mucus produced by the loop can be used to study its physicochemical nature and protecting properties in health and disease.

For many microorganisms the ability to adhere to gastrointestinal mucosa is an essential requirement for colonization and in some cases for later invasion of the tissue. Bacterial adherence to the intestine has been studied mainly *in vitro* (e.g. tissue culture, mucosal scraping, inverted intestinal sections). The inaccuracy of most of the *in vitro* studies is due to the sensitivity of the intestinal mucosal layer to any minor ischaemia that occurs during preparation or incubation of the system and this usually leads to rapid desquamation of cells.

We have developed a method for preparing an isolated colonic loop (Thiry-Vella) in a living rabbit which should be suitable for various *'in vivo'* studies of bacterial and amoebic adherence. The obvious advantage of these loops is that, though isolated, they still maintain their mesenteric attachment and thus their neurovascular supply. An additional advantage of this system is that the protective properties and the function of the colonic mucus produced by these loops can also be investigated.

Albino rabbits (weight 2–3 kg) were starved for 24 h before surgery. The animals were anaesthetized by intramuscular injection of 1–1.5 ml of a mixture of Rompon, Combelen and Vetalar, 2:2:1 (Rompon, Bayer: 2% 2(2,6-xylidino)-5, 6-dihydro-4H-1,3-thiazine HCl; Combelen, Bayer: 1% *N*-(3′-dimethyl-aminopropyl)-3-propionylphenothiazine; Vetalar, Parke Davis: 10% ketamine HCl). A mid-line

1981 Adhesion and microorganism pathogenicity. Pitman Medical, Tunbridge Wells (Ciba Foundation symposium 80) p 94-97

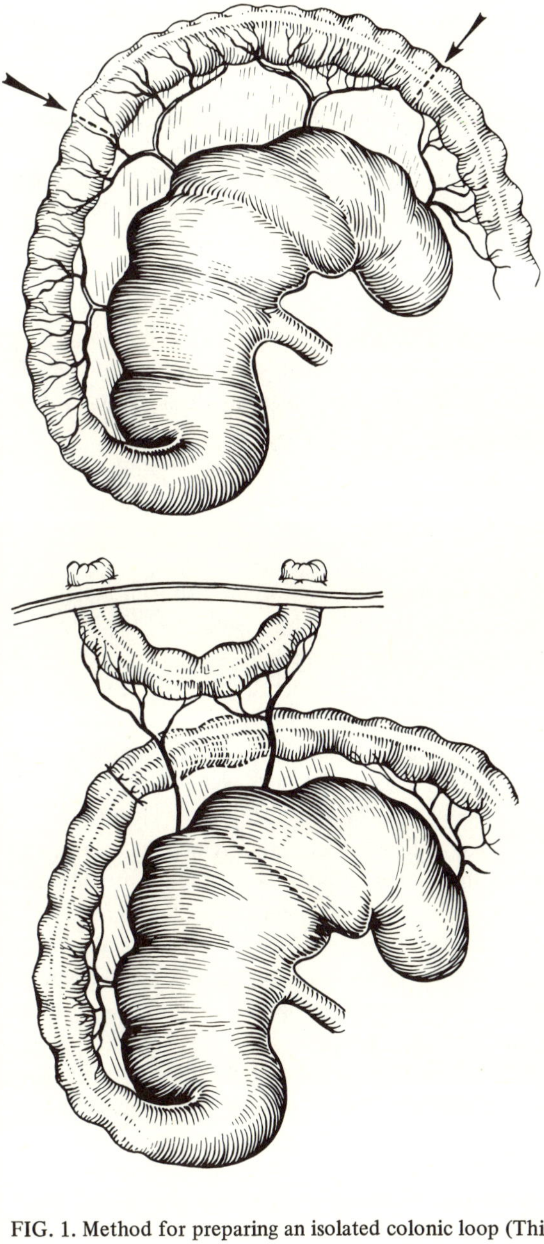

FIG. 1. Method for preparing an isolated colonic loop (Thiry-Vella) in a rabbit.

incision was made and the ascending colon was transected at 6–8 cm and 16–18 cm from the caecum, thus creating an isolated colonic segment of 10 cm. The mesentery was incised to permit mobilization and exteriorization of the two edges of the segment through separate stab wounds in the right abdomen. The colonic continuity was re-established by end-to-end anastomosis with interrupted silk sutures (4-0) (Fig. 1).

The mucosa was everted and sutured to create a matured 'nipple-like' colostomy. When silastic Foley catheters (with inflated balloons) are placed in both ostomies, the isolated colonic loop can be used as a living test tube for adherence experiments. The water-tight system could be easily perfused and collected in fractions.

The loop is flushed daily with distilled water or saline to prevent accumulation of mucus. The mucus is collected and analysed and can be used in incubation media for various adherence studies.

Despite the histological and functional changes, such as shortening of the villi and decreased mucus secretion, that occurred in the isolated colonic loops, this *in vivo* model can be used daily for at least four to six weeks.

DISCUSSION

Choppin: That is obviously a beautiful system, but when you gain something you lose something else. You are losing all the proteolytic enzymes, the bile salts and everything else that comes through the intestinal tract.

Mirelman: This is the colon region and very small amounts of those secretions are meaningful in the colon.

Choppin: Perhaps I picked the wrong examples, but a lot that comes down the tract will not be in that environment. It is a beautiful system, but I am not sure it is much closer to reality.

Taylor-Robinson: The other thing you have to consider is whether the rabbit model, despite the advantages of the functional colonic loop, is relevant to human disease.

Mirelman: The next step is to do these loops in primates. This is just the beginning of our efforts to find a more relevant system for studying the initial stages of intestinal infection. Until now we have used isolated mucosal scrapings and things that sometimes have no real relevance to what happens in the body.

Sussman: In the dog, ileal Thiry-Vella fistulae are unstable unless they are carefully handled. Is any special handling needed to keep the mucosa in the rabbit in a stable, normally functioning state for, say, six months?

Mirelman: The loop has to be made by a highly trained surgeon. One of the problems of older loops is that the amount of mucus secreted goes down with time. We have now done histology studies of sections from a three-month-old animal.

There is atrophy of the glands that are secreting the mucus. This is apparently due to the lack of stimulus from the regular colon stools. We could section it back and then it will regenerate. Apparently stimulus from the normal colon contents keeps it healthier. On the other hand I think these atrophied loops have an advantage: once the colon has degenerated or atrophied it may be even better as a model for infection and disease. With the mucus not secreting well, we may be able to find out whether it has a role in preventing infections.

Freter: Keren et al (1980) have published some work on small intestinal loops. One of the main difficulties was that a bacterial flora became established. I suspect that in your loops you get the opposite state. You probably lack the anaerobic conditions and the extensive bacterial flora normally associated with the large intestine. In the large intestine of normal rabbits we found very little that we could identify as mucus by direct observation.

As we know from germ-free animals, the endogenous mucopolysaccharides that accumulate are degraded in the large intestine. Is there any flora there at all in your loops?

Mirelman: There is bacterial flora. We haven't yet identified which is the flora or followed whether the flora changes in the older loops. However in the initial studies that we did by incorporating radioactively labelled bacteria, they seemed to bind well to the intestinal loop and remained there for quite a while. The animals kept on removing radioactive bacteria with every mucus secretion.

Freter: If you injected the same bacteria into the colon of a normal animal, would they stay there for quite a while? I suspect they would be rejected because of the antagonistic effect of the indigenous flora.

Mirelman: We haven't done that.

REFERENCE

Keren DF, Holt PS, Collins HH, Gemski P, Formal SB 1980 Variables affecting local immune response in ileal loops: role of immunization schedule, bacterial flora and postsurgical inflammation. Infect Immun 28:950-956

Adhesion of mycoplasmas to eukaryotic cells

S. RAZIN, I. KAHANE, M. BANAI and W. BREDT*

*Department of Membranes and Ultrastructure, The Hebrew University–Hadassah Medical School, Jerusalem, Israel and *Institute for General Hygiene and Bacteriology, Albert-Ludwigs University, D-7800 Freiburg, Federal Republic of Germany*

Abstract Many pathogenic mycoplasmas are surface parasites, adhering to the epithelial linings of the respiratory and urogenital tracts. Since mycoplasmas lack cell walls their plasma membrane comes in close contact with that of their host, allowing exchange of components between the two membranes and possibly fusion. The tight association of the parasite with its host is illustrated in scanning electron micrographs of *Mycoplasma pneumoniae* and *M. gallisepticum* adhering to human red blood cells. Specialized structures at the tips of the mycoplasma cells appear to function as attachment organelles. Our main aim has been to chemically define the receptors on the host cell and the binding sites on the mycoplasma cells responsible for adhesion. Glycophorin (the major sialoglycoprotein of human red blood cells) serves as the main or sole receptor for *M. gallisepticum* whereas *M. pneumoniae* binds to additional receptors on human red blood cells. Trypsin treatment of *M. pneumoniae* cells abolishes their ability to attach to human red cells, suggesting the protein nature of the binding sites. *M. pneumoniae* membranes solubilized by detergents were subjected to affinity chromatography on glycophorin–Sepharose so that membrane components with high affinity for glycophorin could be isolated. The fraction isolated consisted of several proteins (relative molecular mass 25 000 and 45 000). The binding of this fraction to red cells was relatively low but appeared to be specific, as it was inhibited by glycophorin but not by its hydrophobic moiety. The possibility is discussed that the exposure of the binding sites on the mycoplasma cell surface is influenced by the electrochemical ion gradient across the membrane.

Animal mycoplasmas usually adhere to and colonize the epithelial linings of the respiratory and urogenital tracts, only rarely invading the tissues and bloodstream. Hence, the mycoplasmas may be regarded as surface parasites. Their attachment is firm enough to prevent them from being removed by the action of ciliated epithelium of the respiratory tract or by the flow of urine. The intimate association between adhering mycoplasmas and their host cells gives rise to a situation in which local

1981 Adhesion and microorganism pathogenicity. Pitman Medical, Tunbridge Wells (Ciba Foundation symposium 80) p 98-118

concentrations of metabolites excreted by the microorganisms can accumulate and cause cell damage. Thus, the H_2O_2 excreted by the attached mycoplasmas may attack the host cell membrane without being rapidly destroyed by catalase or peroxidase present in the extracellular body fluids. Another end-product of mycoplasma metabolism is ammonia, which may cause tissue damage when produced in large quantities, such as during urea hydrolysis by ureaplasmas. It has also been suggested, but not proved experimentally, that proteases, nucleases and lipolytic enzymes produced by the attached mycoplasmas damage host cell membranes as well (Razin 1978). (The review on mycoplasmas by Razin (1978) will be referred to frequently in this paper, as it includes the original references on mycoplasma adhesion published before 1978.)

The importance of adhesion in mycoplasma pathogenicity is well exemplified by the human respiratory pathogen *Mycoplasma pneumoniae.* The finding that loss of virulence in this organism was associated with a markedly decreased adhesion capacity led to the conclusion that the adhesion of *M. pneumoniae* to respiratory epithelium is a prerequisite for the development of respiratory illness (Collier & Clyde 1971). Mycoplasma adhesion to eukaryotic cells differs in one important aspect from that of wall-covered bacteria: there is no wall to separate the plasma membrane of the parasite from that of its host. This may allow the two membranes to fuse, or at least exchange components. Obviously, if fusion occurs a wide variety of potentially cytotoxic proteins and lipids can be introduced directly into the host cells.

Mycoplasmas have been shown to adhere to many types of eukaryotic cells, including erythrocytes, macrophages, lymphocytes, HeLa cells and fibroblasts. Extensive use has also been made of tracheal and oviduct organ cultures, which serve as sensitive indicator systems for the study of adhesion and cell injury by respiratory and genito-urinary mycoplasmas *in vitro* (Collier 1972, Razin 1978).

The main aim of the studies of our groups in Jerusalem and Freiburg has been to chemically define the membrane components of the host and the parasite that are responsible for adhesion. The term 'receptor sites' is used to denote those components of the eukaryotic cell membrane that participate in adhesion, while the term 'binding sites' is reserved for the mycoplasma membrane components taking part in this process (Razin 1978). *M. pneumoniae* and the avian respiratory pathogen *M. gallisepticum* were chosen as test organisms and human or sheep red blood cells (RBC) were the eukaryotic cells used in our experimental systems. Both mycoplasmas are known to adhere to RBC, as manifested either by agglutination of RBC by mycoplasmas in suspension (haemagglutination) or by adsorption of RBC to mycoplasma colonies (haemadsorption). Accordingly, the experimental systems that we have developed are of two types. (1) Washed mycoplasmas, in which membrane lipids were labelled with [9,10-^{3}H] palmitate during growth, are incubated in buffer containing RBC. The RBC with attached mycoplasmas are collected by centrifugation at a low *g* value and the amount of mycoplasmas attached is determined by radioactivity measurements (Banai et al 1978, 1980). (2) A haemadsorption technique can be used in which washed RBC are attached to the surface growth of *M. pneumoniae* in cups of Linbro multi-well plates. The amount of RBC attached is determined by photometric measurement of the haemoglobin released by osmotic lysis of the RBC (Feldner et al 1979).

Factors influencing adhesion

Our experimental systems enabled us to define some of the factors affecting mycoplasma adhesion. The adhesion of *M. gallisepticum* to RBC and of *M. pneumoniae* to tracheal epithelial cells followed first-order kinetics (Powell et al 1976, Banai et al 1978). As would be expected from a reaction that depends on the frequency of impact of particulate matter, adherence of mycoplasmas to cells was found to be temperature-dependent (Manchee & Taylor-Robinson 1968, Gorski & Bredt 1977, Banai et al 1978, Feldner et al 1979). Since adhesion appears to be influenced by the energized state of mycoplasmas (see below), the lower level of cell energization at low temperatures may also be associated with the lower adhesion capacity observed in the cold.

The data available are insufficient to allow definite conclusions about the nature of the bonds responsible for mycoplasma adhesion. The data suffice, however, to indicate that at least some of the bonds are neither electrostatic nor ion-bridging, as increasing the pH of the medium from 6 to 8 and chelating divalent cations has little or no effect on the adhesion of *M. gallisepticum* to RBC (Banai et al 1978) or on adsorption of RBC to *M. pneumoniae* sheets (Feldner et al 1979). Increasing the ionic strength decreased *M. gallisepticum* adhesion to RBC (Banai et al 1978) but increased haemadsorption to *M. pneumoniae* sheets (Feldner et al 1979). Obviously, ionic bonds participate in mycoplasma adherence but cannot be the only ones involved.

Energized state and adhesion

The association of metabolic activity with adhesion capacity has been stressed by Hu et al (1976) and Powell et al (1976), who went so far as to conclude that only metabolically active *M. pneumoniae* cells are capable of adhering to respiratory ciliated epithelium. This conclusion is substantiated by the findings that the adhesion capacity of mycoplasmas is decreased or even nullified by the use of energy metabolism inhibitors (Gorski & Bredt 1977), by ageing of cultures (Manchee & Taylor-Robinson 1968, Banai et al 1980), by killing the mycoplasmas (Powell et al 1976), and by keeping the suspensions at low temperatures (Powell et al 1976, Gorski & Bredt 1977, Banai et al 1978, Feldner et al 1979). Most relevant to this discussion are the findings, described in detail elsewhere in this volume (Bredt et al 1981), which show the remarkable increase in adhesion of *M. pneumoniae* to glass in the presence of a metabolizable energy source, and the decrease in adhesion when metabolic inhibitors and uncouplers are added to mycoplasma preparations. We propose a working hypothesis to explain these findings, according to which the degree of exposure of the binding sites on mycoplasma surfaces depends on membrane energization. Factors which decrease the electrochemical ion gradients across the cell membrane, such as ageing, low temperature, growth inhibitors and ionophores, have been shown to decrease the degree of exposure of membrane proteins on the mycoplasma cell surface (Amar et al 1978, 1979). As will be discussed below, the *M. pneumoniae* binding sites are apparently membrane proteins exposed on the cell surface. Hence, conditions which affect the energized state of the mycoplasma membrane may result in less exposure of the binding sites and a consequent decrease in the adhesion capacity. Experiments are now under way to test this hypothesis.

Receptor sites

Adhesion of *M. pneumoniae, M. gallisepticum* and *M. synoviae* is usually affected, and in some cases nearly abolished, by pretreatment of the eukaryotic cells with neuraminidase (Razin 1978). Fig. 1 shows that the progressive decrease in the sialic acid content of neuraminidase-treated human RBC is accompanied by a decrease in the *M. gallisepticum* cells attached to them. Our finding (Banai et al 1978) that glycophorin, the human RBC glycoprotein carrying most of the sialic acid moieties of the RBC, inhibits adhesion of *M. gallisepticum* to human RBC supports the role of sialic acid moieties as receptor sites. Nevertheless, it appears that sialic acid moieties are not the only receptors for these *Mycoplasma* species. Thus, treatment of tracheal epithelial cells with neuraminidase decreased attachment of *M. pneumoniae* by only about 50–65% (Powell et al 1976, Gabridge et al 1979, Gabridge & Taylor-Robinson 1979). The residual 'background attachment' apparently involves receptors other than sialic acid. Recent observations indicate that various eukaryotic cells may differ in the nature of their receptor sites for the same mycoplasma. Extensive neuraminidase treatment of sheep RBC decreased their ability to bind *M. pneumoniae* by about 80–90% but had little effect on the binding capacity of human RBC and none on that of rabbit RBC (Feldner et al 1979). Similarly, neuraminidase treatment of human lung fibroblasts in culture decreased *M. pneumoniae* attachment by 83–96%, compared to a decrease of about 50% in neuraminidase-treated tracheal epithelial cells (Gabridge et al 1979, Gabridge & Taylor-Robinson 1979). It thus appears that eukaryotic cells may have some other, still unidentified, receptors in addition to those containing sialic acid. Nevertheless, there are good indications that glycophorin serves as the main receptor site

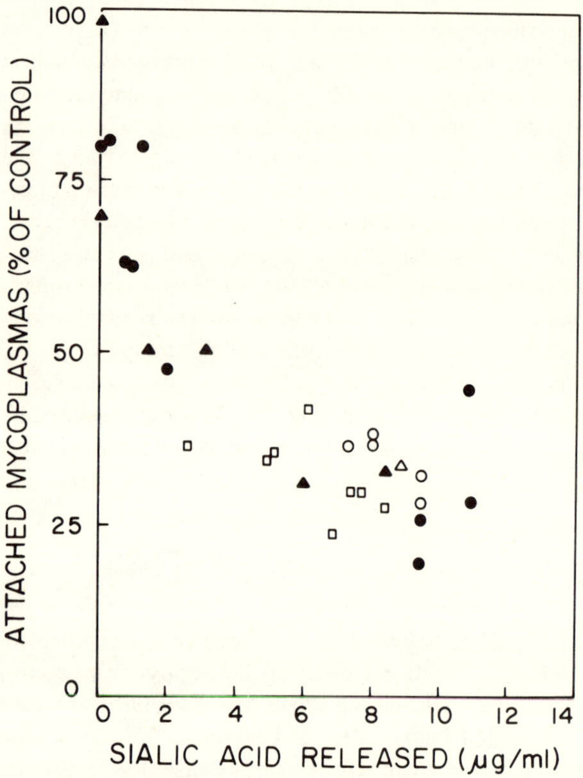

FIG. 1. Effects of removal of sialic acid from human RBC by neuraminidase on the ability of these cells to attach to *M. gallisepticum*. (From Banai et al 1978.)

for *M. gallisepticum* in human RBC (Banai et al 1978), while *M. pneumoniae* binds to other receptors in this membrane as well (Feldner et al 1979). In lung fibroblasts the main receptor sites for *M. pneumoniae* appear to be sialoglycoproteins rather than sialoglycolipids (Gabridge & Taylor-Robinson 1979).

Very little information is available on the chemical nature of the receptors for mycoplasmas other than *M. pneumoniae, M. gallisepticum* and *M. synoviae.* Receptors on HeLa cells for *M. hominis* and *M. salivarium,* though resistant to neuraminidase, were inactivated by proteolytic enzymes, suggesting their protein nature. On the other hand, rabbit RBC receptor sites for *M. dispar* resisted proteolytic treatment as well as treatment with neuraminidase or periodate. Similarly, adhesion of *M. pulmonis* to mouse macrophages was unaffected by pretreatment of the macrophages with neuraminidase, trypsin, chymotrypsin or glutaraldehyde, suggesting that the receptor sites are neither sialic acid residues nor membrane proteins (Razin 1978).

Binding sites

Satisfactory, though largely indirect, evidence is available in support of the protein nature of the binding sites on the surface of *M. pneumoniae* and *M. gallisepticum*. Pretreatment of these mycoplasmas by heat, merthiolate, glutaraldehyde or formalin partially or completely inhibits their attachment to eukaryotic cells (Sobeslavsky et al 1968, Gorski & Bredt 1977, Deas et al 1979). More relevant are the findings that mild trypsin treatment of *M. pneumoniae* cells nullified their ability to adhere to RBC (Gorski & Bredt 1977, Feldner et al 1979, Banai et al 1980) and tracheal cells (Sobeslavsky et al 1968, Hu et al 1977). The ability of *M. pneumoniae* to adhere, abolished by trypsin treatment, could be restored by incubating the treated cells in growth medium for 4 h at 37 °C but not at 4 °C. Chloramphenicol, mitomycin C or ultraviolet radiation inhibited the restoration of adhesion (Gorski & Bredt 1977, Feldner et al 1979). Polyacrylamide gel electrophoresis of *M. pneumoniae* cells showed that a major protein band, designated P_1 (relative molecular mass about 100 000), that was exposed on the cell surface disappeared when the cells were briefly treated with trypsin. Hu et al (1977) proposed P_1 as the *M. pneumoniae* binding site, as the failure of the mycoplasmas to attach correlated with its absence. Moreover, regeneration of P_1 after transfer of the trypsin-treated organisms to fresh growth medium for 6 h renewed the adhesion capacity. Erythromycin inhibited the resynthesis of the protein and the regeneration of adhesion capacity (Hu et al 1977). However, further studies by Hansen et al (1979) were disappointing in showing that an avirulent *M. pneumoniae* strain, with a reduced adhesion capacity, possessed the P_1 protein. Using two-dimensional electrophoresis Hansen et al (1979) detected three proteins that were synthesized by the virulent strain but not by the homologous avirulent one. One of these proteins was located on the cell surface by the lactoperoxidase iodination method and was removable by trypsin; nevertheless this did not prove to be P_1.

Our approach to the isolation of *M. pneumoniae* binding sites is based on using detergents to try to solubilize binding sites from cell membranes in an active form (Banai et al 1980). Our knowledge that *M. pneumoniae* binds to sialic acid receptors suggested the use of affinity chromatography with glycophorin as the ligand to isolate the binding sites from the solubilized membrane material. We expected that the apparent protein nature of the binding sites and their surface localization would allow them to be labelled by lactoperoxidase-mediated iodination of mycoplasma cells, which would facilitate the tracing and quantitation of the binding sites throughout the fractionation procedure.

The detergent sodium deoxycholate was chosen because of its efficiency in solubilizing membranes and its low denaturing activity. However, although about 45–60% of the membrane protein was solubilized when the detergent was used in concentrations of 0.5 to 1.6%, most of the binding activity remained in the insoluble fraction (Table 1), indicating the integral or intrinsic nature of the binding sites.

TABLE 1 Binding of fractions derived from *M. pneumoniae* membranes to RBC and Sepharose beads with and without glycophorin attached

Fraction tested	*Yield (% of total membrane protein)*	*Binding capacity*[a] *(% bound of total fraction in binding mixture)*		
		RBC	*Glycophorin–Sepharose beads*	*Sepharose beads*
Membranes	100.0	7.2	8.0	7.0
Deoxycholate-insoluble fraction	56.2	25.1	17.4	15.6
SDS-soluble fraction	39.0	8.5	22.7	4.0
Fraction eluted from glycophorin–Sepharose column	8.2	5.0	62.2	2.7

[a]Mean of results obtained in five experiments with different batches of membranes. The results represent the % of membranes or membrane fractions bound of the total included in the binding mixture.
(Data of Banai et al 1980.)

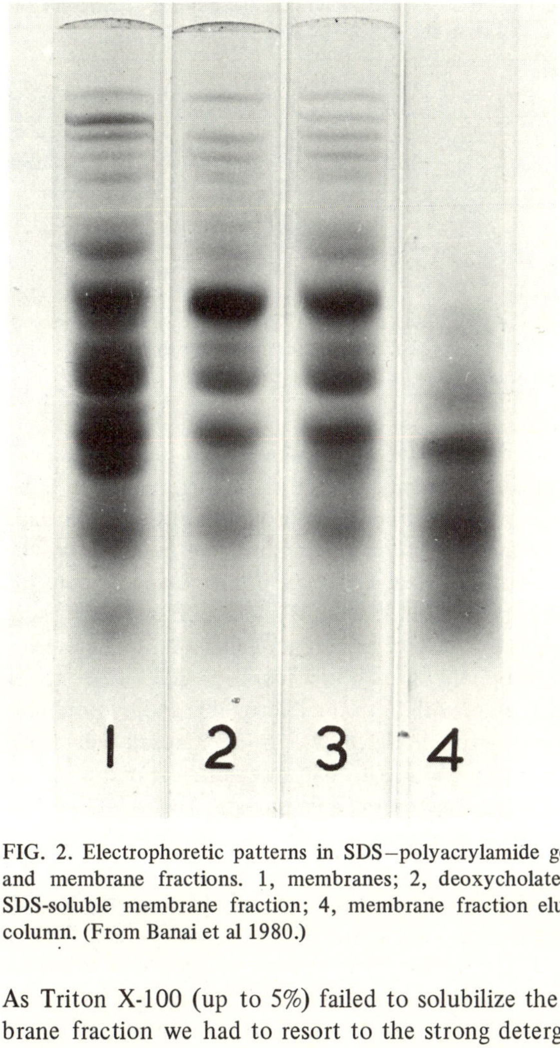

FIG. 2. Electrophoretic patterns in SDS–polyacrylamide gels of *M. pneumoniae* membranes and membrane fractions. 1, membranes; 2, deoxycholate-insoluble membrane fraction; 3, SDS-soluble membrane fraction; 4, membrane fraction eluted from glycophorin–Sepharose column. (From Banai et al 1980.)

As Triton X-100 (up to 5%) failed to solubilize the deoxycholate-insoluble membrane fraction we had to resort to the strong detergent, sodium dodecyl sulphate (SDS). Fortunately, a low SDS concentration (0.1%) sufficed to solubilize a significant part of the deoxycholate-insoluble membrane fraction in an active form suitable for further purification by affinity chromatography (Table 1).

The membrane fraction eluted from the glycophorin–Sepharose column with 0.2% SDS was enriched with two polypeptides with estimated relative molecular masses of 45 000 and 25 000 (Fig. 2). The high affinity of this fraction for glycophorin–Sepharose beads, though not for Sepharose beads themselves, indicated that the fraction consisted of or was enriched with binding sites to glycophorin (Table 1). The almost equal ability of glycophorin and its hydrophobic moiety to

TABLE 2 Effect of glycophorin and its hydrophobic moiety on the binding capacity of the M. pneumoniae membrane fraction eluted from the glycophorin–Sepharose column

Preincubation of the eluted fraction with[a]	*Binding capacity*[b] *(% bound of total fraction in binding mixture)*	
	RBC	*Glycophorin–Sepharose beads*
Buffer alone	4.4	71.0
Intact glycophorin	0.4	41.4
Hydrophobic moiety of glycophorin	4.9	40.3

[a]Glycophorin or its hydrophobic moiety was added to the binding mixture, at a protein concentration 10 times higher than that of the eluted fraction, 30 min before the addition of the fixed RBC or glycophorin–Sepharose beads. (Data of Banai et al 1980.)

inhibit binding of the eluted fraction to glycophorin–Sepharose beads (Table 2) indicates that this fraction contains components capable of binding to both moieties of the glycophorin molecule. This would support the assumption that both sialic acid residues and the hydrophobic moiety of glycophorin are exposed on the Sepharose beads, unlike the disposition of glycophorin in membranes, where the hydrophobic moiety is immersed in the lipid bilayer. Similarly, if the binding sites are integral membrane proteins, as our results suggest, solubilization by detergents is presumed to expose their hydrophobic moieties. Thus, these solubilized mycoplasma membrane proteins can interact with both the hydrophilic and hydrophobic moieties of glycophorin on contact with the glycophorin–Sepharose beads. However, when the isolated binding sites are added to RBC they interact only with the hydrophilic moiety of glycophorin exposed on the RBC surface. This would explain why intact glycophorin, but not its hydrophobic moiety, can inhibit the binding of the eluted fraction to RBC (Table 2). A major drawback of our study is that the eluted fraction had a relatively low capacity for binding to RBC, probably reflecting denaturation of the binding sites by the harsh detergent (SDS) used for the fractionation procedure. We are now seeking a milder agent capable of solubilizing the binding sites from *M. pneumoniae* membranes. Preliminary experiments in our laboratory (I. Kahane, unpublished) indicate that sodium deoxycholate effectively solubilizes human RBC binding sites from *M. gallisepticum* membranes. Thus it may be easier to isolate binding sites from *M. gallisepticum* membranes than from those of *M. pneumoniae.*

Attachment organelles and fusion

An intriguing question is whether the binding sites are evenly distributed throughout the mycoplasma cell surface. Both the filamentous *M. pneumoniae* and the

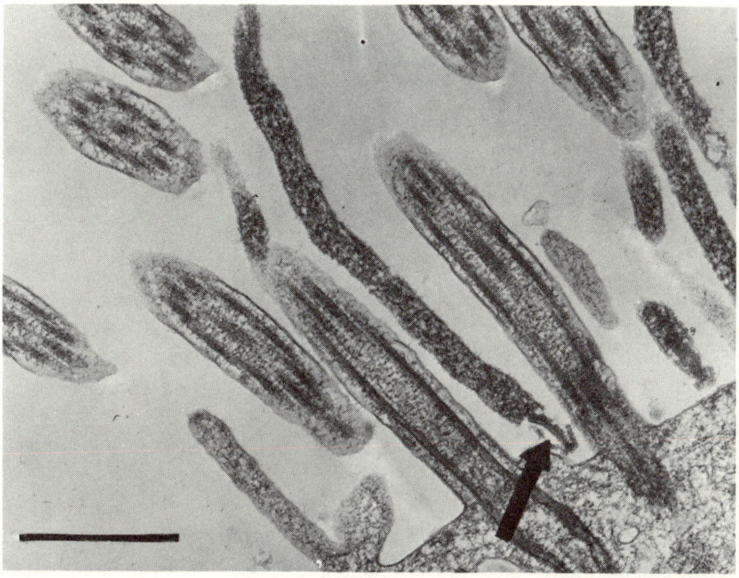

FIG. 3. Orientation of an *M. pneumoniae* filament to ciliated epithelial cells of human fetal trachea infected in organ culture for 48 h. The thin section shows the tip structure containing the dense rod-like element (arrow) adjacent to the epithelial cell surface. Scale bar, 0.5 μm. (From Collier 1972.)

fusiform *M. gallisepticum* possess specialized tip structures that may be associated with adhesion. Phase-contrast and scanning electron microscopy (SEM) show that the filamentous *M. pneumoniae* has a bulb-like 'neck' and a tapered tip. In thin sections the tip is seen to consist of a dense central rod-like core, surrounded by a lucent space and enveloped by the cell membrane. A peculiar bleb has been observed by a variety of electron microscopic techniques at one or both poles of the fusiform *M. gallisepticum* cells (Razin 1978). We have essentially no information on the chemical composition of the tip structures, as they have not yet been isolated.

Zucker-Franklin et al (1966) were the first to associate tip structures with adhesion. Their electron micrographs showed *M. gallisepticum* cells clustered around leucocytes, resembling iron filings on a magnet or flukes attached to their host; the bleb was the contact site more frequently than chance could account for. More recently Uppal & Chu (1977) demonstrated the proximity of the bleb to the surface of tracheal epithelium from fowls infected with *M. gallisepticum.* The attachment of *M. pneumoniae* to the respiratory epithelium of tracheal organ cultures by the tip structure is well documented (Fig. 3). The polarity shown by the adhesion of *M. gallisepticum* and *M. pneumoniae* to cells by their tip structures gives rise to the notion that a high concentration of binding sites exists on the surface of these structures. One way to test this hypothesis might be based on

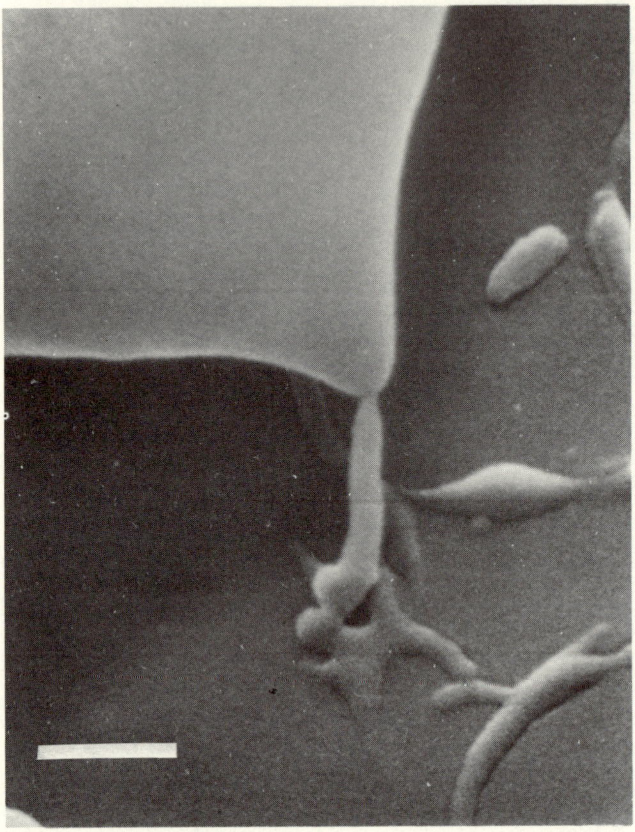

FIG. 4. Attachment of an erythrocyte to the tip of an *M. pneumoniae* filament observed by SEM. The distorted shape of the erythrocyte, caused by its tight attachment to the mycoplasma, can be seen. Scale bar, 0.5 μm. (From Razin et al 1980.)

interaction of the mycoplasma cells with ferritin-labelled antibodies to the binding sites or with labelled receptors such as glycophorin, and on localization of these reagents on the mycoplasma cell surface by electron microscopy. No such tests have been done yet, but labelling of *M. pneumoniae* and *M. gallisepticum* cells with less specific reagents (polycationized ferritin and lectins) failed to show any concentration of groups reacting with these reagents in the tip areas (Brunner et al 1979). There is evidence implying that binding sites are not restricted to the tip structures. Brunner et al (1979) showed by SEM that *M. pneumoniae* adheres to RBC via membrane sites apart from the tip structure. Organisms having spherical, pear-shaped or irregular form with no visible tip structures were found adhering to RBC surfaces. Our recent SEM studies (Razin et al 1980) show that the interaction between individual *M. pneumoniae* filaments and RBC can best be studied

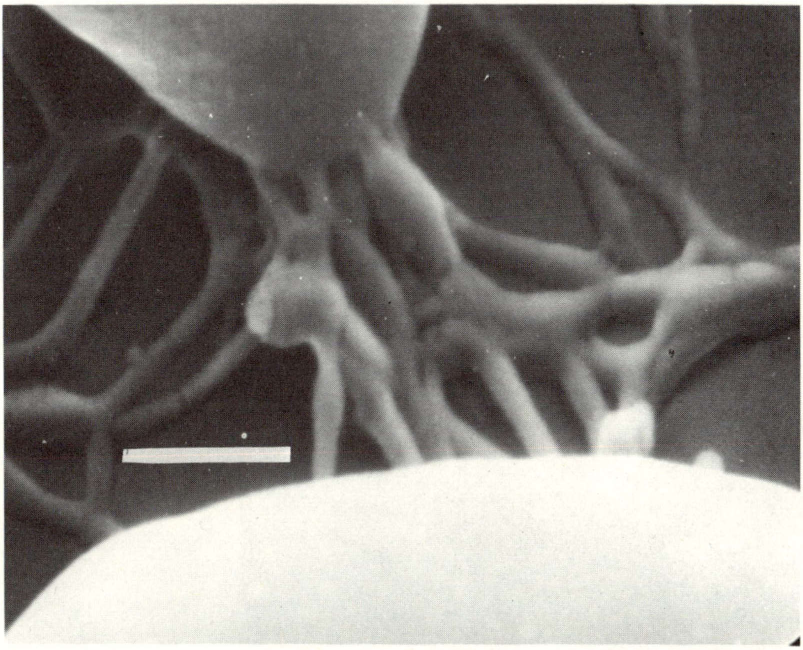

FIG. 5. An erythrocyte attached to *M. pneumoniae* filaments. The possibility of fusion at the site of contact is suggested by this scanning electron micrograph. Scale bar, 0.5 μm. (From Razin et al 1980.)

with RBC attached to the surface growth of *M. pneumoniae* on plastic coverslips of Leighton tubes. Our micrographs frequently show a single filament attaching to the RBC by its tip – an attachment sufficiently tight to distort the shape of the RBC considerably (Fig. 4). Our electron micrographs thus appear to confirm the role of the tip structure in adhesion. In this case, however, it seemed that most of the *M. pneumoniae* filament was attached to the plastic, leaving the tip free to interact with the RBC. Brunner et al (1979) speculated that the perpendicular position of *M. pneumoniae* filaments attached to ciliated respiratory epithelium by their tips (Fig. 3) is due to the gliding movement of the organisms (Razin 1978). The tip is perhaps the part of the organism that leads the way along the cilia, to reach and adhere to the epithelial cell surface. Perhaps the densely packed cilia permit the thin and flexible filamentous organisms to progress but obstruct spherical organisms on their way to epithelial cell surfaces. Accordingly, the perpendicular orientation of mycoplasmas in tracheal organ cultures is supported by the cilia. When there is no physical support, as with fibroblasts or on inert smooth surfaces, the mycoplasma filaments attach along their entire length (Razin 1978).

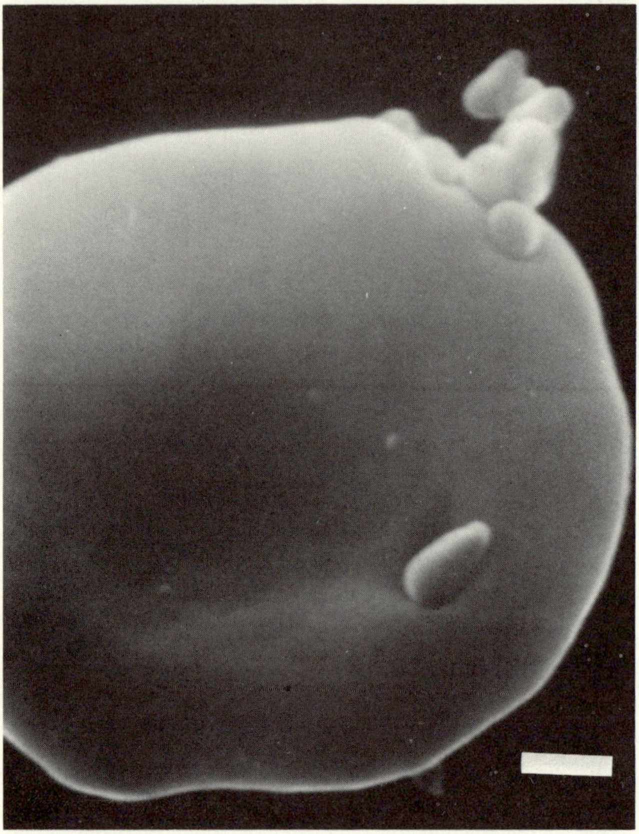

FIG. 6. *M. gallisepticum* cells adhering to human RBC, seen by SEM. The tight association of the mycoplasmas with the RBC is evident from the indentations of the erythrocyte surface at the sites of contact with the mycoplasmas. Scale bar, 0.5 μm. (From Razin et al 1980.)

Since there is no barrier (i.e. no cell wall) separating the plasma membrane of the adhering mycoplasmas from that of the host cells, fusion between the two membranes is a possibility. Using transmission electron microscopy Apostolov & Windsor (1975) provided some evidence for fusion of *M. gallisepticum* with RBC. Recent observations of *M. hyorhinis* capping on the surface of mouse lymphocytes (Stanbridge & Weiss 1978) and of exchange of membrane antigens between murine T lymphoblastoid cells and *M. hyorhinis* (Wise et al 1978) reflect the intimate association between the membrane of the parasite and its host. Our recent SEM study (Razin et al 1980) reveals the tight nature of the attachment of *M. gallispecticum* to RBC (Fig. 6). In *M. pneumoniae* the tips of the filaments appear to fuse with the RBC membrane (Fig. 5). Fusion cannot be proved on the basis of

SEM only, but these observations encourage further studies of the possibility of fusion between eukaryotic and mycoplasma cell membranes.

Concluding remarks

Animal mycoplasmas may be regarded as 'membrane parasites', firmly adhering to and colonizing the membranes of the epithelial lining of the respiratory and genital tracts of infected animals. They appear to be most successful as parasites, for they frequently cause chronic infections which are difficult to eradicate but rarely kill their hosts. Mycoplasmas seldom produce highly potent toxins; the damage they cause appears to depend on their close association with host-cell surfaces, so that even mildly toxic metabolic by-products, such as H_2O_2 and NH_4^+, have a toxic effect on the host cell membrane. Moreover, since mycoplasmas lack cell walls and adhere closely to the animal cell surface, the possibility of fusion between the membranes of the parasite and host has been suggested, but experimental evidence for this is insufficient. Recent studies suggest that the parasite and host membranes exchange antigenic components which perhaps trigger immunological responses of serious consequence to the host. Clearly, the issue of fusion and exchange of membrane components after adhesion of mycoplasmas must be investigated more thoroughly before definite conclusions can be drawn.

Sialic acid moieties on the host cell surface serve as specific, if not exclusive, receptors for the respiratory pathogens *M. pneumoniae, M. gallisepticum* and *M. synoviae.* Very little is known about the chemical nature of eukaryotic cell receptors for other mycoplasmas. There are sound indications that the binding sites in *M. pneumoniae* and *M. gallisepticum* membranes are protein in nature, but these proteins remain to be isolated and characterized and it is still unclear whether they are concentrated on the surface of the specialized tip structures characteristic of these mycoplasmas.

The use of affinity chromatography for the isolation of binding sites from solubilized mycoplasma membranes shows some promise. In this way we obtained an *M. pneumoniae* membrane fraction enriched with proteins having high affinity for glycophorin – the sialic acid-containing receptor of human RBC. Nevertheless, the failure to find a solubilizing agent that causes minimal denaturation still hampers progress by this approach. Little is known about the nature of the bonds responsible for mycoplasma adhesion. Ionic bonds appear to participate in adhesion but they are not the only ones involved. Elucidation of the chemical nature of the receptor and binding sites may clarify this issue as well. We are now testing our hypothesis that the degree of exposure of the binding sites on mycoplasma cell surfaces is influenced by the electrochemical ion gradient across the membrane. If this proves to be correct, as suggested by our preliminary results, conclusions drawn from mycoplasma studies may be useful for a better understanding of interactions among other cell types.

Acknowledgement

Our studies were supported by grant Br 296/12 from the Deutsche Forschungsgemeinschaft, Federal Republic of Germany.

REFERENCES

Amar A, Rottem S, Razin S 1978 Disposition of membrane proteins as affected by changes in the electrochemical gradient across mycoplasma membranes. Biochem Biophys Res Commun 84:306-312

Amar A, Rottem S, Razin S 1979 Is the vertical disposition of mycoplasma membrane proteins affected by membrane fluidity? Biochim Biophys Acta 552:457-467

Apostolov K, Windsor GD 1975 The interaction of *Mycoplasma gallisepticum* with erythrocytes. 1. Morphology. Microbios 13:205-215

Banai M, Kahane I, Razin S, Bredt W 1978 Adherence of *Mycoplasma gallisepticum* to human erythrocytes. Infect Immun 21:365-372

Banai M, Razin S, Bredt W, Kahane I 1980 Isolation of binding sites to glycophorin from *Mycoplasma pneumoniae* membranes. Infect Immun, in press

Bredt W, Feldner J, Kahane I 1981 Attachment of mycoplasmas to inert surfaces. In this volume, p 3-11

Brunner H, Krauss H, Schaar H, Schiefer H-G 1979 Electron microscopic studies on the attachment of *Mycoplasma pneumoniae* to guinea pig erythrocytes. Infect Immun 24:906-911

Collier AM 1972 Pathogenesis of *Mycoplasma pneumoniae* infection as studied in the human fetal trachea in organ culture. In: Pathogenic mycoplasmas. Excerpta Medica, Amsterdam (Ciba Found Symp 6), p 307-320

Collier AM, Clyde WA Jr 1971 Relationships between *Mycoplasma pneumoniae* and human respiratory epithelium. Infect Immun 3:694-701

Deas JE, Janney FA, Lee LT, Howe C 1979 Immune electron microscopy of cross-reactions between *Mycoplasma pneumoniae* and human erythrocytes. Infect Immun 24:211-217

Feldner J, Bredt W, Kahane I 1979 Adherence of erythrocytes to *Mycoplasma pneumoniae.* Infect Immun 25:60-67

Gabridge MG, Taylor-Robinson D 1979 Interaction of *Mycoplasma pneumoniae* with human lung fibroblasts: role of receptor sites. Infect Immun 25:455-459

Gabridge MG, Taylor-Robinson D, Davies HA, Dourmashkin RR 1979 Interaction of *Mycoplasma pneumoniae* with human lung fibroblasts: characterisation of the *in vitro* model. Infect Immun 25:446-454

Gorski F, Bredt W 1977 Studies on the adherence mechanism of *Mycoplasma pneumoniae.* FEMS (Fed Eur Microbiol Soc) Lett 1:265-267

Hansen EJ, Wilson RM, Baseman JB 1979 Two-dimensional gel electrophoretic comparison of proteins from virulent and avirulent strains of *Mycoplasma pneumoniae.* Infect Immun 24:468-475

Hu PC, Collier AM, Baseman JB 1976 Interaction of virulent *Mycoplasma pneumoniae* with hamster tracheal organ culture. Infect Immun 14:217-224

Hu PC, Collier AM, Basemen JB 1977 Surface parasitism by *Mycoplasma pneumoniae* of respiratory epithelium. J Exp Med 145:1328-1343

Manchee RJ, Taylor-Robinson D 1968 Haemadsorption and haemagglutination by mycoplasmas. J Gen Microbiol 50:465-478

Powell DA, Hu PC, Wilson M, Collier AM, Baseman JB 1976 Attachment of *Mycoplasma pneumoniae* to respiratory epithelium. Infect Immun 13:959-966

Razin S 1978 The mycoplasmas. Microbiol Rev 42:414-470

Razin S, Banai M, Gamliel H, Polliack A, Bredt W, Kahane I 1980 Scanning electron microscopy of mycoplasmas adhering to erythrocytes. Infect Immun, in press

Sobeslavsky O, Prescott B, Chanock RM 1968 Adsorption of *Mycoplasma pneumoniae* to neuraminic acid receptors of various cells and possible role in virulence. J Bacteriol 96:695-705
Stanbridge EJ, Weiss RL 1978 Mycoplasma capping in lymphocytes. Nature (Lond) 276:583-587
Uppal PK, Chu HP 1977 Attachment of *Mycoplasma gallisepticum* to the tracheal epithelium of fowls. Res Vet Sci 22:259-260
Wise KS, Cassell GH, Acton RT 1978 Selective association of murine T-lymphoblastoid cell surface alloantigens with *Mycoplasma hyorhinis.* Proc Natl Acad Sci USA 75:4479-4483
Zucker-Franklin D, Davidson M, Thomas L 1966 The interaction of mycoplasmas with mammalian cells. I. HeLa cells, neutrophils and eosinophils. J Exp Med 13:521-531

DISCUSSION

Taylor-Robinson: Apostolov & Windsor (1975) used electron microscopy to study the association of *Mycoplasma gallisepticum* and red cells, and reported evidence of fusion.

Razin: Electron microscopy by itself probably is not sufficient to prove fusion. Usually there is a distance of about 10 nm separating the triple layer of the mycoplasma membrane from that of the eukaryotic cells in culture (Boatman et al 1977). One needs to label mycoplasma cell contents (i.e. DNA) and check their incorporation into the eukaryotic cell to prove fusion.

Bredt: I believe even less in fusion than you do. When mycoplasmas are crawling on glass they quite often attach to each other but they always separate again as individuals. If mycoplasmas could fuse with eukaryotic cells they should also fuse with each other. Fusion cannot be ruled out but the organisms apparently keep their integrity unless something like a virus produces fusion.

Taylor-Robinson: Has anybody really tried to fuse mycoplasmas?

Bredt: Certainly not with a virus.

Razin: I. Kahane and A. Loyter tried long ago to fuse *M. gallisepticum* with Sendai virus. They failed, apparently because of the lack of sialic acid receptors for the virus on the mycoplasma surface (unpublished work).

Taylor-Robinson: Most studies of *M. pneumoniae* attachment have been done in tracheal organ cultures in which there is ciliated epithelium. The organisms tend to be propped up vertically by the cilia and the specialized terminal structure may be directed fortuitously at the cell membrane and only give the appearance of adhering preferentially. However, when Mike Gabridge (Gabridge et al 1979) looked at *M. pneumoniae*-infected fibroblasts in culture, he found that the organisms were lying flat but their terminal structures were stuck to the cell membranes, indicating that this structure is the attachment organelle.

Razin: The peculiar attachment organelles are important in the motility of mycoplasmas on inert surfaces (Bredt 1979). We believe that mycoplasmas are motile not only on glass or plastic but also *in vivo*. The suggestion is that the *M. pneumoniae* filaments move with their tip forward, penetrating the mucus layer along the cilia and finally reaching the epithelial cell surface. Sections of ciliated epithelium contaminated with *M. pneumoniae* show that all the tip organelles face the epithelial cell surface, indicating the polarity of the mycoplasma filaments.

Hughes: I was interested in the mechanical effects on the erythrocyte after binding of the mycoplasma. Is this a purely physical deformation or is there some sort of enzymic or other metabolic effect of the mycoplasma on the red blood cell?

Razin: What I showed you was a physical effect. The attachment is very tight and when we try to wash away the attached erythrocytes this changes their shape. However, if erythrocytes are kept for long with the mycoplasmas the erythrocytes may lyse. Lysis is caused apparently by hydrogen peroxide excreted by the attached mycoplasmas in the vicinity of the erythrocyte membrane. In this area of attachment one would expect the peroxidation of the erythrocyte membrane lipids, resulting in lysis. We also know that when heavy suspensions of *M. gallisepticum* are incubated with red blood cells for 4 h at 37 °C the red blood cells are digested and disappear (Ziv et al 1967).

Hughes: But in the short term is the binding reversible? If you can dissociate them in some way does the mycoplasma go back to its original shape?

Razin: We could not dissociate them at all, and the attachment seems to be irreversible. All we could do was to interfere with attachment if we first added glycophorin to the mycoplasmas and then added the erythrocytes. When we added glycophorin after attachment it didn't detach them (Banai et al 1978).

Beachey: Is binding of glycophorin to mycoplasma reversible?

Razin: Labelled glycophorin binds very well to *M. gallisepticum* and we couldn't remove it even with neuraminidase. However, with purified glycophorin there is a problem as it contains a hydrophobic moiety that can bind in a non-specific manner (Banai et al 1978).

Helenius: Does trypsin remove glycophorin?

Razin: We didn't try it but it should digest part of the glycophorin.

Howard: Does the tryptic fragment released from *M. pneumoniae* retain binding activity to erythrocytes?

Razin: We have tried this approach. We used low concentrations of trypsin to digest *M. pneumoniae* and collected the released peptides, hoping that trypsin would dissociate the binding sites exposed on the mycoplasma surface in a biologically active form. The proteins on the cell surface were first labelled by lactoperoxidase-mediated iodination. The labelled peptides released by the trypsinization failed to bind to red blood cells (Banai et al 1980).

Wallach: I am a little concerned about the hydrophobic moieties of glycophorin and their role here. Once one solubilizes glycophorin, replacing lipid with detergents or attaching it to solid phases, the hydrophobic segment changes in complex ways. This may foster formation of dimers or even larger aggregates which will influence binding studies. One may be dealing with not very specific hydrophobic interactions between two proteins rather than with specific binding sites. How do other sialoproteins behave in terms of protein association or in this binding? There are proteins with equivalent sugar moieties. How do gangliosides behave?

Razin: One would expect isolated glycophorin to form micelles in water. The micellar glycophorin interferes with the binding of *M. gallisepticum* to human erythrocytes. When sialic acid residues were removed by acid hydrolysis of glycophorin the resulting product retained some of the inhibitory activity of intact glycophorin (Banai et al 1978). So you are quite right that the isolated glycophorin may show also hydrophobic attachment, which is not the specific attachment due to the sialic acid moieties of this molecule. We tried α_1 acid glycoprotein, a soluble serum glycoprotein, which was also found to interfere with *M. gallisepticum* attachment to erythrocytes (M. Banai, I. Kahane and S. Razin, unpublished work). It would perhaps be more convenient to use a true water-soluble sialoglycoprotein for the purification of the binding sites. Gabridge & Taylor-Robinson (1979) found that gangliosides containing sialic acid did not bind to *M. pneumoniae* and were not effective in inhibiting attachment of the mycoplasmas to fibroblasts. However, glycoproteins which were isolated by lithium diiodosalicylate from the fibroblast membranes were bound effectively to *M. pneumoniae.*

Feizi: The problem with gangliosides is that they themselves may be incorporated into cell membranes and therefore may even enhance the attachment of the mycoplasmas. Haemagglutination inhibition assays with glycosphingolipids can be difficult even when they are tested in the presence of carrier lipids (Feizi et al 1978).

Wallach: They can be effective in competing with receptors for a variety of peptide hormones. Did you try binding of mycoplasmas to En(a$^-$) erythrocytes?

Razin: The En(a$^-$) erythrocytes still have quite a lot of sialic acid, about 50% of the normal (Tanner & Anstee 1976).

Svanborg Edén: One function of sialic acid may be to help in the presentation of receptor structures in the cell membrane. With Sendai virus it has been shown that two sialic acid residues at the end of a disaccharide increase the affinity of binding 50–100 times compared to one residue (Holmgren et al 1980). The neuraminidase treatment removing sialic acid may be enough to block binding although the proper receptor sequence is still present.

Razin: We tried to measure the sialic acid remaining after extensive neuraminidase treatment of human erythrocytes. We couldn't find any colorimetrically but the technique is not very sensitive. Gas–liquid chromatography indicated that about 5% of the original sialic acid remained in the membrane (Banai et al 1978). Since mycoplasmas also bind non-specifically to surfaces such as plastic and glass, one wouldn't expect the total removal of sialic acid to abolish attachment completely. We may conclude that at least 80% of the receptor sites for *M. gallisepticum* in the red blood cell are sialic acid moieties. The other 20% are either residual sialic acid moieties which are hard to detect, or may consist of other receptors.

Taylor-Robinson: I want to emphasize the diversity of receptors on cells for different mycoplasmas. Dr Razin concentrated on *M. gallisepticum* and *M. pneumoniae* but the order Mycoplasmatales consists of three different families, of which the family Mycoplasmataceae contains the genus *Mycoplasma* which encompasses at least 50 different species that infect humans and animals. Richard Manchee and I (Manchee & Taylor-Robinson 1969a) examined a number of these and instead of mixing suspensions of mycoplasmas with cells, we added cells to organisms, the organisms being in the form of colonies on agar. We could demonstrate haemadsorption, spermadsorption and tissue-culture cell adsorption with some mycoplasma species and the appropriate cell suspension. Human sperm, for example, attached to *M. pneumoniae* colonies very tenaciously by their heads. In fact, the attachment of these or other cells could not usually be broken by vigorous washing. The ability to demonstrate the phenomenon turns on the fact that the mycoplasma colonies themselves are tenaciously stuck to or within the agar and do not separate on washing, as do bacterial colonies.

For the most part there is little specificity. Thus, erythrocytes from a variety of mammalian or avian species will adhere to colonies of a particular mycoplasma species. However, occasionally one sees some specificity. For example, we saw no attachment of human sperm to colonies of *M. bovigenitalium* but there was attachment of bovine sperm.

By treating the cells or the colonies with different chemicals and enzymes so that attachment is prevented, it is possible to come to some conclusions about the chemical nature of the cell receptors and the mycoplasma binding sites. The receptors for many, but not all, of the mycoplasmas are glycoproteins and we agree with Shmuel Razin that the receptor for *M. pneumoniae* is a sialoglycoprotein. For *M. gallisepticum,* also *N*-acetylneuraminic acid-mediated, we were able to demonstrate a 'receptor gradient' in the same way that such gradients have been shown for myxoviruses. Thus, the amount of neuraminidase required to prevent attachment of red cells to one strain of *M. gallisepticum* was 100-fold greater than the amount required to prevent attachment to another strain (Manchee & Taylor-Robinson 1969b). That led us to wonder whether different strains of this avian mycoplasma might show differences in pathogenicity according to the avidity with which they adhered to cells of the respiratory tract. This is an appropriate thought in the context of this conference although we have not had an opportunity of testing its validity.

Our results also show that the binding sites on the membranes of many of the mycoplasmas are proteins, often glycoproteins, and that they are not often identical. Thus, within a large group of different microorganisms that we simply refer to as mycoplasmas, there are variations in binding sites and different receptors on the cells for those binding sites.

Elbein: Neuraminidase of course acts on these things and takes off sialic acid but there are different neuraminidases with different specificities for the various linkages of sialic acids. For example, *V. cholerae* neuraminidase attacks different linkages than clostridial neuraminidase (Ashwell & Morell 1974). The use of neuraminidase is dangerous because there is more than one sialic acid in these receptor sites. Maybe what it is linked to is what we should be looking at. This is probably also true with other glycosidases such as mannosidase.

Feizi: I want to raise some questions and make some comments on the relationship between *M. pneumoniae* and a kind of autoantibody which develops after infection with this microorganism. Among patients with *M. pneumoniae* infection, about 30% develop serum autoantibodies (cold agglutinins) directed against I antigen of erythrocytes. Why do red cell antibodies develop after this infection? Does *M. pneumoniae* itself have I antigen? Are the anti-I antibodies directed against an antigen on the mycoplasma and do they cross-react with the antigen on red cells? There is no concrete evidence that laboratory-cultured strains of *M. pneumoniae* have I antigen. However, it is possible that the mycoplasmas growing in an environment containing I antigen may acquire this antigen. Alternatively the anti-I antibodies may arise because *M. pneumoniae,* in interacting with red cells, modifies the I antigen or somehow renders it autoimmunogenic. In 1969 we immunized rabbits with human red cells which had been pretreated with live *M. pneumoniae* and about 30% of the rabbits developed anti-I cold agglutinins (Feizi et al 1969).

We have heard that *M. pneumoniae* binds to sialic acid receptors on red cell membranes and we now know that a proportion of I-active oligosaccharides of red cells occur in sialylated form (Feizi et al 1978). I antigen is a branched carbohydrate structure carried on glycosphingolipids (Watanabe et al 1979, Feizi et al 1979) and on band 3 protein of erythrocytes (Childs et al 1978). The same carbohydrate sequence also occurs in secreted glycoproteins (Feizi et al 1971, Hounsell et al 1980). In all of these situations I antigen can become further glycosylated and converted into blood group H, A and B active structures, or it can become sialylated. We have recently described gangliosides of human and bovine erythrocyte membranes which have I antigens as their core structure. The mycoplasma binds to glycophorin but it may also bind to these gangliosides. Thus the mycoplasma may come into close contact with I-active oligosaccharides on erythrocyte membranes, perhaps on bronchial epithelial cells also. The increased immunogenicity of I antigen may be related to its desialylation by the mycoplasma or may be due to an adjuvant effect.

It is interesting that cold agglutinins directed against i antigen develop after infectious mononucleosis. The i antigen is a linear oligosaccharide structure related to I (Niemann et al 1978). Peripheral blood lymphocytes are extremely rich in i antigen. It is interesting that infection with Epstein-Barr virus, which has special affinity for B lymphocytes, should elicit anti-i autoantibodies, while infection with *M. pneumoniae* (an agent potentially able to react with I structures) elicits anti-I autoantibodies.

REFERENCES

Apostolov K, Windsor GD 1975 The interaction of *Mycoplasma gallisepticum* with erythrocytes: I. Morphology. Microbios 13:205-215

Ashwell G, Morell AG 1974 The role of surface carbohydrates in the hepatic recognition and transport of circulating glycoproteins. Adv Enzymol 41:99-128

Banai M, Kahane I, Razin S, Bredt W 1978 Adherence of *Mycoplasma gallisepticum* to erythrocytes. Infect Immun 21:365-372

Banai M, Razin S, Bredt W, Kahane I 1980 Isolation of binding sites to glycophorin from *Mycoplasma pneumoniae* membranes. Infect Immun, in press

Boatman E, Cartwright F, Kenny GE 1977 Morphology, morphometry and electron microscopy of Hela cells infected with bovine *Mycoplasma.* Cell Tissue Res 170:1-16

Bredt W 1979 Motility. In: Barile MF, Razin S (eds) The mycoplasmas. Academic Press, New York, vol 1:141-155

Childs RA, Feizi T, Fukuda M, Hakomori S 1978 Blood group I activity associated with band 3, the major intrinsic membrane protein of human erythrocytes. Biochem J 173:333-336

Gabridge MG, Taylor-Robinson D 1979 Interaction of *Mycoplasma pneumoniae* with human lung fibroblasts: role of receptor sites. Infect Immun 25:455-459

Gabridge MG, Taylor-Robinson D, Davies HA, Dourmashkin RR 1979 Interaction of *Mycoplasma pneumoniae* with human lung fibroblasts. I. Characterization of the in vitro model. Infect Immun 25:446-454

Feizi T, Taylor-Robinson D, Shields MD, Carter RA 1969 Production of cold agglutinins in rabbits immunized with human erythrocytes treated with *Mycoplasma pneumoniae.* Nature (Lond) 222:1253-1256

Feizi T, Kabat EA, Vicari G, Anderson B, Marsh W 1971 Immunochemical studies on blood groups XLIX. The I antigen complex: specificity differences among anti-I sera revealed by quantitative precipitin studies; partial structure of the I determinant specific for one anti-I serum. J Immunol 106:1578-1592

Feizi T, Childs RA, Hakomori S, Powell ME 1978 Blood group Ii active gangliosides of human erythrocyte membranes. Biochem J 173:245-254

Feizi T, Childs RA, Watanabe K, Hakomori S 1979 Three types of blood group I specificity among monoclonal anti-I autoantibodies revealed by analogues of a branched erythrocyte glycolipid. J Exp Med 149:975-980

Holmgren J, Svennerholm L, Elwing H, Fredman P, Strannegård O 1980 Sendai virus receptor: Proposed recognition structure based on binding to plastic adsorbed gangliosides. Proc Natl Acad Sci USA, in press

Hounsell EF, Fukuda M, Wood E, Powell ME, Feizi T, Hakomori S 1980 Structural analysis of hexa- to octasaccharide fractions isolated from blood group I and i active sheep gastric glycoproteins. Biochem Biophys Res Commun 92:1143-1150

Manchee RJ, Taylor-Robinson D 1969a Studies on the nature of receptors involved in attachment of tissue culture cells to mycoplasmas. Br J Exp Pathol 50: 66-75

Manchee RJ, Taylor-Robinson D 1969b Utilization of neuraminic acid receptors by mycoplasmas. J Bacteriol 98:914-919

Niemann H, Watanabe K, Hakomori S, Childs RA, Feizi T 1978 Blood group i and I activities of 'lacto-N-*nor*hexaosyl-ceramide' and its analogues: the structural requirements for i-specificities. Biochem Biophys Res Commun 81:1286-1293

Tanner MJA, Anstee DJ 1976 The membrane change in En(a-) human erythrocytes. Biochem J 153:271-277

Watanabe K, Hakomori S, Childs RA, Feizi T 1979 Characterization of a blood group I-active ganglioside. Structural requirements for I and i specificities. J Biol Chem 254:3221-3228

Ziv R, Perek M, Razin S 1967 The action of *Mycoplasma gallisepticum* upon chicken, rabbit and cow erythrocytes. Avian Dis 11:370-377

Bacterial adherence to cell surface sugars

NATHAN SHARON, YUVAL ESHDAT, FREDRIC J. SILVERBLATT* and ITZHAK OFEK†

Department of Biophysics, The Weizmann Institute of Science, Rehovoth and †Department of Human Microbiology, Sackler School of Medicine, Tel Aviv University, Tel Aviv, Israel

Abstract Bacterial adherence to animal cell surfaces is of interest because of its relation to pathogenicity and the insight it provides into determinants of intercellular recognition. The attachment of various strains of *Escherichia coli* and *Salmonella* spp. to epithelial cells and phagocytes is inhibited by D-mannose, and the adherence of other bacteria is inhibited by sugars such as L-fucose and D-galactose, suggesting that sugar-mediated adherence is widespread. This intercellular recognition is thought to be mediated by sugar residues (e.g. D-mannose) on the surface of animal cells, to which bacteria attach by a sugar-binding substance on their surface. The nature of the receptors on the animal cells is unknown. There is evidence that *E. coli* produces lectin-like substances specific for D-mannose, by which it binds to the cells. The most common form of these lectin-like substances appears to be the bacterial pili, which can be reversibly dissociated into their protein subunits. The lectin can also be in the form of bacterial flagella or tightly attached to the outer membrane of the bacteria. Mannose-specific attachment may assist bacteria in colonizing and invading their hosts: methyl α-D-mannoside (but not methyl α-D-glucoside) significantly reduced infection of the urinary tract of mice by virulent strains of *E. coli*. Once bacteria penetrate the host their ability to bind sugars on phagocytes may impair their virulence by facilitating phagocytosis. Further studies of the sugar-mediated bacterial adherence by organisms growing *in vivo* and the structural identification of the host cell receptors may lead to the design of more effective adherence inhibitors that may help to prevent certain bacterial infections.

The first indication that sugar-specific interactions may play a key role in relationships between animals and their microbial pathogens came with the discovery in the 1940s of the binding of influenza virus to sialic acid residues on the surface of erythrocytes. The involvement of sugars in the attachment of bacteria to cells was originally suggested by the findings of Collier & de Miranda (1955), who showed

*On leave of absence from Veterans Administration Sepulveda Medical Center and Department of Medicine, University of California at Los Angeles.

1981 Adhesion and microorganism pathogenicity. Pitman Medical, Tunbridge Wells (Ciba Foundation symposium 80) p 119-141

that D-mannose, alone of many sugars tested, inhibited the agglutination of human erythrocytes by *Escherichia coli.* In a series of investigations, begun in 1955, Duguid and his co-workers (see Duguid et al 1979 and references therein) have demonstrated the adherence of piliated enteric bacteria to erythrocytes, leucocytes and epithelial cells, and have found that this attachment is inhibited specifically by low concentrations of D-mannose. Despite these early reports little attention was paid to the possible role of sugar-mediated recognition in the colonization of mucosal surfaces by bacteria. Research in this area has been greatly stimulated by the realization that cell surface sugars play a central role in biological recognition (Sharon 1979). Evidence has been obtained in our laboratories and elsewhere that strongly suggests that sugar-binding proteins, akin to lectins (Lis & Sharon 1977) and present on bacterial surfaces, may serve to attach certain pathogens to sugar residues on a variety of animal cells. Indeed, the sugar-specific interaction between bacteria and epithelial cells is perhaps the best example of carbohydrates acting as determinants of intercellular recognition. However, bacteria may attach to cell surfaces not only by sugar-specific interactions but also by other means. Moreover, a given bacterial strain may display both sugar-mediated and non-sugar-mediated adherence (Duguid et al 1979).

When studying bacterial attachment to cell surface sugars we should consider the following questions:

(1) How widespread is the occurrence of sugar-binding activity among pathogenic bacteria?
(2) What is the nature of the bacterial surface constituents that bind sugars?
(3) What is the structure of the corresponding receptors on animal cells?
(4) What is the role of sugar-specific adherence in natural infection?
(5) Can the knowledge gained in studies of sugar-mediated bacterial adherence be applied to devise new approaches to the prevention and treatment of bacterial infections?

Before we answer these questions we shall give a brief description of the methods used.

Methodology

Hapten inhibition of cell agglutination (e.g. haemagglutination) by bacteria or of bacterial attachment to animal cells (Table 1) is the method of choice for examining whether sugars are involved in these interactions. However, from such experiments it is not possible to deduce which of the two types of cell contains the sugar-binding lectin or 'adhesin', since all cells are coated by sugars and the occurrence of lectins both in bacteria and in animal cells is well documented (Lis & Sharon 1977). Table 2 summarizes the type of experiments required to ascertain that bacteria do indeed bind to sugars on animal cells surfaces via bacterial lectins. Only with a few strains

TABLE 1 Mammalian cells that bind bacteria via surface sugars

Species	*Cell type*
Man	Epithelial: buccal
	tracheal
	intestinal
	urogenital
	Blood: erythrocytes
	lymphocytes
	polymorphonuclear leucocytes
Mouse	Urinary tract
	Lymphocytes
	Macrophages
Rabbit	Intestinal
	Erythrocytes
Dog	Erythrocytes
Horse	Erythrocytes
Guinea-pig	Erythrocytes
Monkey	Erythrocytes
Pig	Erythrocytes

TABLE 2 Evidence for cell surface sugars as attachment sites for bacteria

1. Inhibition by
   (a) Monosaccharides and derivatives
   (b) Lectins
   (c) Sugar oxidants (periodate, galactose oxidase)
   (d) Glycosidases
2. Binding of bacteria to immobilized sugars
3. Isolation of bacterial lectin
4. Isolation of host cell surface receptor

of *E. coli* have many of the required experiments been done, but even with this organism the corresponding receptors have not yet been isolated.

Studies in our laboratory using *E. coli* K12 have shown that, out of about a dozen sugars tested, only D-mannose and its derivatives inhibited at low concentrations the attachment of the bacteria to human buccal epithelial cells (Fig. 1) or displaced the preattached bacteria from the cells (Ofek et al 1977, 1978). Yeast mannan, a polymer of D-mannose, was also a strong inhibitor, whereas glycogen, a polymer of D-glucose, was not inhibitory. The inhibition by methyl α-D-mannoside was dose-related and linear in the range of 2–6 mg/ml. Treatment of the epithelial cells with concanavalin A, a lectin which binds to D-mannose (or D-glucose) residues on cell surfaces, also inhibited bacterial attachment; no agglutination of *E. coli* K12

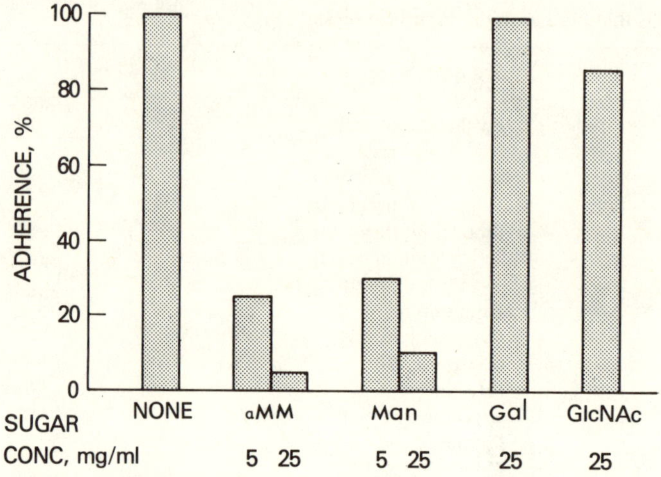

FIG. 1. Inhibition by different sugars of the adherence of *E. coli* 3092 to human oral epithelial cells (αMM, methyl α-D-mannoside). (Data from Ofek et al 1977.)

by concanavalin A was observed. Other lectins with different sugar specificities (wheat germ agglutinin specific for *N*-acetyl-D-glucosamine and penut agglutinin specific for D-galactose) were not inhibitory. Furthermore, epithelial cells treated with sodium metaperiodate were no longer able to bind *E. coli.* Evidence for a mannose-specific lectin on *E. coli* has also been obtained (see p 126).

Detailed information on the specificity of the sugar-combining site of the bacteria that mediates their adherence can be obtained by comparing the inhibition of adherence by a large number of structurally related compounds, as is done in mapping the combining sites of enzymes, antibodies or lectins. Perhaps the only example of such a study with bacteria is by Old (1972), who examined the effect of many monosaccharides, oligosaccharides and their derivatives on the mannose-sensitive agglutination of guinea-pig or horse erythrocytes by certain strains of *Salmonella typhimurium* or *Shigella flexneri.* Modification of any of the hydroxyl groups at the C-2, C-3, C-4 or C-6 positions in the D-mannopyranosyl molecule resulted in loss of inhibitory activity, showing that these groups are necessary for binding to the sugar-combining site on the bacteria. The α-configuration at the C-1 position in D-mannose is important since carbohydrates containing α-linked D-mannose were much stronger inhibitors than those containing β-linked D-mannose. Old (1972) also drew attention to the remarkable degree of similarity between the specificity of the mannose-sensitive bacterial haemagglutination and that of concanavalin A.

Among the criteria listed in Table 2 the demonstration of the binding of bacteria to immobilized sugars is most useful, especially for screening large numbers of isolates for their ability to adhere to cell surface sugars. The immobilized sugar may

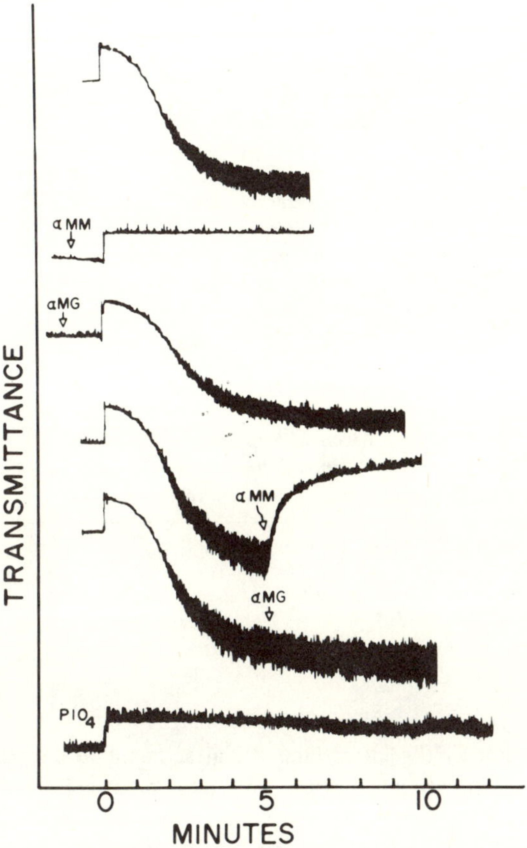

FIG. 2. Aggregation of mannan-containing yeast cells by *E. coli* (top curve) was completely inhibited by preincubation of the yeast cells with methyl α-D-mannoside (αMM, second curve), but not with methyl α-D-glucopyranoside (αMG, third curve). When added 5 min after the addition of *E. coli* to the yeast cells, αMM (fourth curve), but not αMG (fifth curve), reversed aggregation. Yeast cells treated with sodium metaperiodate lost their ability to aggregate with *E. coli* (bottom curve). (From Ofek & Beachey 1978.)

be either a synthetic product, for example a sugar bound to insoluble particles such as agarose beads, or a natural one, such as yeast cells which are covered with mannans. Interaction of bacteria with yeast cells, as well as the effect of inhibitory sugars, can be readily assessed by agglutination experiments, either microscopically on glass slides or in microtitre plates, or by aggregometry, according to the method of Ofek & Beachey (1978) (Fig. 2). In this system, too, periodate oxidation of the yeast cells has been found to abolish their ability to bind *E. coli*. Also, excellent correlation has been found between the attachment of *E. coli* to yeast cells, as

TABLE 3 Sugar-binding specificity of bacteria

Sugar	*Organism*
D-Mannose[a]	*Escherichia coli*
	Salmonella spp.
	Shigella spp.
	Citrobacter freundii
	Klebsiella spp.
	Proteus spp.
	Serratia marcescens
	Vibrio cholerae
L-Fucose[b]	*Vibrio cholerae*
D-Galactose[c]	*Actinomyces viscosus*
	Actinomyces naeslundii
Sialic acid[d]	*Mycoplasma*
N-Acetyl-D-glucosamine[e]	*Chlamydia psittaci*
	Chlamydia trachomatis
Galβ(1→3)GalNAcβ(1→3)Gal[f]	*Neisseria gonorrhoeae*

[a]Duguid et al 1979, Mirelman et al 1980, Jones & Freter 1976.
[b]Jones & Freter 1976.
[c]Costello et al 1979.
[d]Razin et al, this volume.
[e]Levy 1979.
[f]Buchanan & Pearce 1979.

measured by the rate of aggregation of the latter, and the attachment to buccal epithelial cells, as measured by the number of bacteria that adhere per epithelial cell (Ofek & Beachey 1978).

A similar approach may be useful for studying the adherence of bacteria with specificities for sugars other than mannose, using cells or particles which possess well-defined sugar structures containing, for example, D-galactose or *N*-acetyl-D-glucosamine.

Occurrence of sugar-binding activity

Studies in various laboratories have revealed the existence of bacteria with specificities for six different sugars (Table 3). These sugars are common constituents of glycoproteins and glycolipids found in animal cell membranes. It appears that the mannose-binding activity is most widely distributed among various species of Gram-negative bacteria, as shown both by inhibition of haemagglutination (Duguid et al 1979) and by yeast agglutination (Mirelman et al 1980). The latter workers examined a total of 393 routine hospital isolates for their ability to agglutinate yeast cells, either *Saccharomyces cerevisiae* or *Candida albicans*. Positive agglutination

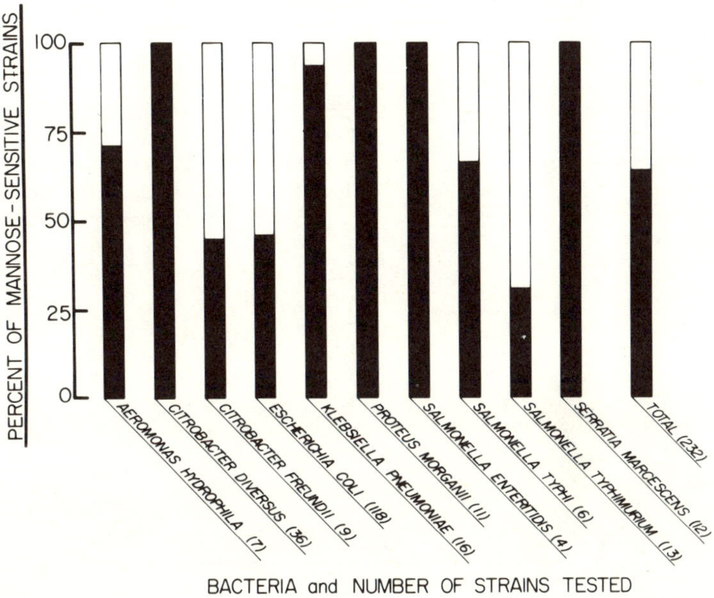

FIG. 3. Distribution of mannose-binding activity among bacterial isolates from clinical materials. (Data from Mirelman et al 1980.)

which could be inhibited by methyl α-D-mannoside (0.125 M) was taken to indicate the presence of a mannose-specific lectin on the surface of the bacteria tested. The results summarized in Fig. 3 show that all isolates of *Serratia marcescens, Salmonella enteritidis, Proteus morganii* and *Citrobacter diversus,* as well as most of those of *Klebsiella pneumoniae,* had mannose-binding activity. On the other hand, only about half the strains of *E. coli* and *Citrobacter freundii,* and two-thirds of *Salmonella typhi* and *Aeromonas hydrophila,* were mannose binders.

Among the bacterial strains in which no isolate was found to possess mannose-binding activity were *Haemophilus influenzae, Neisseria meningitidis* and *Staphylococcus aureus.* Of particular interest is the finding that none of the 12 isolates of *Pseudomonas aeruginosa,* a Gram-negative rod, were mannose-binding, since extracts of this organism have been shown by Gilboa-Garber et al (1977) to contain large quantities of a mannose-specific lectin. However, this lectin is located in the periplasm (Glick & Garber 1980) and apparently does not reach the surface of the bacteria. It would appear that besides being able to produce a lectin, bacteria must also possess the genes required for anchoring the lectin on the outer surface of the cells. Even with the right genetic constitution the amount of sugar-binding activity formed varies among different isolates (Table 4). The phenotypic expression of the cell-surface lectin depends on a variety of environmental factors. Thus, *E. coli* cells

TABLE 4 Variations in the mannose-binding activity of different E. coli strains (adapted from Ofek & Beachey 1978)

	No. adherent bacteria/human buccal epithelial cell in the presence of		
Strain	*Phosphate-buffered saline*	*Methyl α-D-mannoside (0.02 M)*	*Methyl α-D-glucoside (0.02 M)*
M_1	79	4	86
M_2	51	7	56
M_3	79	4	75
M_4	23	5	23
M_5	30	4	28
M_6	41	3	44
M_7	56	5	60
M_8	76	6	70
M_9	89	4	91
M_{10}	25	3	26
M_{11}	77	7	77
M_{12}	70	6	68

in the logarithmic phase of growth were non-adherent whereas stationary-phase bacteria were strongly adherent, the adherence being mannose-sensitive (Ofek et al 1977). Recently it has been shown that *E. coli* cells grown in the presence of subinhibitory doses of β-lactam or aminoglycoside antibiotics were incapable of forming the mannose-binding activity (Ofek et al 1979, Beachey et al, this volume).

Nature of the carbohydrate-binding constituent

The participation of bacterial pili (or fimbriae) in the mannose-specific adherence of *E. coli* and other Gram-negative bacilli to cell surfaces was first pointed out by Duguid & Gillies (1957), who found a striking correlation between piliation and mannose-sensitive haemagglutination by the bacteria. In fact, agglutination of guinea-pig erythrocytes reversed by D-mannose has been used, together with electron microscopy, to define type 1 pili. Ofek & Beachey (1978) fractionated cultures of *E. coli* into non-adherent and adherent populations by adsorbing the latter on buccal epithelial cells followed by elution with methyl α-D-mannoside. The eluted cells showed a high degree of mannose-binding activity and were heavily piliated, whereas the non-adherent cells lacked mannose-binding activity and were devoid of pili.

Direct evidence for the role of pili in mannose-sensitive bacterial adherence was obtained in studies with isolated and purified type 1 pili. Such pili agglutinated guinea-pig erythrocytes (Rivier & Darekar 1975, Salit & Gotschlich 1977a) and adhered to monkey kidney cells grown *in vitro* (Salit & Gotschlich 1977b); the haemagglutination was inhibited by low concentrations of D-mannose or methyl α-D-mannoside. Moreover, purified *E. coli* type 1 pili strongly agglutinated yeast cells (Eshdat et al 1980a, Korhonen 1979).

Very little is known about the molecular mechanisms by which pili attach to mannose-containing constituents of the cell surface. Since pili are made up of protein subunits known as 'pilin' (M_r 17 000) the question may be asked whether all the subunits have the capacity to bind D-mannose. Moreover, one cannot exclude the possibility that a minor protein component associated with the pili is the mannose-binding constituent, in particular as pili are highly hydrophobic and thus may adsorb protein contaminants. A major difficulty in studying structure–function relationships in pili is their high stability. So far, type 1 pili have been dissociated into their subunits only in conditions known to cause irreversible protein denaturation (Salit & Gotschlich 1977a, McMichael & Ou 1979).

In a recent study (Y. Eshdat, F. J. Silverblatt & N. Sharon, unpublished observations) we found that pili were completely dissociated on incubation in saturated guanidine–HCl at 37 °C, as shown by electron microscopy (Fig. 4). Even in concentrations of guanidine-HCl as high as 7.5 M the pili were only partially dissociated. Gel filtration of completely dissociated pili on a Sepharose CL-6B column in the presence of saturated guanidine-HCl yielded a single protein peak of M_r 17 000, identical with that of pilin (Fig. 5). The material obtained after removing guanidine-HCl by dialysis was fractionated on a mannan-Sepharose 4B column; 15% of the protein was bound to the column and was eluted with 0.1 M-methyl α-D-mannoside. We do not know whether the unbound protein consists of pilin originally devoid of mannose-binding activity, or whether the lack of binding is due to denaturation during the preparative procedure. The dissociated protein gave a single band on polyacrylamide gel electrophoresis in 0.1% dodecyl sulphate or 10 M-urea, and penetrated completely into 7% gels in the absence of denaturants. After dialysis against 5 mM-Tris buffer (pH 8.0) containing 5 mM-$MgCl_2$ the protein reassembled into pili that were similar to native pili, as we could see by gel filtration and electron microscopy (Fig. 4). Thus, the procedure described permits the preparation of a non-denatured soluble form of pili subunits of low molecular mass.

Studies in our laboratory have also shown that sugar-specific constituents may be present on the bacterial surface in forms other than pili. We have isolated a preparation of flagella (Fig. 6) with mannose-specific lectin activity from *E. coli* 7343, a urinary tract isolate grown at 37 °C for 40 h in broth in static cultures (Eshdat et al 1978, 1980b). The lectin was adsorbed to yeast cells and could be partially eluted with methyl α-D-mannoside. Its sugar specificity was similar to that of the intact bacteria (Table 5). Of special interest is the strong inhibition by *p*-nitrophenyl α-D-mannoside, which implies that the lectin has a hydrophobic site close to its sugar-binding site. In this respect it is similar to plant lectins such as

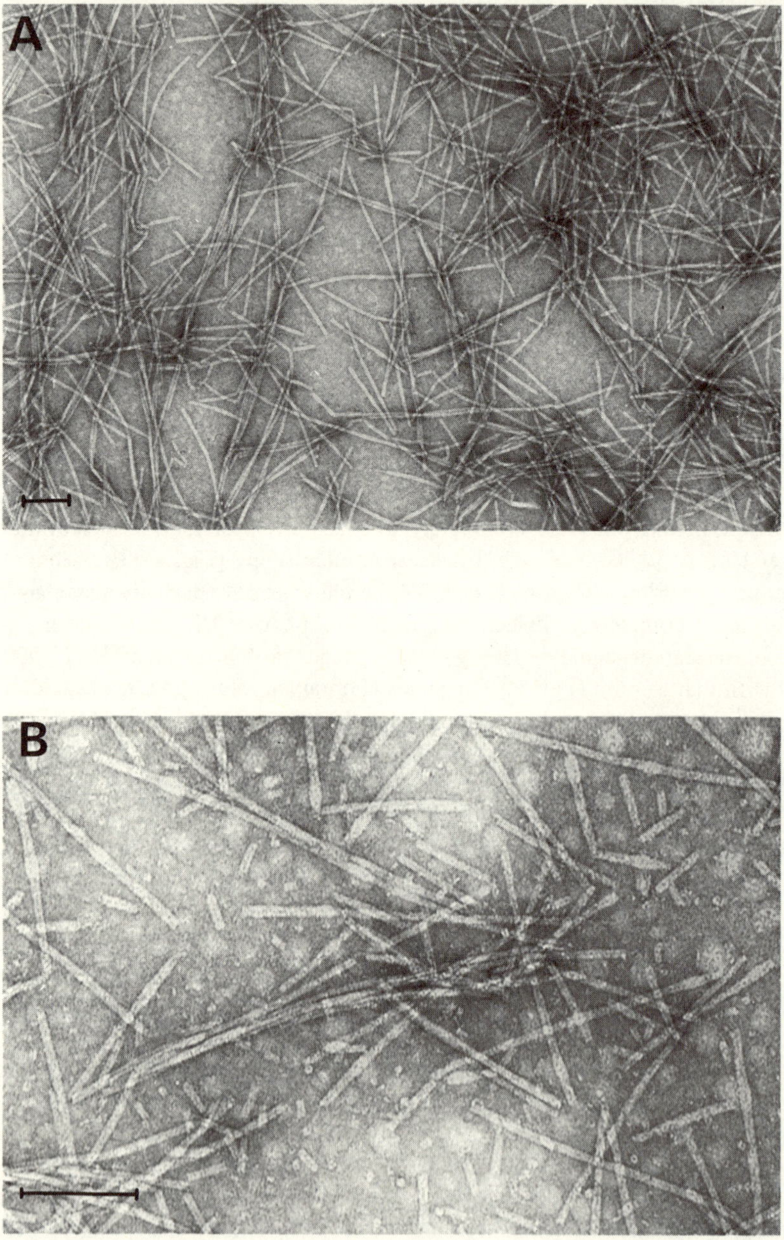

FIG. 4. Electron micrographs of pili from *E. coli* 346: (A) purified pili; (B) pili incubated for 2 h at 37 °C in 5 mM-Tris buffer (pH 8.0) and 7.5 M-guanidine hydrochloride; (C) pili incubated for 2 h at 37 °C in 5 mM-Tris buffer (pH 8.0) and saturated guanidine hydrochloride;

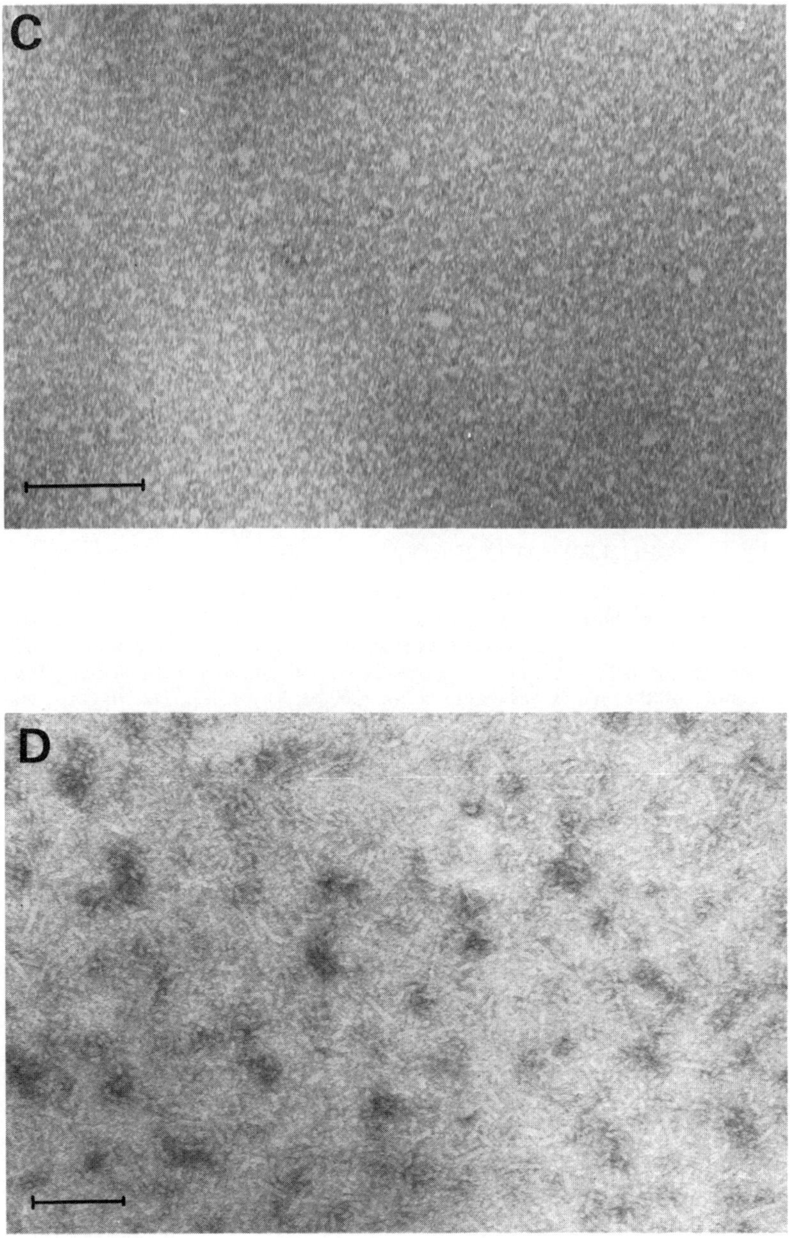

(D) reassembled pili in 5 mM-Tris buffer (pH 8.0) and 5 mM-$MgCl_2$. The pili were negatively stained with uranyl acetate. Scale bars, 0.1 μm.

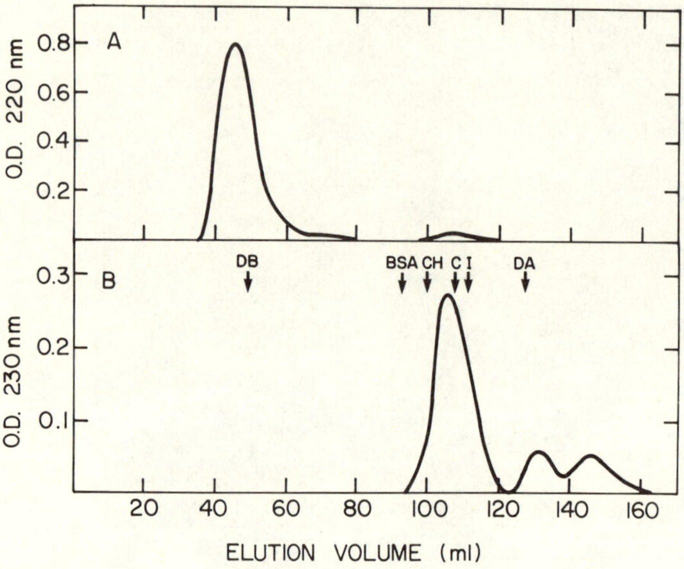

FIG. 5. Gel filtration on a Sepharose CL-6B column (1.5 × 84 cm). Elution pattern of: (A) purified pili in 5 mM-Tris buffer, pH 8.0; (B) dissociated pili in same buffer saturated with guanidine hydrochloride. Protein markers used: dextran blue (DB), bovine serum albumin (BSA), α-chymotrypsinogen (CH), cytochrome *c* (C), insulin (I) and dinitrophenyl-L-alanine (DA). The optical density of the eluates was continuously monitored by a Beckman DB-G spectrophotometer. Fractions of 2 ml were collected.

concanavalin A and soybean agglutinin (Lis & Sharon 1977). The flagellar lectin consists of protein subunits with an M_r of 36 500 and its amino acid composition is different from that reported for the type 1 pili protein, the K99 and K88 antigens, colonization factor antigen I, and the major outer membrane protein Ia of *E. coli*, which has a similar molecular weight (Eshdat et al 1978).

We have obtained additional indications that the lectin-like activity on bacterial surfaces is not necessarily in the form of pili. For example, homogenization for up to 90 min, a procedure known to release type 1 pili (as well as flagella) from bacteria, did not release mannose-binding activity from a mannose-specific strain of *Salmonella typhimurium* 1826, which is only slightly piliated (T. J. Pistole et al, unpublished results).

Receptors on animal cells

The availability of receptors for bacterial lectins, as for other adhesins, depends on the particular host cell under study (Ofek & Beachey 1980a). For example, erythrocytes from different species, or epithelial cells from various sites of the same

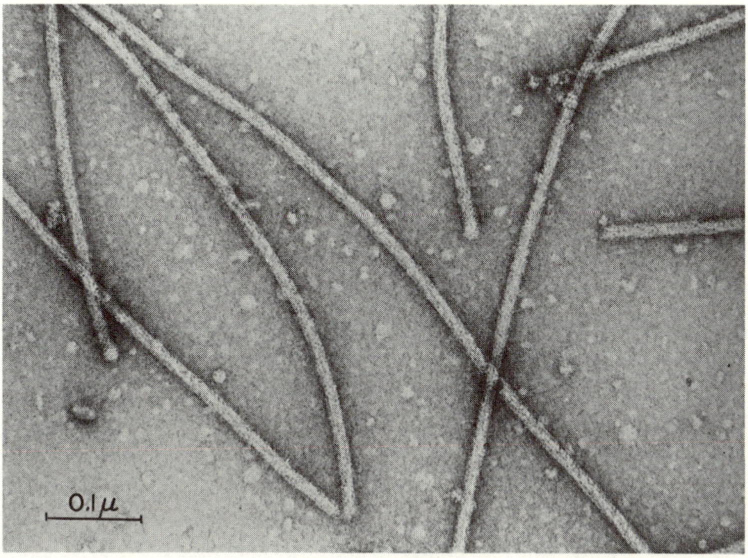

FIG. 6. Electron micrographs of *E. coli* 7343 flagella with mannose-specific lectin activity; negative staining with uranyl acetate.

animal, differ markedly in their ability to interact with the same bacterial strain. In addition, the sensitivity of this interaction to sugars (e.g. D-mannose) may also differ markedly from one type of cell to another. It follows, therefore, that the surfaces of animal cells differ in the structure and availability of sugar residues that can serve for the attachment of bacteria. This is in accord with the extensive body of evidence obtained in studies with plant lectins (Lis & Sharon 1977). Information on the chemical structure of sugar receptors for bacteria is, however, scanty.

Glycophorin, the major sialoglycoprotein of the human erythrocyte membrane, has been shown to serve as the receptor for *Mycoplasma gallisepticum* specific for sialic acid (Razin et al, this volume). In studies of the attachment of *E. coli* to human uroepithelial cells Svanborg Edén et al (this volume) have found that although monosaccharides do not inhibit this attachment, glycolipids isolated from uroepithelial cells were effective inhibitors. They concluded that globoside, GalNAcβ(1→3)Galα(1→4)Galβ(1→4)Glcβ-ceramide, and possibly substances with internal sugar sequences identical to globoside, fulfil the criteria of a receptor. Receptors for other sugar-specific bacteria have not yet been isolated. It may now be possible to isolate such receptors by affinity techniques with the aid of the purified lectins from the corresponding bacteria. An alternative is to use readily available plant lectins with a specificity similar to that of the bacterial lectin. For example, the mannose-containing receptors for *E. coli* and other Gram-negative bacteria (Table 3) can be isolated with the aid of concanavalin A.

TABLE 5 Inhibition by carbohydrates of yeast agglutination by E. coli 7343 and its lectin

Compound	*Lowest inhibitory concentration, mM* *Intact bacteria*	*Isolated lectin*
D-Mannose	2.1	1.25
Methyl α-D-mannoside	1.05	0.625
p-Nitrophenyl α-D-mannoside	0.05	0.010
p-Nitrophenyl α-D-glucoside	6.0	2.5

No inhibition by D-glycose 6-phosphate, D-mannose 6-phosphate, D-galactose, or D-arabinose, at concentrations up to 100 mM. (Results from Eshdat et al 1978 and Y. Eshdat et al, unpublished experiments.)

Function of sugar-binding activity

There is a general consensus that bacterial adherence contributes to the virulence of the organisms. This is based on the assumption that bacteria must adhere to the mucosal surfaces of the host cells in order to colonize host tissues and is supported by extensive data showing a positive correlation between the ability of bacteria to adhere to cell surfaces and the expression of virulence (Ofek & Beachey 1980a, b). The role of sugar-mediated attachment in pathogenicity is, however, not clear, since there is no evidence for such a correlation between the two.

An indication of the possible role of the sugar-binding activity in infectivity was obtained in studies of a mouse model of genitourinary tract infection (Aronson et al 1979). In these experiments infective strains of *E. coli* were injected into the bladders of mice in the absence or presence of different sugars. It was found that methyl α-D-mannoside, but not methyl α-D-glucoside, caused a marked reduction in the number of bacteriuric mice (Fig. 7). In this system, methyl α-D-mannoside and methyl α-D-glucoside were both inactive against a strain of *Proteus mirabilis,* which agrees with their inability to inhibit the adherence of this organism to epithelial cells. Similar preliminary results were obtained in a rabbit model of gastrointestinal infection where colonization of *E. coli* was specifically blocked by D-mannose but not by other sugars (Hirschberger et al 1977).

When the role of sugar-mediated attachment in virulence is considered, the ability of bacteria such as *E. coli* and *Salmonella typhi* to bind to surface sugars of phagocytic cells is of special significance (Bar-Shavit et al 1977, Silverblatt et al 1979). Such binding occurs in the absence of opsonins on the surface of the bacteria and is followed by their ingestion and killing by the phagocytes. This means that once the bacteria penetrate into the host, their ability to bind to sugars on the phagocytes may, in fact, impair virulence by facilitating phagocytosis.

It would appear, therefore, that the sugar-binding ability of the bacteria has a dual role in the host–parasite relationship: it may enable the organisms to establish

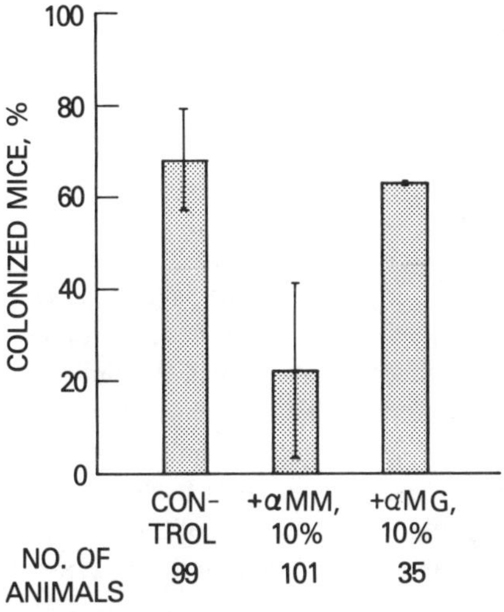

FIG. 7. Effect of sugars on urinary tract *E. coli* infection in mice: α-MM, methyl α-D-mannoside; αMM, methyl α-D-glucoside. (Data from Aronson et al 1979.)

colonization of mucosal surfaces by adherence to sugar residues on epithelial cells, and it may also cause them to be recognized by such residues on the surface of phagocytes and as a result be ingested and digested by the latter cells.

The studies on bacterial adherence to cell surface sugars described so far have all been done with strains subcultured and grown in the laboratory. Since bacteria display remarkable phenotypic variations in their sugar-binding activity it is crucial to examine this activity in bacteria grown *in vivo*. Such a study was recently initiated in our laboratories (I. Ofek & N. Sharon, unpublished results). Bacteria were obtained from the urine of patients with *E. coli* urinary tract infections. The bacteria were sedimented within three hours of urine collection by centrifugation in the cold and resuspended in phosphate-buffered saline to a concentration of 10^9 cells/ml. Preliminary results show that, among the 12 cases so far examined, bacteria from only one urine specimen exhibited weak mannose-binding activity, as assayed by yeast agglutination; bacteria from three other specimens agglutinated human erythrocytes (type A) and the haemagglutination was not inhibited by mannose (0.125 M). Interestingly, when subcultured and grown in the laboratory for 48 hours in liquid medium, about half the strains acquired potent mannose-binding activity; one of these also exhibited haemagglutinating activity, which was, however, mannose-resistant.

It is not known whether bacteria growing *in vivo* can produce mannose-binding substances, even when they possess the required genes. It is also possible that the shed bacteria found in the urine represent a fraction of organisms lacking the mannose-binding lectin, or other adhesins. In any event, our results clearly emphasize the importance of studying bacterial adherence with organisms grown *in vivo*.

Future prospects

Obviously, more work is needed for the role of bacterial adherence to cell surface sugars in natural infection to be assessed. Special attention should be paid to bacteria grown *in vivo* rather than those subcultured and grown *in vitro*. A better understanding of the mechanism of colonization of mucosal surfaces will undoubtedly be gained from investigations leading to the isolation and chemical characterization of bacterial surface lectins and of the corresponding sugar-containing receptors from animal cells. More importantly, perhaps, such studies may lead to the preparation of lectin vaccines and of compounds which mimic the structure of the receptors and may be useful as effective inhibitors of adherence and infection by bacterial pathogens.

Acknowledgement

This study was supported in part by a grant from the United States–Israel Binational Science Foundation (2123/80) to I. Ofek.

REFERENCES

Aronson M, Medalia O, Schori L, Mirelman D, Sharon N, Ofek I 1979 Prevention of colonization of the urinary tract of mice with *Escherichia coli* by blocking of bacterial adherence with methyl α-D-mannopyranoside. J Infect Dis 139:329-332

Bar-Shavit Z, Ofek I, Goldman R, Mirelman D, Sharon N 1977 Mannose residues on phagocytes as receptors for the attachment of *Escherichia coli* and *Salmonella typhi.* Biochem Biophys Res Commun 78:455-460

Beachey EH, Eisenstein BI, Ofek I 1981 Sublethal concentrations of antibiotics and bacterial adhesion. In this volume, p 288-300

Buchanan TM, Pearce WA 1979 Pathogenic aspects of outer membrane components of Gram-negative bacteria. In: Inouye M (ed) Bacterial outer membranes: biogenesis and functions. Wiley-Interscience, New York, p 475-514

Collier WA, de Miranda JC 1955 Bacterien-hemagglutination. I. Versuche mit einem haemagglutinerenden Stamm von *E. coli.* Antonie van Leeuwenhoek J Microbiol Serol 21:133-140

Costello AH, Cisar JO, Kolenbrauder PE, Gabriel O 1979 Neuraminidase-dependent hemagglutination of human erythrocytes by human strains of *Actinomyces viscosus* and *Actinomyces naeslundii.* Infect Immun 26:563-572

Duguid JP, Gillies RR 1957 Fimbria and adhesive properties in dysentery bacilli. J Pathol Bacteriol 74:397-411

Duguid JP, Clegg S, Wilson MI 1979 The fimbrial and non-fimbrial haemagglutinins of *Escherichia coli.* J Med Microbiol 12:213-227

Eshdat Y, Ofek I, Yashouv-Gan Y, Sharon N, Mirelman D 1978 Isolation of a mannose-specific lectin from *Escherichia coli* and its role in the adherence of the bacteria to epithelial cells. Biochem Biophys Res Commun 85:1551-1559

Eshdat Y, Speth V, Jann K 1980a The role of pili and outer membrane proteins in bacterial adherence. Israel J Med Sci 16:479 (abstr.)

Eshdat Y, Sharon N, Ofek I, Mirelman D 1980b Structural association of the outer surface mannose-specific lectin of *Escherichia coli* with bacterial flagella. Israel J Med Sci 16:479 (abstr.)

Gilboa-Garber N, Mizrahi L, Garber N 1977 Mannose-binding hemagglutinins in extracts of *Pseudomonas aeruginosa*. Can J Biochem 35:975-981

Glick J, Garber N 1980 Periplasmic localization of the D-mannose-binding lectin of *Pseudomonas aeruginosa*. Israel J Med Sci 16:481 (abstr.)

Hirschberger M, Mirelman D, Thaler MM 1977 The effects of mannose and its derivatives on interactions of pathogenic *E. coli* with newborn intestinal mucosa. Gastroenterology 72:1069

Jones GW, Freter R 1976 Adhesive properties of *Vibrio cholerae:* nature of the interaction with isolated rabbit brush border membranes and human erythrocytes. Infect Immun 14:240-245

Korhonen TK 1979 Yeast cell agglutination by purified enterobacterial pili. FEMS (Fed Eur Microbiol Soc) Lett 6:421-425

Levy NJ 1979 Wheat germ agglutinin blockage of *Chlamydial* attachment sites: antagonism by *N*-acetyl-D-glucosamine. Infect Immun 25:946-953

Lis H, Sharon N 1977 Lectins: their chemistry and application to immunology. In: Sela M (ed) The antigens. Academic Press, New York, vol 4:429-529

McMichael JC, Ou JT 1979 Structure of common pili from *Escherichia coli.* J Bacteriol 138: 969-975

Mirelman D, Altmann G, Eshdat Y 1980 Screening of bacterial isolates for mannose-specific lectin activity by agglutination of yeasts. J Clin Microbiol 11:328-331

Ofek I, Beachey EH 1978 Mannose binding and epithelial cell adherence of *Escherichia coli.* Infect Immun 22:247-254

Ofek I, Beachey EH 1980a Bacterial adherence. Adv Intern Med 25:503-538

Ofek I, Beachey EH 1980b General concepts and principles of the adherence of bacteria to animal cells. In: Beachey EH (ed) Bacterial adherence. Chapman & Hall, London (Receptors and recognition, series B) vol 6:1-27

Ofek I, Mirelman D, Sharon N 1977 Adherence of *Escherichia coli* to human mucosal cells mediated by mannose receptors. Nature (Lond) 265:623-625

Ofek I, Beachey EH, Sharon N 1978 Surface sugars of animal cells as determinants of recognition in bacterial adherence. Trends Biochem Sci 4:159-160

Ofek I, Beachey EH, Eisenstein BI, Alkan ML, Sharon N 1979 Suppression of bacterial adherence by subminimal inhibitory concentrations of β-lactam and aminoglycoside antibiotics. Rev Infect Dis 1:832-837

Old DC 1972 Inhibition of the interaction between fimbrial hemagglutinins and erythrocytes by D-mannose and its derivatives. J Gen Microbiol 71:149-157

Razin S, Kahane I, Banai M, Bredt W 1981 Adhesion of mycoplasmas to eukaryotic cells. In this volume, p 98-113

Rivier DA, Darekar MR 1975 Inhibitors of the adhesiveness of enteropathogenic *E. coli.* Experientia (Basel) 31:662-664

Salit IE, Gotschlich EC 1977a Hemagglutination by purified type I *Escherichia coli* pili. J Exp Med 146:1169-1181

Salit IE, Gotschlich EC 1977b Type I *Escherichia coli* pili: characterization of binding to monkey kidney cells. J Exp Med 146:1182-1194

Sharon N 1979 Some biological functions of cell surface sugars. In: Yagi K (ed) Structure and function of biomembranes. Japan Scientific Societies Press, Tokyo, p 63-82

Silverblatt FJ, Dreyer JS, Schauer S 1979 Effect of pili on susceptibility of *Escherichia coli* to phagocytosis. Infect Immun 24:218-223

Svanborg Edén C, Hagberg L, Hanson LA, Korhonen T, Leffler H, Olling S 1981 Adhesion of *Escherichia coli* in urinary tract infection. In this volume, p 161-178

DISCUSSION

Svanborg Edén: May I suggest that the bacteria we should look at are not the ones that are excreted but those sitting on the mucosa.

Sharon: I would welcome any suggestions or comments on that from people who know more about it than I do.

Sussman: One of my students looked at her own excreted epithelial cells and isolated adhering bacteria. When she was symptomatic she confirmed what you say, Dr Sharon. That is, the organisms were mannose-resistant (MR), not mannose-sensitive, and the appearance of MR bacteria sticking to her uroepithelial cells was associated with symptomatic episodes. When the symptoms resolved, the MR organisms disappeared and MS organisms returned.

Sharon: In the urine of some patients we found epithelial cells which were heavily covered with bacteria but we have not yet removed these bacteria. One thing which worried us for a while was that since these patients were all from one hospital they might have been infected by a single strain. However, after specimens had been subcultured in brain–heart infusion broth for 48 h, over half of them developed mannose-sensitive adherence to completely different extents, making infection by one strain unlikely. We also know that the bacteria in the urine have the capacity to produce mannose binding activity, but either we do not find it (e.g. because it is masked) or the bacteria do not express it.

Svanborg Edén: Were those patients untreated?

Sharon: They were not treated with antibiotics and the infection was asymptomatic.

Svanborg Edén: That certainly agrees with our results, which show that in contrast to the patients with acute pyelonephritis a large proportion of asymptomatic infections occur in patients who carry strains that do not attach to the mucosa. They may have underlying defects such as residual urine, or binding to the mucus may be enough for them to remain in the urinary tract.

Levine: What you see is clearly analogous to diarrhoeal infections with enterotoxigenic *E. coli*, many of which don't have mannose-resistant haemagglutinins but do have mannose-sensitive haemagglutinins. The same question arises: what is the role *in vivo* of mannose-sensitive haemagglutinin? Although there are antibody rises to mannose-resistant haemagglutinins when individuals are infected with such strains, we have not yet found rises in antibody to purified type 1 pili in patients who are infected with strains that have type 1 pili but don't have mannose-resistant haemagglutinins. Perhaps the type 1 pili are not playing a major role *in vivo* and therefore are not stimulating an antibody response.

Silverblatt: We had the opportunity of looking at five patients with classical pyelonephritis with bacteraemia. All had a significant rise in anti-type 1 pili antibody. However no antibodies were detected in the urine (Rene & Silverblatt 1980).

Svanborg Edén: It is known that patients with acute pyelonephritis produce antibodies to Tamm-Horsfall protein (Fasth et al 1979). The Ørskovs have suggested that the Tamm-Horsfall protein is identical with the mucus of the urinary tract (Ørskov et al 1980), and that strains carrying type 1 fimbriae bind to this mucus. There may be a linkage between all these phenomena.

Sharon: Has anybody tested the ability of purified Tamm-Horsfall glycoprotein to inhibit adherence?

Svanborg Edén: Ørskov et al (1980) gave no figures and we have only preliminary results.

Sharon: It is not difficult to prepare it. Each of us excretes daily in our urine about 45 mg of Tamm-Horsfall glycoprotein.

Choppin: Didn't the Ørskovs say that Tamm-Horsfall protein inhibited binding?

Svanborg Edén: Yes.

Sharon: Was this a mannose-sensitive strain?

Svanborg Edén: Yes.

Feizi: Has anyone tested brain–heart broth for the presence of mannose-binding lectin? Since mannose-sensitive adhesive property is so commonly acquired after culturing the bacteria in this broth, it would be important to exclude the passive uptake of lectin by the bacteria.

Sussman: If organisms are grown in nutrient broth, you will tend to select type 1 pili. Under these conditions you wouldn't get full expression of MR pili. Brain–heart infusion broth will tend to select for MR pili.

Pearce: What happens if bacteria are incubated in urine for a few hours and then tested?

Svanborg Edén: Urine is not a defined and stable medium.

Silverblatt: I have looked at the question *in vivo*. I collected urine from rats experimentally infected with *E. coli* bearing type 1 pili and after a week I looked for pili on bacteria from the urine. Electron microscopy showed that every infected animal had heavily piliated organisms in their urine, so apparently growth in urine is not inimical to the expression of pili.

Sussman: MR pili don't like rat cells particularly, so you would be selecting out MS pili in the rats. We have some pure MR strains and they show poor attachment, if any, to rat cells.

Taylor-Robinson: Is treatment in the way you have mentioned with the rats a likely possibility in humans?

Sharon: We have been toying with this idea but physicians seem to be too busy curing patients to collaborate with us.

Taylor-Robinson: Is there an ethical problem?

Sharon: No.

Mirelman: The Talmud reports that the cherob fruit (St John's bread), which is known to be rich in α-mannans, is beneficial for many types of diarrhoea. It is possible that many of the old wives' tales of treatment may have some scientific basis if substances like mannan remove bacteria that have mannose-binding activity.

Freter: If mannose-sensitive fimbriae are important to adhesion *in vivo* why can't that be shown directly on bladder cells coated with organisms, serologically or by electron microscopy?

Sharon: We don't know. We have to proceed very carefully now. The ones that are eluted may be the ones that are coated with Tamm-Horsfall glycoprotein.

Freter: That is what I am saying. If you can collect bacteria on bladder cells and test them, perhaps serologically, for fimbriae, it should not be too difficult to determine whether fimbriae are synthesized *in vivo.*

Sharon: We should do that. Using competing sugars, it may perhaps be possible to remove mucins that are coating the bacteria and preventing their attachment to the urinary tract.

Vosbeck: I have been working on the selection of models for measuring bacterial adhesion to human tissues, in order to study the molecular mechanisms of adhesion. We felt that isolated human tissue cells were too cumbersome to use because of the individual variability of the donors and because of differences in sampling procedures. We therefore chose to risk losing some biological relevance but to gain reproducibility and operational simplicity by using the following system (Vosbeck et al 1979).

We use tissue culture monolayers of a human epithelioid cell line, Intestine 407 (ATCC CC1 6), which is related to HeLa cells. We add a suspension of *E. coli* labelled with [^{14}C]acetate. After incubation we wash the preparation to remove unbound bacteria. We solubilize the monolayer together with the attached bacteria with sodium dodecyl sulphate and count the radioactivity, which is proportional to the number of bound bacteria.

One of our standard *E. coli* strains came from Professor Richmond's laboratory, where it was isolated from the urine of a patient with pyelonephritis. This strain adheres well to three human epithelioid cell lines we tested, Intestine 407, Wish, and HEp-2, which are all related to HeLa cells. It adheres less well to human fibroblastic cell lines such as HEL (human lung fibroblasts), and still less well to animal cell lines, whether epithelioid (Y-1, mouse adrenal, PK_1, pig kidney) or fibroblastic (Vero, BHK, 3T3, SV 101).

These results, to my mind, show that we are measuring a quite specific interaction between the Intestine 407 cells and the bacteria, not just a physicochemical interaction such as hydrophobic binding. The principal question remains, of course, how relevant this binding is *in vivo*. A survey of *E. coli* strains showed that many strains appear to bind in this system. Most of them do not have known adhesion factors. K88 strains adhere and K99 strains usually do too, but strains carrying the human intestinal adhesion factors CFA/I and CFA/II unfortunately do not.

Binding is nearly always mannose-resistant. If we grow bacteria at 20 °C instead of 37 °C adhesion is lost, and loss of adhesion is associated with loss of piliation. In other words the behaviour of this adhesion assay system is similar to that of other systems measuring adhesion due to plasmid-mediated adhesion factors. We therefore believe it is worth investigating this system further. It should not be too difficult to find a similar tissue culture system for measuring binding mediated by CFA/I, CFA/II or any other adhesion factor. Such simple, easily manipulable systems are in my opinion much better suited for studying the molecular mechanism of adhesion to human epithelia than the very complex and variable systems that use freshly isolated material from human surgical or biopsy specimens or exfoliated material.

Mirelman: The bacteria used in most experiments are grown in a rich medium. The problem of how disease is caused will have to be solved by studying other systems for growth of bacteria, mainly for prolonged incubation periods and in poorly defined media that provide conditions which are more like those bacteria encounter in a host. Have you tested whether bacteria found in urine or faeces bind to your monolayers?

Vosbeck: No, we have tested bacteria only after subculturing them at least twice on agar.

Richmond: The interaction of a eukaryotic cell and the bacterial cell is such a complicated thing that one has to find systems where as many of the variables are controlled as possible. To examine many of these phenomena at the molecular level it is important to be able to examine the kinetics of the system. So one needs a microorganism in which good kinetics can be done and which is as well defined as possible, for example *E. coli* K12. For the eukaryotic cell system I don't know what one would choose. I would be interested in proposals. In such a relatively defined system, once it was agreed, you could then start to insert genes from various sources into the *E. coli* K12 and look at their effect in the defined system. The objective must be to clarify how many distinctly different types of binding phenomena are at issue. When that has been done one can try to define simple indicators for which system is operating, and then try to fit that to the clinical picture.

Taylor-Robinson: You are talking in terms of molecular biology, but some of us are not yet at that stage of sophistication.

Richmond: Are you saying that it is too difficult?

Taylor-Robinson: It may be in some organism–cell situations, and a defined system may appear to have little clinical relevance.

Richmond: E. coli K12 is a bacterial strain in which the expression of characters can be achieved, and you must have some system where the various aspects of the situation can be examined systematically one by one.

Tramont: If I understand you correctly, you want a well-defined and understood system in which you can examine and define what the attachment factors are: how much is mannose, what factor 1 is, and so on. Once you have found that at a molecular level, you can try to fit that with what is observed clinically. That is how a molecular biologist would approach the problem but not how a clinician would approach it. The clinician will try to make a correlation with a broader phenomenon, such as adhesion of whole bacteria to cells, and then work backwards to determine the exact adhesive antigen. Both will come to the same point or conclusions if it turns out to be something that is really relevant to the human disease state.

Svanborg Edén: The organisms we have studied have many binding properties present at the same time. So unless you also guarantee your receptor sites on the host cell you risk ending up with a system that illustrates binding beautifully but doesn't say anything about binding to the host cell in relation to disease.

Taylor-Robinson: In that context shouldn't one be more concerned with the use of organ cultures in which the cells remain physiologically similar to their state *in vivo* than with cells isolated from each other or cells established in tissue culture?

Svanborg Edén: We have compared our system with that of Dr Vosbeck and exchanged strains. About 50% of the strains that attach in his HeLa cell system also attach in my system. Unless you can somehow go back to the source of the infection and demonstrate that your system parallels that to some extent, the *in vitro* results may be misleading.

Richmond: But I don't see how you can do it. When you say go back to the nature of the infection, what eukaryotic cell line are you going to use?

Svanborg Edén: I don't use a cell line but single cells exfoliated into the urine.

Vosbeck: Systems like yours may be useful for studying what is happening at the bacterial cell surface. But we are also interested in the eukaryotic cell surface and the receptors. Especially in the urinary tract system, where there are a lot of dead cells and a mixture of cells, you run into trouble if you want to examine the cell side. You have to take the risk that what you find may not be relevant.

Tramont: So your system is better suited to determining what the receptors are, Dr Vosbeck, and Dr Svanborg Edén's system is better for looking at organisms that are causing disease in an individual patient. Again it comes down to what question you are asking from the start.

Taylor-Robinson: Obviously we can't lay down strict criteria and say that only one system or approach should be used. If we use a variety and find that there is a correlation in the sense that each produces a similar answer, then we are more likely to get the right answer.

Sussman: In the early days of bacteriophage studies, great advances were made when people had agreed on the strains and the hosts that they were going to use.

Sharon: Maybe they were lucky in choosing the right system in the first place.

Pearce: Among the chlamydiae a large number of strains appear to attach poorly to tissue cell cultures. Could one use buccal or other mucosal epithelial cells, if attachment to these is greater, in order to define possible receptors for such strains?

Candy: There are certain experiments that you can only do in cell cultures and we use cultured cells as well as cells directly obtained from human donors. But if you don't go back to the human tissue which the bacteria can colonize *in vivo* you can get surprises. For example, *E. coli* CFA/I binds to red cells in a mannose-resistant fashion. In the two human biopsies that we have studied the adhesion phenomenon has been mannose-sensitive. The number of CFA/I *E. coli* that bound decreased from 57 in 20 fields to 14. In another patient it went from 89 without mannose to 15 with mannose. This was an entirely unexpected finding. In this system the jejunal biopsy comes straight out of the patient and into the experiment; hence tissue viability is less of a problem. A further advantage of using tissue or cells from patients is that the adhesiveness of different strains of bacteria to material from various groups of patients can be compared.

Choppin: In the myxoviruses and paramyxoviruses it is known precisely which protein is binding to the receptor, the nature of the receptor is known, and quite a lot is known about the chemistry of both components. But if you take many strains of influenza virus and many types of red blood cells, you find that some viruses agglutinate some cells and some agglutinate others; a sheep red blood cell is agglutinated by only a very few viruses, whereas a chicken cell is agglutinated by many. If you knew nothing about the chemistry in such a system, you might conclude – completely erroneously – that all these viruses had different receptors, and were binding to them by different proteins. Furthermore, with neuraminidase you can produce a receptor gradient that would lead you to conclude that there was a whole family of different receptors. In fact, in every case a specific glycoprotein on the virus is reacting with a neuraminic-acid-containing glycoprotein or, in some cases, glycolipid. Within that formula there is a lot of variation, based on the stereochemistry of the various glycoproteins. Thus it may be misleading to say that there are a whole lot of different binding affinities within one family of bacteria. It may just be variation within a central theme. My bet would be that if you understand any one of these systems completely, that will help you in understanding all the others, but you have to start somewhere. One has to keep probing from both sides. Once one understands how *E. coli* organisms attach to tissue culture cells, that helps one to understand how they infect the urinary epithelium.

REFERENCES

Fasth A, Hanson LÅ, Jodal U, Peterson H 1979 Autoantibodies to Tamm-Horsfall protein associated with urinary tract infections in girls. J Pediatr 95:54-60

Ørskov I, Ferencz A, Ørskov F 1980 Tamm-Horsfall protein or uromucoid is the normal urinary slime that traps type I fimbriated *Escherichia coli.* Lancet 1:887

Rene P, Silverblatt FJ 1980 Immune response to pilus antigen in pyelonephritis. In: Current chemotherapy and infectious disease. American Society for Microbiology, Washington, in press

Vosbeck K, Handschin H, Menge EB, Zak O 1979 Effects of subminimal inhibitory concentrations of antibiotics on adhesiveness of *Escherichia coli* in vitro. Rev Infect Dis 1:845-851

Adhesion of enterotoxigenic *Escherichia coli* in humans and animals

MYRON M. LEVINE

Center for Vaccine Development, Division of Infectious Diseases, University of Maryland School of Medicine, Baltimore, Maryland 21201, USA

Abstract Enterotoxigenic *Escherichia coli* (ETEC), an important cause of diarrhoea in humans and animals, require accessory virulence properties in addition to enterotoxin to manifest virulence. Several classes of pili (hair-like protein surface organelles) promote adhesion of ETEC to small intestinal mucosa. Antibody directed against adhesion pili interferes with colonization of the small intestine and prevents disease. This paper reviews studies with purified K88, K99 and 987-type pili used as parenteral vaccines in pregnant pigs and cattle. Infant animals suckled on immunized mothers were significantly protected against fatal disease.

Colonization factor antigen (CFA) I and II pili, and type 1 somatic pili, promote adhesion of human ETEC pathogens to epithelial cells *in vitro* and are generally recognized as accessory virulence factors. CFA/I and II were found in only 25% of 36 human ETEC infections; positive strains were usually LT^+/ST^+ (LT: heat-labile; ST: heat-stable). Strains lacking CFA/I and II are virulent; other factors must be responsible for adhesion in such strains. While none of 14 LT^+/ST^- strains elaborated CFA/I or II, 10 (71%) possessed type 1 somatic pili.

An initial ETEC diarrhoeal infection in volunteers stimulated protective immunity against diarrhoea on re-challenge with the same strain. Despite clinical protection healthy 'veterans' excreted the ETEC strain to the same degree as ill controls. Thus the mechanism of immunity was not bactericidal. Disease-induced LT antitoxic immunity failed to protect volunteers against challenge with a heterologous (LT^+/ST^-) strain. One explanation of these observations is that the mechanism of protection was anti-adhesive with antibody directed against adhesive factors on the bacterial surface preventing attachment of bacteria to receptors on small intestinal mucosal cells. Immunoprophylaxis against ETEC in humans with purified pili vaccines appears feasible.

Enterotoxigenic *Escherichia coli* (ETEC) is an important pathogen causing travellers' diarrhoea, infant diarrhoea in less-developed areas, and neonatal enteric colibacillosis of piglets and calves. ETEC can elaborate either a heat-labile (LT) or a heat-stable (ST) enterotoxin, or both, and genes dictating enterotoxin production

1981 Adhesion and microorganism pathogenicity. Pitman Medical, Tunbridge Wells (Ciba Foundation symposium 80) p 142-160

typically reside in transferable plasmids. LT and ST induce diarrhoea by activating adenylate cyclase or guanylate cyclase, respectively, in enterocytes of small intestinal mucosa. As a consequence intracellular cyclic AMP or GMP accumulates, resulting in decreased absorption by villous cells and net secretion by crypt cells. When the two types of enterotoxin and their transferability by plasmids were first recognized, it was naively assumed that virtually any *E. coli* strain could become an enteric pathogen if enterotoxin genes were acquired by plasmid transfer. This proved not to be true. ETEC exert the pathogenic effect in the proximal small intestine, where they must reside, in some manner counteracting the clearing effects of gut peristalsis, and where they must release toxin within a critical proximity to mucosal cells. Studies in piglets first demonstrated that ETEC must have accessory virulence properties in addition to LT or ST in order to cause disease (Smith & Linggood 1971). The best characterized accessory virulence properties, adherence or colonization factors, enable ETEC to attach to mucosa of the small intestine, thereby overcoming the potent peristaltic defence as well as allowing release of toxin closer to reactive sites. The antigens responsible for colonization in several animal ETEC have been identified (Burrows et al 1976, Nagy et al 1977, Ørskov & Ørskov 1966, Ørskov et al 1975, Smith & Linggood 1971). In each instance they have proved to be hair-like organelles (pili, fimbriae) on the surface of the bacterium (Nagy et al 1977, Ørskov et al 1975, Stirm et al 1967). These pili have been purified and used as vaccines in animals (Acres et al 1979, Morgan et al 1978, Nagy et al 1978, Rutter & Jones 1973); by stimulating antibody that binds the adhesion pili (Rutter et al 1976) these veterinary vaccines have prevented colonization of the upper small bowel and diarrhoeal disease.

The purpose of this paper is (1) to review the pertinent work describing colonization antigens and vaccines in animal ETEC strains that served as a background for studies of human ETEC, and (2) to review several classes of pili found in human ETEC that exhibit adhesive properties for mucosal epithelial cells.

Pili

The term pili (Latin for hairs) was coined by Brinton (1959) to describe the rod-like, non-flagellar bacterial appendages assembled from protein subunits. The term fimbriae (Latin for fibres) was used by Duguid (1955) in identifying the same organelles. Conjugal or sex pili are plasmid-mediated organelles responsible for the transfer of DNA between bacteria. Other classes of pili are not involved in nucleic acid transfer. The earliest of these to be well characterized were type 1 somatic pili (also referred to as 'common' pili) (Brinton 1959, 1967). These pili, hollow fibres 7 nm in diameter and 0.5–2.0 μm in length, and 50–400 per bacterial cell, are coded for by chromosomal genes (Brinton 1959, 1967, 1978). Type 1 somatic pili confer upon bacteria many properties lacking in the unpiliated state, including surface translocation, enhanced growth in limiting oxygen concentrations, tight colonial association of bacterial growth, and the ability to agglutinate guinea-pig erythrocytes and to adhere to epithelial surfaces (Brinton 1978). The adhesive and haemagglutinating properties are abolished by the monosaccharide D-mannose, leading to the terms mannose-sensitive haemagglutination (MSHA) and adhesion, typical of type 1 somatic pili (Duguid et al 1955).

K88 and K99 antigens

In the early 1960s Ørskov & Ørskov (1966) described a new capsular antigen, K88, in *E. coli* strains isolated from piglets with diarrhoea. This protein antigen was found to be a pilus-like surface structure (Stirm et al 1967) morphologically distinguishable from type 1 somatic pili in electron photomicrographs. K88 pili were found to be plasmid-mediated (Ørskov & Ørskov 1966, Smith & Linggood 1977) and preferentially expressed after growth on solid medium at 37 °C but not after growth at 18 °C or in broth (Ørskov & Ørskov 1966, Stirm et al 1967). K88 pili confer on ETEC the ability to agglutinate guinea-pig erythrocytes in the presence of D-mannose (mannose-resistant haemagglutination) (Stirm et al 1967). The K88 antigen was shown by Smith & Linggood (1971) to be an essential virulence determinant of K88-positive strains of ETEC that cause diarrhoea in newborn piglets. K88-positive strains manifest adhesive properties allowing them to attach to proximal intestinal mucosa and proliferate to high numbers there. K88-negative mutants of K88-positive strains, despite being comparable in enterotoxin production, are avirulent in piglets because they cannot attach and propagate in the upper small bowel. Rutter et al (1975) observed that K88-positive ETEC strains did not adhere to intestinal tissue of all piglets. They noted two host phenotypes, 'adhesive' and 'non-adhesive'. The gene for the adhesive phenotype behaved as an autosomal dominant inherited in a simple Mendelian manner (Rutter et al 1975).

Ørskov et al (1975) identified an analogous pilus-like, protein surface antigen originally associated with ETEC diarrhoea in calves and lambs. Initially called Kco, its official designation became K99 (Ørskov et al 1975); further work demonstrated that K99 was also found in many piglet ETEC. K99 was found to be analogous to K88: (1) they appeared similar in electron photomicrographs; (2) K99, like K88, was preferentially expressed after cultivation on solid agar at 37 °C, and K99 did not appear after incubation at 18 °C (Burrows et al 1976); (3) genes for both are contained in transferable plasmids (Ørskov et al 1975); and (4) K99-containing bacteria cause mannose-resistant agglutination of sheep erythrocytes (Burrows et al 1976).

E. coli 987 pili in porcine ETEC

Many ETEC strains isolated from piglets possess K88 or K99 antigen and mutants of these strains lacking K88 or K99 also lack adhesiveness and pathogenicity. On the other hand, many ETEC strains isolated from piglets with diarrhoea lack K88 or K99 antigens, yet are clearly pathogenic; furthermore, these strains adhere to porcine intestinal epithelium *in vivo* and *in vitro* and cause diarrhoea when fed to piglets. One such strain, *E. coli* 987, has been intensively studied (Isaacson et al 1978, Nagy et al 1977). *E. coli* 987 apparently adheres to intestinal epithelium by a class of pili distinct from K88 or K99 (Isaacson et al 1978). In electron photomicrographs *E. coli* 987 pili resemble type 1 somatic pili as defined by Brinton yet they differ in that they do not possess haemagglutinating properties against guinea-pig or other erythrocytes and are distinct biochemically and antigenically (Isaacson et al 1978, Nagy et al 1977). It is unclear whether genes for 987 pili are located in chromosomal or plasmid DNA.

Adhesion of E. coli pili in vitro

An assay was developed by Isaacson et al (1978) whereby attachment of *E. coli* and purified pili to porcine epithelial cells could be evaluated. Purified *E. coli* 987, K99 or type 1 somatic pili adhered to epithelial cells *in vitro*, as did bacteria possessing these antigens (Isaacson et al 1978). Purified pili competitively inhibited the attachment of *E. coli* bearing the homologous pilus antigen; similarly, specific anti-pilus antibody prevented attachment to epithelial cells of purified pili or of *E. coli* bearing the homologous pilus antigen (Isaacson et al 1978).

Veterinary vaccine studies

Because colibacillosis in piglets occurs in the neonatal period, ETEC disease is controlled by active immunization of pregnant gilts to stimulate colostral antibodies which passively protect piglets suckled by successfully immunized mothers (Rutter et al 1976). Several controlled studies have been done using purified pili as parenteral vaccines followed by experimental challenge of suckling piglets with ETEC that have the homologous pilus antigen.

Results of these trials in pigs and cattle are summarized in Table 1. The offspring were protected against diarrhoea and the death rate from this disease was greatly reduced. Several groups demonstrated that protection correlated with anti-pili antibodies in the colostrum and milk of immunized mothers (Acres et al 1979, Morgan et al 1978, Nagy et al 1978, Rutter & Jones 1973, Rutter et al 1976). Small amounts of O antigen contaminated the purified pili preparations, and small rises in serum and colostral O antibody were seen after immunization (Morgan et al 1978, Rutter et al 1976). The antibody levels were low in comparison with anti-pili antibody levels. Piglets born to control and immunized dams have been challenged with an ETEC strain bearing the homologous pilus antigen and O and H antigens

TABLE 1 Prevention of neonatal diarrhoea in pigs and cattle by inoculation of pregnant animals with purified pili antigens

Purified pilus antigen (*E. coli* strain of origin)	Reference	Animal actively immunized	Dosage schedule (time pre-partum)	Challenge strain (inoculum)	Disease in challenged newborns: Attack rate of diarrhoea		Mortality rate due to diarrhoea		Vaccine efficacy against fatal disease (%)
					Offspring of immunized mothers	Offspring of control mothers	Offspring of immunized mothers	Offspring of control mothers	
K88 (O8:K27,K88ab:NM O8:K27,K88ac:NM)	Rutter & Jones 1973	Pregnant gilts	1st: 15 mg intra-mammary, 11 wk; 2nd: 30 mg s.c., 10 d	O149:K91,K88ac:H10 (10^{10})	–	–	4/31[a]	20/29	81
987 (O9:K103,987P:NM)	Nagy et al 1978	Pregnant gilts	1st: 5 mg s.c.,21–27 d; 2nd: 5 mg s.c.,7–13 d	O9:K103,987P:NM (5×10^8)	0/69[b]	11/39[c]	0/69	17/56	100
987 (O9:K103,987P:NM)	Morgan et al 1978	Pregnant gilts	1st: 9 mg s.c., 16–28 d; 2nd: 9 mg s.c., 5–17 d	O9:K103,987P:NM (5×10^8)	–	–	0/38	19/85	100
987 (O9:K103,987P:H⁻)	Morgan et al 1978	Pregnant gilts	1st: 9 mg s.c., 16–29 d; 2nd: 9 mg s.c., 5–17 d	O20:K101,987P:NM (10^{10})	–	–	1/37	12/74	83
K99 (*E. coli* K12 with K99)	Morgan et al 1978	Pregnant gilts	1st: 9 mg s.c., 16–29 d; 2nd: 9 mg s.c., 5–17 d	O101:K30,K99:NM (10^{10})	–	–	0/41	23/65	100
K99 (*E. coli* K12 with K99)	Acres et al 1979	Pregnant cows	1st: 10 mg s.c., 6 wk; 2nd: 10 mg s.c., 3 wk	O9:K30,K99:NM (10^{11})	2/9	9/10	0/9	9/10	100

[a]No. dead/No. challenged. [b]No. ill/No. challenged [c]No. ill/No. survivors at 6th day after challenge.

distinct from those of the strain from which the vaccine was prepared; the results show that vaccine efficacy was indeed mediated by anti-pili antibody (Morgan et al 1978, Rutter & Jones 1973).

The exciting and encouraging results of these preliminary studies suggest that up to 80 or 90% of ETEC diarrhoea of piglets can be prevented by immunizing with a trivalent pilus vaccine containing K88, K99 and 987-type pili. Furthermore, these studies serve as a guide and encouragement for studies with human ETEC.

Adhesion pili in human ETEC

Early studies by DuPont, Hornick and co-workers (Dupont et al 1971) in adult volunteers clearly demonstrated that more than enterotoxigenicity was required in order for an *E. coli* strain to cause diarrhoea in humans. Volunteers who ingested up to 10^{10} organisms of *E. coli* 263 (a highly enterotoxigenic, $K88^+$ pig pathogen) failed to develop diarrhoeal disease. Similarly, when volunteers ingested a normal colonic flora strain of *E. coli* (HS) to which an enterotoxin plasmid had been transferred, diarrhoea was not induced.

While working with ETEC strain H10407 (O78:H11), originally isolated from a case of watery diarrhoea in Bangladesh, Evans et al (1975) noted that their laboratory strain suddenly ceased to cause accumulation of fluid in the infant rabbit assay despite continued production of heat-labile enterotoxin. They also noted that the laboratory-derived mutant was less effective in colonizing rabbit intestine than the fully virulent parent strain, and that the mutant strain (H10407 P) lacked a plasmid present in the parent strain. Using H10407 P organisms to absorb antiserum prepared against the fully virulent parent strain Evans et al (1975) produced an antiserum to the surface antigen present on H10407 but missing on H10407 P. Subsequent work showed this antigen to be a pilus-like organelle distinct from the type 1 somatic pili also present on H10407 (Evans et al 1978). In several respects this pilus, currently referred to as Colonization factor antigen I (CFA/I), is analogous to the K88 pilus antigen of porcine ETEC strains. CFA/I resembles K88 morphologically, is plasmid-mediated, is preferentially expressed after growth on solid agar at 37 °C, is not found after incubation at 18 °C, and confers upon bacteria the property of mannose-resistant agglutination of human type A and B and bovine erythrocytes. This last observation has provided a simple tool for screening agar-grown ETEC strains for CFA/I pili.

Independently, Brinton also observed that *E. coli* H10407 elaborates a second pilus, antigenically, morphologically and biochemically distinct from type 1 somatic pili (Brinton 1978). These pili have been purified and their haemagglutinating properties confirmed. The pili, referred to as non-mannose-sensitive pili by Brinton (1978), are apparently identical to CFA/I.

Evans et al (1978) and Ørskov & Ørskov (1977) noted that CFA/I appeared commonly in only certain serogroups, including O15, O20, O25 and O78 (Ørskov & Ørskov 1977). Evans & Evans (1978) have identified a second pilus-like antigen, termed CFA/II, which is also plasmid-mediated, resembles CFA/I morphologically, is not expressed when cultured at 18 °C, and confers on bacteria the ability to cause mannose-resistant agglutination of bovine but not human erythrocytes. Haemagglutination is maximal at 4 °C. CFA/II is frequently found in ETEC strains of serogroups O6 and O8 (Evans et al 1978).

Prevalence of CFA/I, CFA/II and type 1 somatic pili in human ETEC

The concept of the immunological control of ETEC diarrhoea in humans by multivalent purified pili vaccines is attractive. If only a few antigenic types of colonization factors were common to human ETEC, and if they could be identified and purified, immunoprophylaxis of human ETEC disease, particularly travellers' diarrhoea, would be feasible.

Controversy exists over the prevalence of CFA/I and II in human ETEC strains and over whether these particular pili antigens are required by all (or most) ETEC strains in order for pathogenicity, intestinal colonization and immunological responses to be manifested in humans. For example, Evans et al (1978) reported that 25 of 29 (86%) of their ETEC strains from travellers' diarrhoea possessed CFA/I, and that 98% of ETEC belonging to serogroups O6, O8, O15, O25, O63 and O78 produced CFA/I or II (Evans & Evans 1978). In contrast, while Ørskov & Ørskov (1977) found CFA/I commonly in O78 serogroup isolates from human diarrhoea, it occurred in none of 49 ETEC pathogens from five other serogroups. Similarly, Gross et al (1978) examined 89 ETEC strains from patients with diarrhoea and found that only nine strains (10%) exhibited MRHA and only six strains (7%) reacted with CFA/I antibody (Gross et al 1978). Since CFA/I and II pili are plasmid-mediated, one explanation for the divergent observations is that, in the surveys recording a low prevalence, the strains may have lost their plasmids before the time of testing. An alternative explanation is that not all ETEC strains may need CFA/I or II in order to cause disease in man; in strains lacking CFA/I or II other factors may promote adherence and colonization (Levine et al 1980).

In an attempt to resolve this fundamental controversy we examined the ETEC and enteropathogenic *E. coli* (EPEC) strains that have been used at the University of Maryland in volunteer challenge studies for the presence of CFA/I, CFA/II and type 1 somatic pili (Levine et al 1980). The virulence or avirulence of these strains is indisputable (Table 2). Strains were tested after growth on CFA agar as well as in broth. Only one strain, H10407, produced CFA/I, and another had CFA/II. Four ETEC strains and two EPEC strains which lack CFA/I or II nevertheless caused clear-cut diarrhoeal illness in volunteers. Furthermore, ingestion of these strains by volunteers resulted in intestinal colonization and in vigorous immune responses to

TABLE 2 Colonization factor antigens I and II and type 1 somatic pili in Escherichia coli strains fed to volunteers and in 15 normal flora control strains

Strains	*Serotype*	*Enterotoxin*	*Caused diarrhoea in volunteers*	*CFA/I*[a]	*CFA/II*[b]	*Type 1 somatic pili*[c]
ETEC						
H10407	078:H11	LT^+/ST^+	+	+	–	+
B_2C	06:H16	LT^+/ST^+	+	–	+	–
B7A	0148:H28	LT^+/ST^+	+	–	–	+
TD225-C4	078:H9	LT^+/ST^-	+	–	–	+
E2528-C1	025:NM	LT^+/ST^-	+	–	–	+
214-4	Non-typable	LT^-/ST^+	+	–	–	–
H10407P	078:H11	LT^+/ST^-	–	–	–	+
EPEC						
E851/71	0142:K86:H6	EPEC	+	–	–	+
E2348/69	0127:K63:H6	EPEC	+	–	–	+
E74/68	0125:K67:H2	negative	–	–	–	–
Normal flora controls						
15 strains	–	Negative	NA[d]	– in 15	+ in 1 – in 14	+ in 11 – in 4

[a]Identified by mannose-resistant agglutination of human type A erythrocytes and confirmed by agglutination with specific antiserum.
[b]Identified by mannose-resistant agglutination of bovine but not human erythrocytes at 4 °C.
[c]Identified by mannose-sensitive agglutination of guinea-pig erythrocytes.
[d]Not applicable.

LT or O antigens. The virulent ETEC strains lacking CFA/I or II included all enterotoxigenic phenotypes: LT^+/ST^+, LT^+/ST^-, and LT^-/ST^+ (Levine et al 1980).

Our interpretation of these findings is that in human *E. coli* enteropathogens there exist classes of adhesion pili other than CFA/I and II, or other surface structures such as polysaccharides, slime layers or lectins that promote adhesion to small intestinal mucosa.

To further explore the prevalence of CFA/I and II in ETEC we collaborated with Dr R. Bradley Sack of Baltimore City Hospital. Dr Sack provided a collection of ETEC isolates from cases of travellers' diarrhoea in Peace Corps workers in Africa, Asia and Latin America. The isolates had been subcultured for from one to three passages before their arrival at the Center for Vaccine Development laboratories. The ETEC strains were cultured on both CFA agar and in broth before being tested by haemagglutination and with CFA/I and II antisera. Thirty-six ETEC strains from 36 patients with pure ETEC diarrhoeal infections were examined. As seen in Table 3, of nine persons with LT^+/ST^+ infections, four (44%) had CFA/I (two strains) or CFA/II (two strains). LT^-/ST^+ strains occurred in 13 patients; one of these strains

TABLE 3 Prevalence of CFA/I, CFA/II and type 1 somatic pili in enterotoxigenic Escherichia coli isolated from patients[a] with acute travellers' diarrhoea: relationship with toxin type

	Class of pili		
Enterotoxin phenotype	*CFA/I*	*CFA/II*	*Type 1 somatic*
LT^+/ST^+	2/9[b] (22%)	2/9 (22%)	0/9
LT^-/ST^+	1/13 (8%)	2/13 (15%)	4/13 (31%)
LT^+/ST^-	0/14	0/14	10/14 (71%)
Totals	3/36 (8%)	4/36 (11%)	14/36 (39%)

[a] Patients with infections from which enterotoxigenic *E. coli* of only a single serotype and toxin type were isolated.
[b] Infections due to ETEC with pilus/number of infections.

had CFA/I and two had CFA/II. Neither CFA/I nor II were found in any of the 14 LT^+/ST^- strains that infected 14 patients; in contrast 10 of these strains (71%) had type 1 somatic pili.

Increasingly LT^-/ST^+ strains have been found to be as important as LT^+/ST^+ strains in causing diarrhoea. The collection of travellers' diarrhoea isolates is consistent with this finding. While CFA/I or CFA/II occurred in about half the LT^+/ST^+ isolates (still well below the finding of Evans et al 1978), LT^+/ST^+ strains appear to represent only a minority of ETEC pathogens. Thus, while purified CFA/I and II pili may be needed in a polyvalent pili vaccine to protect against LT^+/ST^+ strains, common antigens must be identified in the remaining LT^+/ST^- and LT^-/ST^+ strains for inclusion in a future vaccine.

As noted in Tables 2 and 3, type 1 somatic pili occur commonly in ETEC, particularly LT^+/ST^- strains. Brinton (1978) has shown that about 40% of ETEC have type 1 somatic pili antigenically closely related to those of *E. coli* H10407.

Immunity to enterotoxigenic Escherichia coli in volunteer studies

Challenge studies were done in volunteers (college students and other groups of adults) to investigate the mechanisms of disease-induced immunity (Levine et al 1979). Seventeen volunteers ingested (with $NaHCO_3$) 10^6 or 10^8 organisms of ETEC strain B7A (O148:H28). This strain produces LT and ST and has type 1 somatic pili but lacks CFA/I or II. Ten volunteers developed diarrhoeal illness closely mimicking natural travellers' diarrhoea; among these 10 persons, rises in titre of serum LT antitoxin and anti-O antibody occurred in eight (80%). Eight of the volunteers who developed diarrhoea in the first challenge agreed to be rechallenged nine weeks later with 10^8 B7A organisms. Only one of these eight

TABLE 4 Clinical, serological and bacteriological response of eight volunteers to initial challenge with E. coli B7A and subsequent re-challenge nine weeks later in comparison with 12 control volunteers

							4 × or > rise in antibody			
Volunteer group	*Number of volunteers*	*Diarrhoea*	*Incubation (h)*	*Total diarrhoeal stool volume (ml)*	*Total loose stools*	*Nausea or vomiting*	*Malaise*	*Antitoxin (%)*	*Antisomatic (%)*	*Positive stool culture (%)*
Initial challenge of veterans[a]	8	8 (100%)	44[b] (19–124)	7035	30	3 (38%)	6 (75%)	75	75	100
		$P = 0.002$								
Re-challenge[a]	8	1 (13%)	52	557	9	0	0	50	13	100
		$P = 0.05$								
Controls[a]	12	7 (58%)	45 (17 – 68)	6355	33	2 (17%)	7 (58%)	92	83	100

[a]Challenged orally with 10^8 *E. coli* B7A after ingestion of $NaHCO_3$.
[b]Mean (range).

TABLE 5 Heterologous challenge with E. coli E2528-C1 of volunteers who previously had diarrhoeal infection due to E. coli B7A in comparison with controls

Volunteer group	*Number of volunteers*	*Number with diarrhoea*	*Incubation (h)*	*Mean total diarrhoeal stool volume (ml) per ill volunteer*	*Mean total number loose stools per ill volunteer*
B7A veterans[a]	4	3	19.7[b] (13–24)	514 (301–622)	2.7 (1–5)
Controls[a]	6	2	16 (9.5–22.5)	482 (415–549)	4.5 (4–5)

[a]Challenged orally with 10^9 *E. coli* E2528-C1 after ingestion of 2.0 g $NaHCO_3$.
[b]Mean (range).

'veterans' developed diarrhoea, compared to seven of 12 controls (Table 4) who also participated in the re-challenge (P = 0.05). Despite clinical protection, all 'veterans' excreted B7A after rechallenge. These observations showed that clinical protection was mediated by a non-bactericidal mechanism, possibly anti-adhesive or antitoxic in nature and probably operating at the local intestinal mucosal level.

Four controls who developed diarrhoea during the homologous B7A re-challenge test were re-challenged nine weeks later with 10^9 organisms of ETEC strain E2528-C1 (O25:H^-), which produces only LT and possesses type 1 somatic pili but not CFA/I or II. The O,H and type 1 somatic pilus antigen composition of E2528-C1 is distinct from B7A, while LT is a common antigen shared between the two strains. Three of these four 'veterans' and two of six concomitantly challenged controls developed comparable diarrhoea (Table 5). Heterologous protection was not conferred where the only common antigen was heat-labile enterotoxin, indicating that infection-derived LT antitoxin is apparently not protective. On the other hand, these studies demonstrate that prior disease due to ETEC confers homologous immunity against subsequent challenge and that the operative mechanism is not bactericidal and is not mediated by serum anti-O antibodies. Additional homologous re-challenge studies were done in volunteers with ETEC strain H10407 (O78:H11), which produces LT, ST, type 1 somatic pili and CFA/I. Despite clinical protection in re-challenged volunteers, quantitative stool cultures revealed no difference in the level of faecal excretion of H10407 between protected re-challenged volunteers and ill controls; both groups excreted 10^6–10^8 H10407 organisms/g (or ml) of stool for several days.

The demonstration of homologous immunity to ETEC in volunteers mediated by non-bactericidal mechanisms has convinced us that the efficacy of purified pili vaccines should be evaluated as a rational approach to the immunoprophylaxis of ETEC diarrhoea in humans.

Purified pili vaccines: the near future

Preliminary studies are under way at the University of Maryland's Center for Vaccine Development to evaluate in volunteers the safety, immunogenicity and efficacy of *E. coli* purified pili vaccines prepared by Dr Charles C. Brinton, Jr and his co-workers at the University of Pittsburgh. Vaccines to be tested include oral and parenteral purified preparations of type 1 somatic pili from *E. coli* H10407 and CFA/I from the same strain.

Acknowledgements

These studies were supported by research contracts NO1A142553 from the National Institute of Allergy and Infectious Diseases and DAMD-17-78-C-8011 from the US Army Research and Development Command.

REFERENCES

Acres SD, Isaacson RE, Babiuk LA, Kapitany RA 1979 Immunization of calves against enterotoxigenic colibacillosis by vaccinating dams with purified K99 antigen and whole cell bacterins. Infect Immun 25:121-126

Brinton CC Jr 1959 Non-flagellar appendages of bacteria. Nature (Lond) 183:782-786

Brinton CC Jr 1967 Contributions of pili to the specificity of the bacterial surface and a unitary hypothesis of conjugal infectious heredity. In: Davis BD, Warren L (eds) The specificity of cell surfaces. Prentice-Hall, Englewood Cliffs, NJ, p 37-70

Brinton CC Jr 1978 The piliation phase syndrome and the uses of purified pili in disease control. In: Proceedings of the XIIIth joint US–Japan conference on cholera. National Institutes of Health, Bethesda, Md (DHEW publ. No. NIH 78-1590) p 33-70

Burrows MR, Sellwood R, Gibbons RA 1976 Haemagglutinating and adhesive properties associated with the K99 antigen of bovine strains of *Escherichia coli.* J Gen Microbiol 96:269-275

Duguid JP, Smith IW, Dempster G, Edmunds PN 1955 Non-flagellar filamentous appendages ("fimbriae") and haemagglutinating activity in Bacterium coli. J Pathol Bacteriol 70:335-348

DuPont HL, Formal SB, Hornick RB, Snyder MJ, Libonati JP, Sheahan DG et al 1971 Pathogenesis of *Escherichia coli* diarrhea. N Engl J Med 285:1-9

Evans DG, Evans DJ Jr 1978 New surface-associated heat-labile colonization factor antigen (CFA/II) produced by enterotoxigenic *Escherichia coli* of serogroups O6 and O8. Infect Immun 21:638-647

Evans DG, Silver RP, Evans DJ Jr, Chase DG, Gorbach SL 1975 Plasmid-controlled colonization factor associated with virulence in *Escherichia coli* enterotoxigenic for humans. Infect Immun 12:656-667

Evans DG, Evans DJ Jr, Tjoa WS, DuPont HL 1978 Detection and characterization of colonization factor of enterotoxigenic *Escherichia coli* isolated from adults with diarrhea. Infect Immun 19:727-736

Gross RJ, Cravioto A, Scotland SM, Cheasty T, Rowe B 1978 The occurrence of colonization factor (CF) in enterotoxigenic *Escherichia coli.* FEMS (Fed Eur Microbiol Soc) Lett 3:231-233

Isaacson RE, Fusco PC, Brinton CC, Moon HW 1978 *In vitro* adhesion of *Escherichia coli* to porcine small intestinal epithelial cells: pili as adhesive factors. Infect Immun 21:392-397

Levine MM, Nalin DR, Hoover DL, Bergquist EJ, Hornick RB, Young CR 1979 Immunity to enterotoxigenic *Escherichia coli.* Infect Immun 23:729-736

Levine MM, Rennels MB, Daya V, Hughes TP 1980 Hemagglutination and colonization factors in enterotoxigenic and enteropathogenic *Escherichia coli* that cause diarrhea. J Infect Dis 141:733-737

Morgan RL, Isaacson RE, Moon HW, Brinton CC, To CC 1978 Immunization of suckling pigs against enterotoxigenic *Escherichia coli*-induced diarrheal disease by vaccinating dams with purified 987 or K99 pili: protection correlates with pilus homology of vaccine and challenge. Infect Immun 22:771-777

Nagy B, Moon HW, Isaacson 1977 Colonization of porcine intestine by enterotoxigenic *Escherichia coli:* selection of piliated forms *in vivo,* adhesion of piliated forms to epithelial cells *in vitro,* and incidence of a pilus antigen among porcine enteropathogenic *E. coli.* Infect Immun 16:344-352

Nagy B, Moon HW, Isaacson RE, To CC, Brinton CC 1978 Immunization of suckling pigs against enterotoxigenic *Escherichia coli* infection by vaccinating dams with purified pili. Infect Immun 21:269-274

Ørskov I, Ørskov F 1966 Episome-carried surface antigen K88 of *Escherichia coli.* I. Transmission of the determinant of the K88 antigen and influence on the transfer of chromosomal markers. J Bacteriol 91:69-75

Ørskov I, Ørskov F 1977 Special O:K:H serotypes among enterotoxigenic *E. coli* strains from diarrhea in adults and children. Occurrence of the CF (colonization factor) antigen and of hemagglutinating abilities. Med Microbiol Immunol 163:99-110

Ørskov I, Ørskov F, Smith HW, Sojka WJ 1975 The establishment of K99, a thermolabile, transmissible *Escherichia coli* K antigen, previously called "Kco", possessed by calf and lamb enteropathogenic strains. Acta Pathol Microbiol Scand Sect Microbiol Immunol 83:31-36

Rutter JM, Jones GW 1973 Protection against enteric disease caused by *Escherichia coli* – a model for vaccination with a virulence determinant? Nature (Lond) 242:531-532

Rutter JM, Burrows MR, Sellwood R, Gibbons RA 1975 A genetic basis for resistance to enteric disease caused by *E. coli.* Nature (Lond) 257:135-136

Rutter JM, Jones GW, Brown GTH, Burrows MR, Luther PD 1976 Antibacterial activity in colostrum and milk associated with protection of piglets against enteric disease caused by K88-positive *Escherichia coli.* Infect Immun 13:667-676

Smith HW, Linggood MA 1971 Observations on the pathogenic properties of the K88, Hly and Ent plasmids of *Escherichia coli* with particular reference to porcine diarrhoea. J Med Microbiol 4:467-485

Stirm S, Ørskov F, Ørskov I, Birch-Anderson A 1967 Episome-carried surface antigen K88 of *Escherichia coli.* III. Morphology. J Bacteriol 93:740-748

DISCUSSION

Sharon: When you were studying type 1 pili did you always test for the mannose sensitivity?

Levine: We simultaneously cultured each strain on CFA agar at 37 °C and inoculated it into a test tube containing 15 ml Mueller-Hinton broth, a minimal broth. We incubated the stagnant culture for 48 h and then subcultured into the broth again, to get type 1 pili. Then we haemagglutinated in the presence and absence of mannose with human type A, bovine and guinea-pig cells at 24 °C and at 4 °C. Antisera exist which agglutinate purified type 1 somatic pili of *E. coli* H10407. Some people, including Charles Brinton, have done surveys with these

E. coli strains. If a strain causes mannose-sensitive agglutination of guinea-pig erythrocytes and is negative by agglutination with an antiserum of this kind, that does not mean that type 1 pili are not present. It means that the strain probably has type 1 pili but of an antigenic type distinct from that of H10407. Multiple antigenic varieties of type 1 pili apparently exist.

Sharon: Was the non-piliated isolate that adheres very strongly mannose-sensitive?

Levine: It caused no haemagglutination and had no pili on electron microscopy. Buccal cell adhesion was mannose-resistant.

Sharon: Is the 987 mannose-sensitive?

Levine: No. It causes no haemagglutination. It adheres to epithelial cells in a mannose-insensitive way – it is a different receptor.

Sharon: Is anything known about the subunits of the pili studied by you, which are similar morphologically to type 1 pili? Do they also consist of pili type subunits?

Levine: Charles Brinton (1977) has shown by absorption spectroscopy that 987 pili are quite different from the somatic type 1 pili from H10407. *E. coli* BAM and H10407 are reasonably close; 987 is quite different. Immunologically the antiserum directed against H10407 type 1 pili had very little cross-agglutinability for recognizing antigens on 987. But to the best of my knowledge nobody has broken them down. Morphologically they look alike but they seem to be biochemically and immunologically quite distinct.

Taylor-Robinson: If you really believe that a pili vaccine is going to work, then clearly you must believe that pili constitute the most important virulence factor and that anti-pili antibody will prevent attachment. Yet you judiciously pointed out that there were other factors. Could you expand on that?

Levine: A multivalent vaccine for humans should contain CFA/I, CFA/II, which would cover a small but important percentage of strains, and type 1 pili from perhaps three different antigenic varieties that are very common in the rest of the strains. I think type 1 pili are not a virulence property in the same class or strength as CFA/I and CFA/II but that they are a virulence property in a general way. If one uses them in a vaccine and stimulates antibody against them it turns out that one can protect, and that is what really counts. Antibody directed against them apparently gives sufficient steric hindrance, or whatever the mechanism is, to prevent disease. Mechanistically all we have to do is prevent adhesion in the upper small bowel and the rest doesn't matter.

Taylor-Robinson: Why can't you use the whole organism? Why do you need to use purified pili?

Levine: There are 164 *E. coli* O serogroups, of which perhaps only 20 occur commonly as LT/ST strains. But many other serogroups are found all over the world in strains with LT alone or ST alone. So to have a vaccine giving anti-O immunity one needs perhaps fifty O antigens. That approach has been used in Eastern Europe, for example. Pili are apparently much better antigens to include because many fewer are needed. In his study, Charles Brinton is using his purified

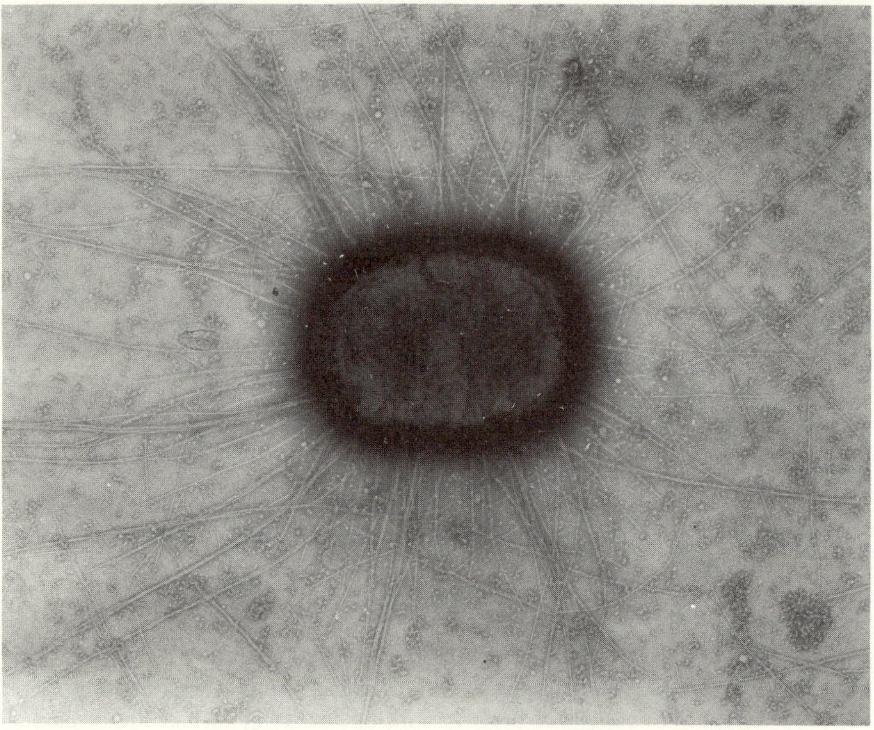

FIG. 1 (Candy). Electron micrograph of negatively stained preparation of *E. coli* showing morphology of 987 pili (x 57 000). Photograph provided by Dr Julian Beasley.

antibodies against the antigenic variety of type 1 pilus found on H10407; 40% of a large number of toxigenic *E. coli* isolates from all over the world strongly agglutinate with that antibody, suggesting that they have an identical or closely related type 1 pilus. B7A has, as far as we can tell, a type 1 pilus antigenically identical to H10407. Their titres against each other are identical.

Candy: Fig. 1, supplied by Dr Julian Beasley (Wellcome Research Laboratories, Beckenham, UK), shows the rather spectacular 987 pili.

I would now like to stress that fimbriae, or pili, are not the only mechanism by which bacteria adhere. An infant admitted to Queen Elizabeth Hospital for Children in the East End of London under the care of Dr John Walker-Smith had protracted diarrhoea following a rotavirus infection and excreted a strain of *E. coli* (O111) for a prolonged period. We found that this was a mannose-resistant strain which agglutinated both bovine and human red cells when grown on solid media; under this growth condition it was not fimbriated. High resolution scanning electron microscopy has shown some relatively large surface projections which seemed to be sticking the *E. coli* to erythrocytes and to each other. Similar structures were seen by negative staining electron microscopy (Fig. 2).

Finally, another child under Dr Walker-Smith's care at Queen Elizabeth Hospital for Children who had severe protracted traveller's diarrhoea showed, on biopsy of the small intestine, *E. coli* adhering to enterocytes by very close bacteria-to-cell apposition (A. D. Phillips, personal communication). Wherever this *E. coli* had stuck the microvilli were destroyed. This is unlike K88-mediated adhesion, where $K88^+$ *E. coli* adhere to structurally normal microvilli with a clear space between bacteria and brush border. A similar phenomenon was described by Ulshen & Rollo (1980).

My colleague Alan Phillips identified the adhering *E. coli* by indirect immunofluorescence as serogroup O128. These *E. coli* were, of course, not subcultured; it was an adhesion 'experiment' performed *in vivo* by nature. What also occurred was adhesion of the same *E. coli* to the mucus layer, very much as in the pictures that Professor Freter showed of *Vibrio cholerae.* Not only does it adhere to and appear to destroy the brush border but it also sticks to the mucus layer.

Rutter: In humans with *E. coli* diarrhoea, what are the numbers per gram of intestinal contents of pathogenic strains that have CFA/I, CFA/II or neither CFA factors?

Levine: Dupont et al (1971) got 10^5–10^7 organisms/ml of jejunal fluid in volunteers with diarrhoea due to *E. coli* B_2C, a strain now known to have CFA/II.

To get positive cultures in humans with strains lacking CFA/I or II one has to put jejunal tubes in very early in illness. We got about 10^5/ml for a short time.

Rutter: That contrasts markedly with the results in animals. K88-positive strains can reach high numbers (about 10^9/g contents) in the anterior small intestine (Rutter et al 1975, Smith & Huggins 1978) and strains with K99 or 987P reach similarly high numbers but lower down the intestine (Smith & Huggins 1978). If the bacteria don't attach to the gut cells they may not be able to reach high numbers, and transient counts of 10^5 are quite low compared with the results in animals. Is there evidence for attachment of the human pathogenic strains to intestinal cells or is there just a close association in the mucus layer?

Levine: Evans et al (1979) have done some fluorescent antibody studies of CFA/I-positive and CFA/II-positive organisms in infant rabbits and shown their close association to the mucosa. A lot of our thinking stems from analogies to animal strains but you have pointed out an important difference.

Taylor-Robinson: One of the advantages of working with animals is that you can easily take specimens at autopsy or before to see what is happening. Can you do jejunal biopsies or other procedures on your human volunteers, Dr Levine?

Levine: Yes. With both cholera and toxigenic *E. coli* infections it was difficult to find organisms, except during a very brief period. We did light and electron microscope studies on biopsies and we did cultures. In cholera we collected serial jejunal fluid for quantitative cultures during incubation. *Vibrio cholerae* from the upper

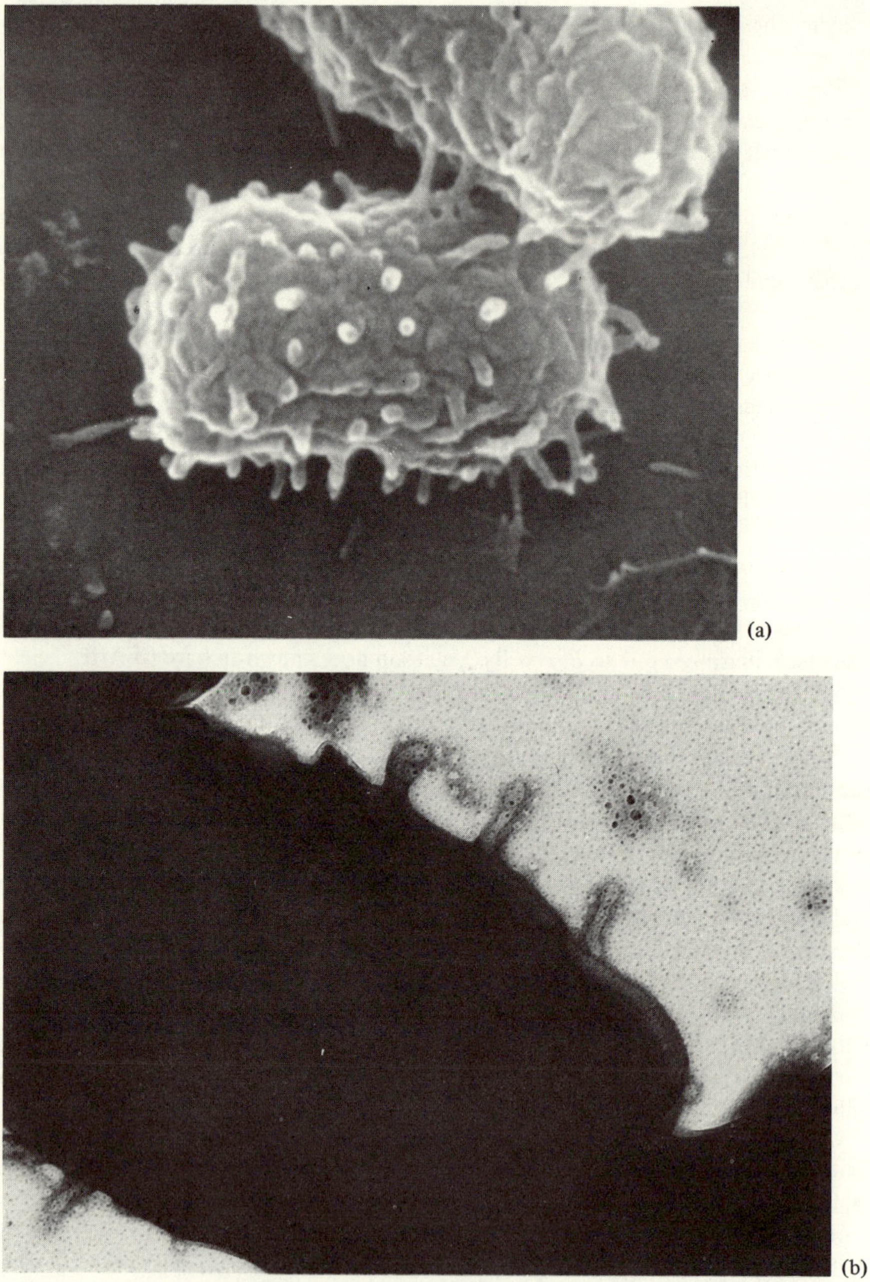

FIG. 2 (Candy). *E. coli* (O111) isolated from repeated stool cultures of an infant with protracted diarrhoea. (a) High resolution scanning electron micrograph showing surface projections. Note bacteria-to-substrate and bacterium-to-bacterium connections (X 80 000). (b) Negative staining preparations to show surface projections (X 94 000).

small bowel, where it is a pathogen, cannot be cultured until about eight hours or less before the onset of diarrhoea. The organisms stay there during early illness and then disappear. This relatively short period of colonization in volunteers surprised us. In contrast Dupont et al's (1971) early studies with US volunteers and Sack et al (1971) and Gorbach et al's (1971) studies with naturally occurring diarrhoeal disease due to toxigenic *E. coli* in India show a lot more colonization, although of short duration. The bowels of Bengali patients are distinctly different from those of American volunteers. The motility is different and the histology of the Bengali upper small bowel often presents a sprue-like picture in the 'normal' state with blunted villi. That may have something to do with the difference.

Wallach: During the course of disease could attachment become limited by factors other than immunological factors, and apart from cellular destruction – for example by a change in the ionic environment of the gut cells? This might move the organisms into a different region of the bowel.

Levine: I think that is correct. There is a fair amount of evidence that the recovery of toxigenic *E. coli* and *V. cholerae* is not related to the immunological response. The bacteria attach for a short time, the toxin is taken in, and with the onset of fluid they reach large numbers for a short time in the upper small bowel. Then they disappear and the diarrhoea continues. They then move to a lower part of the bowel. We think that the disease is sequential, starting proximally and moving its way down the jejunum. The diarrhoea disappears before there is local antibody, when the enterocytes from the crypts make their way up to the tips of the villi and are sloughed off. The toxin continues to work for the life of the cell. It is only when those cells are sloughed that people recover from cholera. Patients can be purging heavily with cholera and when you sterilize them with tetracycline they can purge at a life-threatening rate for 48 h more. This is all due to the toxin. They get better unrelated to antibody. They get better when the cells with the toxin are sloughed off. So I think you are absolutely right.

Svanborg Edén: It has been feared that by inducing immunity to common pili one would also interfere with the normal enterobacterial flora of the intestinal tract. You said earlier that you didn't have any evidence that the normal flora was upset. That is surprising, considering that the type 1 pili are said to mediate attachment important for normal intestinal colonization.

Levine: That is a very important question. The initial studies with the type 1 pilus vaccine was held up because the human volunteer research committee, which included an immunologist, had this fear. That is, if pili are produced *in vivo* they must have some function, and if we produced antibodies to protect against adhesion in the upper small bowel what might we do in the colon? In a few immunized volunteers we looked very intensively at gut motility, at intestinal absorption (by D-xylose absorption and excretion tests), and at the prevalence of colonic *E. coli* bearing H10407 type 1 somatic pilus antigen, before and after immunization. Despite the fact that we generated antibody levels that were protective this did not adversely affect absorption, motility or the prevalence of *E. coli* with that antigen. The reason for this has been obvious to us all along: *E. coli* organisms actually represent a very small amount of the total enteric flora. We tend to overestimate their importance because we do aerobic cultures, but 95% of the dry weight of stools consists of anaerobes, particularly Bacteroides. There is only a film of *E. coli* on the mucosal surface in the large bowel. Most of them are in the lumen, where they are doing important things in terms of bile acid metabolism and interactions etc., so it has not been a problem.

Silverblatt: Another answer to Dr Svanborg Edén's question is that patients who are treated prophylactically with the antibiotic trimethoprim-sulphamethoxyzole for urinary tract infections do very well indeed without any *E. coli* in their stools at all.

Why doesn't the presence of normal type 1 pili in the gut induce spontaneous vaccination, Dr Levine?

Levine: That is the critical question! I think the answer is that type 1 pili are not a firm mechanism for adhesion to cells of the upper small bowel, or receptors may not be there in the upper small bowel. Whatever the reason, they are not stimulating an immune response in the upper small bowel. In the lower bowel they apparently are, because it is not uncommon to have low levels of type 1 pilus antibody. What we do with immunization is essentially to hyperimmunize and put antibodies onto the mucosal surface at such levels that they don't ordinarily exist in the upper small bowel. Clearly those antibodies must also be coming onto the colonic surface, but fortunately with no adverse effect. That critical question bothered us until we started working with the vaccine.

REFERENCES

Brinton CC 1977 The piliation phase syndrome and the uses of purified pili in disease control. In: Proceedings of the 13th joint US–Japan conference on cholera. National Institutes of Health, Bethesda, Maryland (DHEW publ. No. 78-1590) p 33-70

Dupont HL, Formal SB, Hornick RB, Snyder MJ, Libonati JP, Sheahan DG, LaBrec EH, Kalas JP 1971 Pathogenesis of *Escherichia coli* diarrhea. N Engl J Med 285:1-9

Evans DG, Evans DJ Jr Clegg S, Pauley JA 1979 Purification and characterization of the CFA/I antigen of enterotoxigenic *Escherichia coli.* Infect Immun 25:738-748

Gorbach SL, Banwell JG, Chatterjee BD, Jacobs B, Sack RB, 1971 Acute undifferentiated human diarrhea in the tropics. I. Alterations in intestinal microflora. J Clin Invest 50:881-889

Rutter JM, Burrows MR, Sellwood R, Gibbons RA 1975 A genetic basis for resistance to enteric disease caused by *E. coli*. Nature (Lond) 257:135-136

Sack RB, Gorbach SL, Banwell JG, Jacobi B, Chatterjee BD, Mitra RC 1971 Enterotoxigenic *Escherichia coli* isolated from patients with severe cholera-like disease. J Infect Dis 123:378-385

Smith HW, Huggins MB 1978 The influence of plasmid-determined and other characteristics of enteropathogenic *Escherichia coli* on their ability to proliferate in the alimentary tract of piglets, calves and lambs. J Med Microbiol 11:471-492

Ulshen MH, Rollo JL 1980 Pathogenesis of *Escherichia coli* gastroenteritis in man – another mechanism. N Engl J Med 302:99-101.

Adhesion of *Escherichia coli* in urinary tract infection

C. SVANBORG EDÉN, L. HAGBERG, L. Å. HANSON, T. KORHONEN*, H. LEFFLER and S. OLLING

*Departments of Clinical Immunology, Pediatrics and Medical Biochemistry, University of Göteborg, Guldhedsgatan 10, 41346 Göteborg, Sweden and *Department of General Microbiology, University of Helsinki, Finland*

Abstract In individuals prone to urinary tract infection the intestine is colonized by *E. coli* strains that possess a combination of properties determining virulence. Such an *E. coli* strain may colonize the vaginal and periurethral areas and ascend the urinary tract. The ability to attach to the mucosal surface is thought to be essential for *E. coli* to colonize and to remain in the urinary tract.

Most *E. coli* from patients with urinary tract infection show one or both of two adherence properties. One may depend on the recognition by type 1 fimbriae of mannose-containing residues in the urinary slime. It is measured as mannose-sensitive haemagglutination and is found on most *E. coli* strains. The second adherence property is detected as attachment to human urinary tract epithelial cells and as mannose-resistant agglutination of human erythrocytes. This may depend on the recognition of globo-series glycolipids in the epithelial cell surface. Possession of this adherence factor is strongly related to virulence. Most strains from patients with acute pyelonephritis and cystitis have this property but it is rare in strains from patients with asymptomatic bacteriuria and strains from normal faeces.

Local antibodies may interfere with bacterial attachment, thus possibly preventing the colonization that precedes urinary tract infection or modifying an established infection. Vaginal antibodies are known to coat *E. coli* from the stools. Antibodies in the urine of patients with acute pyelonephritis inhibit attachment of the infecting strain to uroepithelial cells. Antibodies directed against several bacterial surface structures, for example O antigen and fimbriae, are likely to inhibit attachment by steric hindrance or agglutination. The role of antibodies in adhesion-mediating structures such as fimbriae in susceptibility to and the outcome of human urinary tract infection remains to be investigated.

Mucous membranes form a barrier between host tissues and infectious agents and other potentially dangerous substances in the external milieu. Several mechanisms

1981 Adherence and microorganism pathogenicity. Pitman Medical, Tunbridge Wells (Ciba Foundation symposium 80) p 161-187

combine to keep out unwanted microorganisms and other harmful agents. First, mucosal surfaces are mechanically cleaned by the flow of secretions such as saliva, intestinal juice and urine. Peristalsis and ciliary movements potentiate this cleansing action and propel loose matter towards the body exits. Secondly, the mucus layer overlying epithelial surfaces hinders contact between epithelial cells and agents in the lumen; it also acts as a carrier of defence factors secreted from epithelial cells, such as secretory IgA antibodies. Thirdly, the epithelial linings participate in immune recognition and defence in a way specific for each host and tissue (for review see Gibbons 1973, Hanson et al 1980).

In spite of this complex barrier most infections occur via mucous membranes. Bacteria have developed special properties for overcoming host defences. To resist the flow of secretions and colonize the tissue surface it may be sufficient for bacteria to associate with the mucus layer, whereas additional properties may be necessary to induce infection (Freter 1978). To penetrate the mucus barrier bacteria may need chemotactic properties (Allweiss et al 1977) and specific adhesins may be needed for binding to the epithelial surface. Symptoms of infection may then be induced directly by the attached bacteria through the action of toxins or by bacterial invasion. The role of these properties may be studied in the *Escherichia coli* that colonize or infect the urinary tract. In this paper we summarize information on attachment as a virulence factor in *E. coli* strains that cause urinary tract infection (UTI).

Methods

In vitro models using human epithelial cells

In studies of the attachment of bacteria to human urogenital epithelium *in vitro,* bacteria and epithelial cells are usually mixed to allow attachment to occur, any unattached bacteria are eliminated, and the bacteria attached to the cells are counted. Epithelial cells can be collected from the vaginal introitus (Mårdh & Weström 1976, Fowler & Stamey 1977, Parsons et al 1979, Botta 1979), from the periurethral area (Källenius & Winberg 1978), or from the urine, which contains transitional epithelial cells representative of the bladder epithelium and squamous epithelial cells from the urethra and external genital region (Svanborg Edén et al 1977, Schaeffer et al 1979). Cultured HeLa cells have also been used (Vosbeck et al 1979). The specificity involved in attachment makes the choice of target cell crucial, since bacteria attaching to urogenital epithelial cells may not bind to HeLa cells and strains attaching to buccal cells may not bind to urinary tract epithelial cells.

After incubation unattached bacteria are eliminated by repeated washing and centrifugation or by filtration. The number of attached bacteria is determined by light microscopy of unfixed samples (Svanborg Edén et al 1977) or of fixed and stained samples (Mårdh & Weström 1976, Fowler & Stamey 1977, Källenius & Winberg 1978), or by determination of counts when ^{14}C-labelled (Parsons et al 1979) or [^{3}H]uridine-labelled bacteria (Schaeffer et al 1979) are used. With radio-labelled bacteria attachment to the total number of epithelial cells in the suspension is measured. Counting by microscopy is laborious and only a limited number of epithelial cells, usually 40-50, can be inspected in each sample. Microscopic inspection, however, also gives qualitative information. For example, bacteria may be attached to the epithelial cells or to slime surrounding the cells (Ørskov et al 1980). *Proteus mirabilis* strains, for instance, attach only to squamous epithelial cells whereas many *E. coli* strains attach to both squamous and transitional epithelial cells (Svanborg Edén et al 1980). Information on these points would be missed if labelled bacteria were used.

Animal models

Many studies indirectly support the hypothesis that bacterial attachment is a prerequisite for colonization and infection. *In vivo* models suitable for studies of attachment are, however, rare. The ability of *E. coli* to produce diarrhoea in pigs depends on attachment to the mucosa of the small intestine. Pigs lacking the receptor for the adhesion-mediating K88 antigen are resistant to both colonization and diarrhoeal disease (Rutter et al 1975). So far, no experimental model for studying the role of attachment in animals naturally prone to UTI has been described. Most existing models of haematogenous (Lehmann et al 1968) and ascending infection (Kaijser et al 1978) were developed for studying kidney damage or abscesses. A model mimicking the initial stages of infection, namely the colonization phase (Uehling et al 1978), with adhesion specificity similar to that of the human urinary tract, is desirable.

Relations between attachment in vitro and severity of UTI in vivo

Escherichia coli

The urinary tract is normally resistant to infection. *E. coli* originate from the stools, colonize the vaginal (Stamey & Howell 1976) and periurethral areas (Bollgren & Winberg 1976), and ascend the urinary tract. Bacteria may colonize the urinary tract without producing overt symptoms (asymptomatic bacteriuria, ABU). Alternatively, colonization may lead to infection of the kidney – acute pyelonephritis – or infection may be limited to the bladder – acute cystitis. The type of infection produced depends partly on the virulence factors in the bacteria, partly on host susceptibility and defence mechanisms.

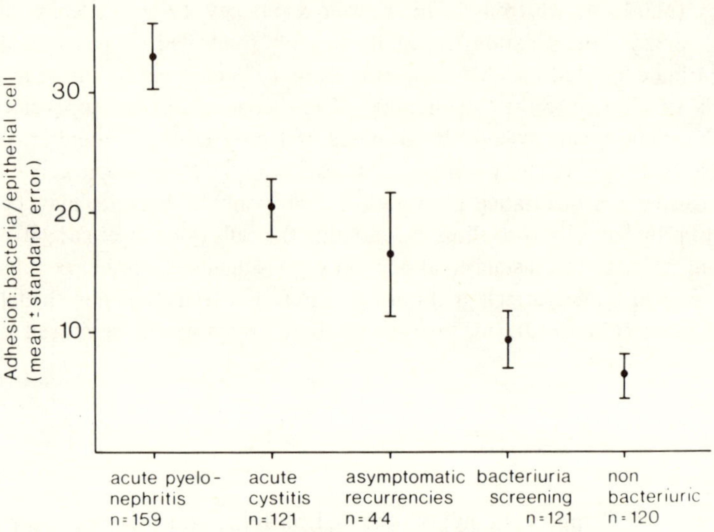

FIG. 1. Relation between clinical origin and capacity to attach to human urinary tract epithelial cells of *E. coli* strains from the urine of patients with various forms of urinary tract infection or from the stools of healthy children. Bacteria/cell, mean and SEM (*n*, number of strains).

Using our adhesion test system with human epithelial cells exfoliated into the urine (Svanborg Edén et al 1977) we compared the adhesive capacities of some 450 *E. coli* strains isolated from patients with various forms of UTI and from the stools of healthy children. A relationship was found between the adhesive capacity *in vitro* and the severity of UTI (Fig. 1) *in vivo* (Svanborg Eden et al 1976, 1979b). Patients with acute pyelonephritis had the highest number of attaching *E. coli* strains, those with acute cystitis had intermediate numbers, and those with asymptomatic bacteriuria had low numbers, as did strains from the stools of healthy children (Svanborg Edén et al 1976, 1979b). No relationship between infection site and bacterial attachment was found in adult women with urological complications (Fowler & Stamey 1978) but such a relationship was confirmed for isolates from adult women without underlying defects of the urine flow (Svanborg Edén et al 1979b).

E. coli strains that cause UTI can be isolated from the faecal flora before or at the onset of symptomatic UTI (Grüneberg 1969). These strains have similar adhesive properties whether they are isolated from the stools or from the urine (Svanborg Edén et al 1979b). Adhesive capacity may thus be a selective factor for *E. coli* strains from the faecal flora, determining their ability to reach and colonize the urinary tract.

How can the poorly adhering strains from patients with ABU colonize and remain in the urinary tract? First, a decrease in adhesive capacity may be induced by host defence mechanisms when strains are carried for a time in the urinary tract. *E. coli* from the urine of patients with ABU of short duration had greater adhesive capacity than strains likely to have been carried for a longer time (Svanborg Edén et al 1979b). Secondly, binding to urinary slime may be sufficient to allow bacteria to colonize the urinary tract and remain there. Over 60% of ABU strains carry type 1 pili thought to mediate binding to the slime (Ørskov et al 1980). Thirdly, bacteria may remain in residual urine without attaching to the mucosa. *E. coli* isolated from ABU patients with residual urine adhered less well than *E. coli* isolated from patients without residual urine. On the other hand why do ABU strains that can adhere to epithelia fail to induce symptomatic infections? The ABU strains rarely possess additional virulence factors (see p 169); alternatively, bacterial virulence factors and host defence mechanisms, such as antibodies that inhibit attachment, may be evenly balanced in some patients with ABU.

Proteus mirabilis

Whereas *E. coli* cause most episodes of UTI in patients without obstruction of urine flow, *P. mirabilis* strains are often found in patients with recurrent UTI, stones or malformations. It has been suggested that *P. mirabilis* is less virulent than *E. coli* and will not produce bacteriuria unless an underlying defect is present. A major difference in the antigenic composition of the cell wall of the two species is that *P. mirabilis* lacks a polysaccharide capsule. The adhesion characteristics of *P. mirabilis* strains from various infectious foci differ from those of *E. coli* (Svanborg Eden et al 1980). Most of the 335 *P. mirabilis* strains tested attached in large numbers to the epithelial cells of human urinary sediment. No difference was noted between strains from patients with UTI and strains from other sources, for example from the stools of healthy controls. However, the *P. mirabilis* strains attached to squamous but not to transitional epithelial cells, whereas *E. coli* bound to both cell types – acute pyelonephritis strains more so than ABU strains. These findings suggest that *P. mirabilis* is less efficient in colonizing the bladder in patients without urinary tract obstruction because these strains do not bind to the transitional epithelium. Residual urine or stones may be required for *P. mirabilis* to remain in the bladder.

Bacterial adherence properties

Adhesins

Pili or fimbriae are believed to mediate the attachment of *E. coli* to various host tissues (Brinton 1965, Duguid 1968). The hydrophobic nature of pili has complicated

TABLE 1 Specificity of epithelial attachment in relation to haemagglutination pattern of whole fimbriated E. coli and their isolated fimbriae (only strains carrying fimbriae inducing mannose-resistant haemagglutination attached to human uroepithelial cells)

	Haemagglutination				*Attachment to epithelial cells*		
	Human		*Guinea-pig*		*Human*		*Rat*
Bacterial strain	*Bacteria*	*Fimbriae*	*Bacteria*	*Fimbriae*	*Buccal*	*Urinary*	*Urinary*
E. coli 3669	MR	MR	–	–	MR	MR	–
E. coli 3048	MR	MR	MS	MS	(MR)	MR	(MR)
E. coli 6013	–	–	MS	MS	(MR)	–	(MR)
S. typhimurium	–	–	MS	MS	MS	–	MS

MS: mannose-sensitive (reversed by D-mannose, 25 mg/ml).
MR: mannose-resistant (unaffected by D-mannose, 25 mg/ml).
(MR): decreased by D-mannose but not significantly.

attempts to isolate and purify them and determine their role in bacterial attachment. Indirect methods such as electron microscopy and haemagglutination have been used. A new technique that separates pili from outer membrane vesicles by solubilization in deoxycholate and pili from flagella with 6 M-urea allows the isolation of immunologically pure pili (Korhonen et al 1980b). Using this technique we have purified pili from three attaching *E. coli* strains isolated from patients with acute pyelonephritis. The pilus preparations were free of detectable amounts of capsular polysaccharide or lipopolysaccharide and retained the binding properties of whole bacteria. Thus, the haemagglutination patterns of whole piliated bacteria and isolated pili were identical (Korhonen 1979). Pili labelled with ^{125}I specifically bound to epithelial cells of the human urinary tract. This binding was inhibited by unlabelled pili and was not affected by D-mannose (Korhonen et al 1980a). Thus, for these strains, pili are likely to participate in the binding reactions.

We originally suggested that fimbriae had a role in the attachment of *E. coli* to epithelial cells of the human urinary tract, since their presence as shown by electron microscopy was related to adhesive capacity (Svanborg Edén & Hansson 1978). Different workers reported, however, contradictory results on the adhesion characteristics of fimbriated *E. coli* and on the effect of D-mannose on attachment (Ofek et al 1977, Schaeffer et al 1979, Källenius & Möllby 1979). The three *E. coli* strains mentioned above, where isolated fimbriae and whole piliated bacteria showed the same binding characteristics, were used to study the relation between fimbriation, haemagglutination pattern and adhesion to different epithelial cells (Table 1). *E. coli* 6013 carried only type 1 fimbriae (which induce agglutination of guinea-pig erythrocytes reversed by D-mannose) (Duguid 1968), *E. coli* 3669 carried only

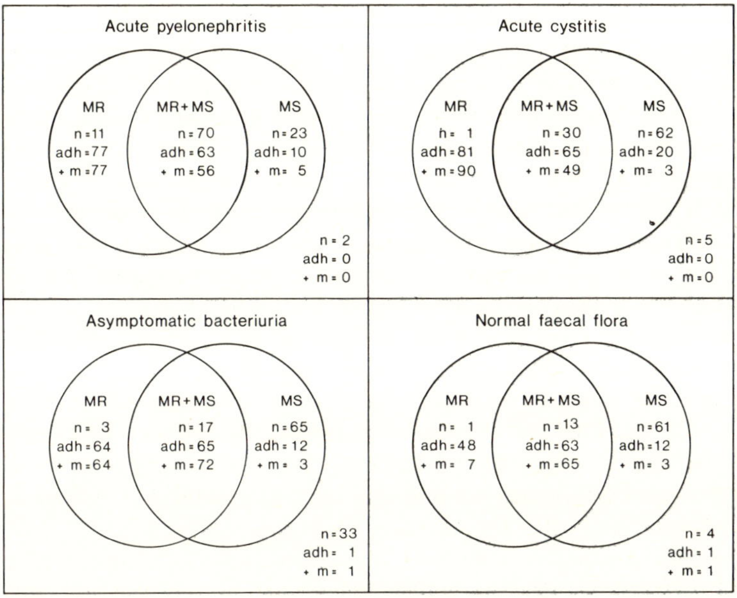

FIG. 2. Relation between ability to attach to human uroepithelial cells and to agglutinate erythrocytes of *E. coli* from patients with UTI or healthy controls. Four main patterns of haemagglutination are shown: MR, mannose-resistant haemagglutination of human erythrocytes (left circle); MS, mannose-sensitive agglutination of guinea-pig erythrocytes (right circle); both patterns simultaneously (area common to both circles); or no agglutination (area outside the circles). n, number of strains; adh, mean adhesion of all strains in the area (bacteria per cell) (tested in PBS). + m, adhesion in 25 mg/ml D-mannose. Regardless of their origin, strains with mannose-resistant haemagglutination attached in large numbers. The highest proportion of such strains was isolated from patients with acute pyelonephritis.

fimbriae inducing mannose-resistant haemagglutination of human erythrocytes, and *E. coli* 3048 simultaneously carried both types of fimbriae. As seen in Table 1, only the two strains carrying the mannose-resistant fimbriae attached to human urinary tract epithelial cells. The attachment was not affected by D-mannose. All strains attached to human buccal cells, but only the two with mannose-sensitive fimbriae attached to rat urinary tract epithelial cells.

The strong relation between capacity to attach to human uroepithelial cells and to induce mannose-resistant agglutination of human erythrocytes was confirmed for the large collection of UTI strains mentioned earlier (p 164). Three main combinations of haemagglutinins were found (Fig. 2): only mannose-resistant agglutination of human erythrocytes (16 out of 453 strains), only mannose-sensitive haemagglutination (199 of the 453 strains) or both haemagglutination patterns together (144 of the 453 strains). The remaining 94 strains did not induce agglutination with the erythrocytes tested (human, guinea-pig, horse and yeast cells).

Regardless of their origin, strains that induced mannose-resistant agglutination of human erythrocytes alone or simultaneously with mannose-sensitive agglutination of guinea-pig erythrocytes adhered well (Fig. 2). Seventy-seven per cent of pyelonephritis strains but only 16% of normal faecal isolates had this property. The strains that induced only mannose-sensitive agglutination of guinea-pig erythrocytes had low mean adhesion. Such strains were abundant in all groups of strains and were found in 91% of cystitis and 64% of normal faecal strains. None of the non-agglutinating strains attached. The attachment of strains with mannose-resistant haemagglutination was weakly decreased and that of strains with mannose-sensitive haemagglutination was more markedly decreased by D-mannose (Fig. 2).

These results suggest that the same group of bacterial surface structures mediate agglutination of human erythrocytes and attachment to epithelial cells of the human urinary tract, and that this may be a factor selecting the *E. coli* from the faecal flora which colonize and infect the urinary tract. On the strains shown in Table 1 this structure may be of a fimbriate nature, but strains free of pili or fimbriae agglutinate erythrocytes and attach to epithelial cells. The nature of this type of 'adhesin' is not known.

What is the significance of type 1 fimbriae – the mannose-sensitive haemagglutinin carried by 343 of the 453 strains we studied? Ørskov et al (1980) suggest that, rather than attaching *E. coli* to the urinary tract cells, type 1 pili mediate binding to the urinary slime. This would explain the high frequency of such fimbriae on normal faecal strains and on the urinary tract pathogens. Whereas type 1 fimbriae may be sufficient for colonization of the mucus layer, the mannose-resistant adhesins/agglutinins may be necessary for attachment to the epithelial cells and additional virulence factors for tissue invasion.

Other bacterial surface structures

Several properties are associated with the virulence of UTI pathogens, for example possession of lipopolysaccharide and capsular polysaccharide, ability to ferment dulcitol, production of haemolysin and resistance to the bactericidal effect of normal human serum (Hanson et al 1977). Table 2 shows that *E. coli* isolated from patients with acute pyelonephritis often simultaneously possess complete lipopolysaccharide (O antigen), capsular polysaccharide (K antigen), the ability to adhere, and resistance to the bactericidal effect of serum. Only a few strains isolated from the stools of healthy children possess all four characteristics simultaneously, although many strains have one or two of them. None of the strains isolated from ABU patients had all four characteristics. During the infection process a bacterial strain may produce different virulence factors at different stages. Formation of fimbriae may be triggered first, if we assume that the ability to adhere is needed for colonization of the urinary tract. Once bacteria are attached and can increase their

TABLE 2 Number of virulence factors (O antigen, K antigen, capacity to attach to epithelial cells of the human urinary tract, and resistance to the bactericidal effect of human serum) on E. coli strains isolated from the urine of patients with acute pyelonephritis, acute cystitis or asymptomatic bacteriuria and from the stools of non-bacteriuric schoolchildren

	Number of virulence factors					
Diagnosis	*Four*	*Three*	*Two*	*One*	*None*	
Acute pyelonephritis	31	50	26	6	0	113
	(28)[a]	(44)	(23)	(5)	(0)	(100)
Acute cystitis	15	44	30	12	1	102
	(15)	(43)	(29)	(12)	(1)	(100)
Asymptomatic bacteriuria	0	12	38	41	26	117
	(0)	(10)	(33)	(35)	(22)	(100)
No bacteriuria	2	21	50	43	4	120
(normal faecal *E. coli*)	(2)	(17)	(42)	(36)	(3)	(100)

[a]Numbers in parentheses: percentages of patients or healthy children.

concentration of nutrients, a polysaccharide capsule may be formed, which may help them to avoid phagocytosis. Secretion of lipopolysaccharides may be important for the inflammatory reaction and onset of symptoms. The relation between capsule and fimbriae is hard to test since the cultivation conditions that promote capsule formation select for non-piliated variants. We have not, however, been able to inhibit attachment with pure lipopolysaccharide or capsular polysaccharide at concentrations up to 40 mg/ml.

Epithelial cell receptors

Two hypotheses currently dominate the discussion about mechanisms of bacterial adhesion. One assumes that bacteria possess a general stickiness due to hydrophobic or other non-specific physicochemical interactions with various surfaces (Brinton 1965). The second hypothesis is that a specific receptor molecule on the target cell is recognized by a specific molecule on the bacteria, possibly a lectin (Eshdat et al 1978). Our results strongly suggest that attachment to epithelial cells of the human urinary tract is a specific process, during which bacterial surface ligands such as mannose-resistant haemagglutination-inducing antigens react with specific epithelial cell receptors (Table 1).

Mannose-containing receptors

The inhibiting effect of D-mannose and L-fucose on adhesion and haemagglutination led to the assumption that sugar residues on the epithelial cell and erythrocyte

TABLE 3 Haemagglutination and attachment to epithelial cells of the human urinary tract by E. coli strains isolated from patients with UTI. Recognition of globoseries glycolipids coated onto guinea-pig erythrocytes or present on human erythrocytes and uroepithelial cells

	Agglutination of erythrocytes					
	Human		*Guinea-pig*		*Guinea-pig coated with globoside*[b]	*Attachment (bacteria/ epithelial cell)*
E. coli strain	*PBS*[a]	*D-mannose*	*PBS*	*D-mannose*		
36692	+++	+++	–	–	+++	90
1445	+++	+++	–	–	+++	98
3976	+++	+++	–	–	+++	65
4283	+++	+++	–	–	+++	69
2737	+++	+++	+++	–	+++	71
364	+++	+++	++	–	+++	64
548	+++	+++	++	–	+++	45
693	+++	+++	+++	–	+++	49
1011	–	–	+++	–	–	6
246	–	–	+++	–	–	0
3550	–	–	+++	–	–	0
3276	–	–	+++	–	–	0
4166	–	–	–	–	–	0
392	–	–	–	–	–	0

Bacteria that bind to and agglutinate human erythrocytes adhere to uroepithelial cells. Globotetraosylceramide acts as a receptor for both these reactions. Guinea-pig erythrocytes which were not agglutinated by the *E. coli* strains became agglutinable after they were coated with globoside. The test was done in D-mannose to avoid reactions due to mannose-sensitive fimbriae on the strains.

[a]PBS, phosphate-buffered saline.
[b]Erythrocytes incubated with globoside as described by Leffler & Svanborg Edén (1980). Agglutination in PBS containing 25 mg D-mannose/ml.

surfaces are recognized by bacteria (for review see Duguid 1968). A mannose-containing receptor is unlikely to be important for the attachment of human UTI pathogens to the epithelial cells of the urinary tract. Strains that agglutinate yeast cells and induce mannose-sensitive agglutination of guinea-pig erythrocytes attached poorly to human uroepithelial cells unless they also possessed mannose-resistant haemagglutinins (Tables 1 and 3, Fig. 2). Mannose also did not decrease the binding of strains showing mannose-resistant haemagglutination to human uroepithelial cells. Mannose did not even always inhibit the attachment of strains with type 1 fimbriae, suggesting that adhesins that do not induce haemagglutination may be

Gal β1→4Glc β1→1CERAMIDE (a)

Gal α1→4Gal β1→4Glc β1→1CERAMIDE (b)

GalNAc β1→3Gal α1→4Gal β1→4Glc β1→1CERAMIDE (c)

GalNAc α1→3GalNAc β1→3Gal α1→4Gal β1→4Glc β1→1CERAMIDE (d)

FIG. 3. Chemical structures of globoseries glycolipids. (a) Lactosylceramide; (b) globotriaosylceramide; (c) globotetraosylceramide; (d) Forssman glycolipid hapten. The region of the molecules possibly recognized by the bacteria is indicated by the bar. (From Leffler & Svanborg Edén 1980.)

present on type 1 piliated strains. Mannose-containing receptors may be present in urinary slime and may be recognized by type 1 fimbriae on *E. coli* strains that colonize mucosal surfaces by binding to the mucus layer (Ørskov et al 1980).

Glycolipid receptors

Recent evidence supports the hypothesis that a special class of glycosphingolipids, the globoseries glycolipids (Fig. 3), act as receptors for uropathogenic *E. coli* that attach to epithelial cells of the human urinary tract or agglutinate human erythrocytes. A fraction of glycolipids isolated from the urinary tract cells inhibited attachment of *E. coli* to these cells from the same donor (Svanborg Edén & Leffler 1980). Chemical analysis of the glycolipid fraction revealed a number of components; we selected analogous glycolipids from other sources for tests of receptor activity. Two test systems were used in parallel:

(a) Inhibition of attachment of *E. coli* to human urinary tract epithelial cells by suspensions of glycosphingolipids.

(b) Bacterial agglutination of erythrocytes coated with glycolipids.

Globotetraosylceramide and to a lesser extent globotriaosylceramide inhibited bacterial attachment. The bacterial agglutination of erythrocytes from various species paralleled their content of globotetraosylceramide or Forssman glycolipid (globopentaosylceramide) (Fig. 3). Furthermore, erythrocytes not normally agglutinated by the *E. coli* strains became agglutinable after being coated with globotetraosylceramide (Leffler & Svanborg Edén 1980) (Table 3). The efficiency of the recognition of globoside also depended on how it is presented to the bacteria (Fig. 4).

Receptor activity could be further linked to the structural features specific for the globoseries glycolipids by genetic evidence (Marcus et al 1976). The total neutral glycolipid fractions from human erythrocytes of blood group P_1 P_2 containing globoseries glycolipids and group p lacking globoseries glycolipids were tested (Marcus et al 1976). Fractions from P_1 and P_2 but not from p erythrocytes inhibited attachment (Leffler & Svanborg Edén 1980). Källenius et al (1980) have reported that *E. coli* from patients with acute pyelonephritis do not agglutinate p but only P_1, P_2 and $P_1{}^k$ erythrocytes. On the basis of this and the inhibition of haemagglutination by the reduced free trisaccharide, the P^k antigen (formula b, Fig. 3) was suggested as the receptor.

Further experiments are necessary to define the exact structural feature and site in the saccharide chain recognized by bacteria. The region indicated in Fig. 3 may explain our results (Leffler & Svanborg Edén 1980). This does not, however, agree entirely with the results of Källenius et al (1980).

The identification of the globoseries glycolipids as bacterial receptors is interesting in several ways. First, receptor analogues may block bacterial attachment and prove efficient in prophylaxis against UTI. Secondly, it may now be possible to identify the nature of the bacterial surface ligands involved in attachment and inducing mannose-resistant haemagglutination. Thirdly, the susceptibility of different individuals to bacterial colonization may depend on the composition and glycoproteins of their mucous membranes. Thus, individuals of blood group p who lack one receptor for attaching bacteria could be resistant to infection by these bacteria, just as pigs lacking receptors for K88 are resistant to infection by *E. coli* carrying the K88 antigen (Rutter et al 1975).

Fourthly, although the sugar chains of most glycolipids including gangliosides have been found in glycoproteins or as free saccharides in milk and urine (Rauvala & Finne 1978), the globotetraosylceramide saccharide chain has been found only in glycolipid form and it is thus bound in direct proximity to the lipid bilayer of the cell membrane. This may be important in bacterial invasion or for other effects on the epithelial cells after attachment.

Lastly, the epithelial cells of the urinary tract are rich in glycolipids but the globoseries glycolipids are minor components. One reason why bacteria do not bind to the other glycolipids may be that bacterial ligand specificity has adapted to avoid competitive inhibition of the binding by free or glycoprotein-bound saccharides of similar structure excreted in the urine.

Susceptibility to urinary tract infection

Children without gross defects in their urinary flow may nevertheless get recurring UTI. In Göteborg 3% of girls and 1% of boys under 11 years contracted symptomatic UTI between 1964 and 1966, and the prevalence of ABU in the population is about 1%. Of the children followed from their first known febrile infection, 30% had a recurrence within a year and in 60% of those UTI kept recurring. Three main reasons for this increased susceptibility to UTI have been suggested and investigated: first, less pronounced defects of urine flow, such as residual urine or low-grade reflux, may play a role; secondly, receptivity for bacterial attachment in patients with UTI may differ from that in healthy controls; thirdly, local immune mechanisms may also differ in patients and controls (for review see Hanson et al 1977).

Susceptibility to bacterial attachment

E. coli bacteria with a full set of virulence factors may be found in the intestinal flora of healthy individuals. Only in those susceptible to UTI do *E. coli* succeed in colonizing the vaginal and periurethral area and in ascending the urinary tract. An increased receptivity for attaching bacteria has been reported for vaginal (Fowler & Stamey 1977), periurethral (Källenius & Winberg 1978) and urinary tract epithelial cells (Svanborg Edén & Jodal 1979) from patients with UTI compared to those from healthy and age-matched controls, when the patients' own infecting strain was used. The differences are, however, small and their existence has been doubted (Parsons et al 1979, Schaeffer et al 1979). Subtle differences in the amount of availability of receptors rather than their presence or absence might explain these findings. A defective capacity for synthesizing globoseries glycolipids, which act as receptors for a number of UTI pathogenic *E. coli*, is known only for individuals of the rare blood group p. Rather, the way the receptor is presented in the membrane may affect the efficiency of its recognition by the bacteria. Fig. 4 shows that smaller amounts of globoside are needed to inhibit attachment when the glycolipid is mixed with lactosylceramide, which by itself is not recognized by the bacteria. Variation in the density of receptors, depending on the maturity of the epithelial cells, might also explain the relation between susceptibility to attachment and the hormonal status of the cell donor. Such a relation has been clearly demonstrated for *P. mirabilis* (Svanborg Edén et al 1980), and streptococci (Botta 1979), and has also been suggested for *E. coli* (Schaeffer et al 1979, Svanborg Eden et al 1980).

Immune mechanisms

No gross defects in cellular or humoral immune mechanisms have been found in patients susceptible to UTI compared to healthy controls. UTI is not considered a major problem in patients with immunodeficiencies. In addition, UTI recurs in spite of the immune response to earlier infection (Hanson et al 1977). Immune defence mechanisms are still thought to be important in determining the onset and course of UTI.

Antibodies in vaginal and periurethral secretions and in urine may prevent the

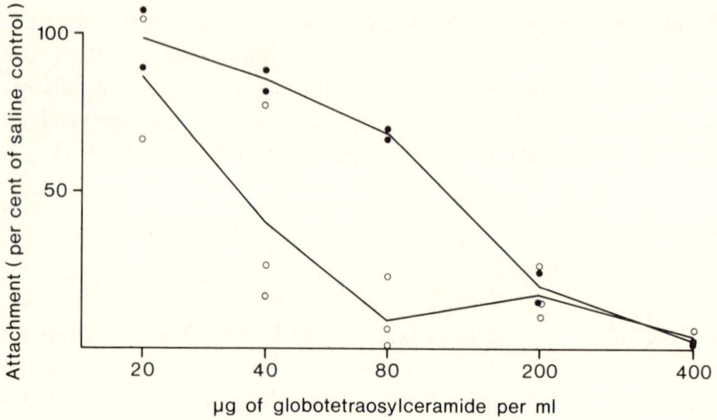

FIG. 4. The recognition of globoside by bacteria depends on how the globoside is presented. Smaller amounts of globotetraosylceramide were needed for inhibition of adhesion when lactosylceramide, which by itself does not affect attachment, was mixed with the globotetraosylceramide. The glycolipids were sonicated in phosphate-buffered saline solution to form micelles. Bacteria were preincubated with the glycolipids, epithelial cells were added, and adhesion was tested as described in text (p 162). ○, globotetraosylceramide + 100 µg lactosylceramide; ●, globotetraosylceramide alone.

bacterial colonization that precedes UTI. Stamey & Howell (1976) have shown that antibodies coating faecal *E. coli* are common in healthy women whereas vaginal antibodies do not coat the UTI strains in patients. Furthermore, the immune response may exert a selective pressure on strains colonizing the urinary tract, since 80% of strains causing recurring UTI differ serologically from the strain causing the preceding attack (for review see Hanson et al 1977). One way in which secretory antibodies act is by preventing attachment (Freter 1969). IgG and secretory IgA antibodies isolated from the urine of patients with acute pyelonephritis prevent attachment to urinary tract epithelial cells of the strain causing the infection (Svanborg Edén & Svennerholm 1978). Absorption studies have shown that antibodies directed against the O antigen efficiently inhibited attachment whereas anticapsule antibodies were less efficient. Human milk and commercial immunoglobulin also inhibit attachment (Svanborg Edén & Svennerholm 1978).

Antibodies to purified pili inhibit, in a dose-related manner, attachment of the strain from which pili were isolated and attachment of strains with cross-reacting pili. Antibodies to purified pili of both mannose-sensitive and mannose-resistant types are found in breast milk and may modulate bacterial colonization in the intestine of babies (Svanborg Edén et al 1979a). Induction of antibodies via vaccination may be a future means of preventing recurrent UTI in patients at risk.

Immune mechanisms may also affect the course of an established urinary tract infection. Antibodies occur in the serum and urine of patients with acute pyelonephritis and reach peak values about a week after the onset of symptoms (for review see Hanson et al 1977). We are now investigating whether these antibodies limit damage to the kidney, shorten the duration of fever or protect the host in any other way, or whether they should be considered as markers of reactions between host and bacteria that have already taken place.

Antibacterial agents and attachment

Since the attachment of *E. coli* is mediated by bacterial surface structures, antibacterial agents altering the bacterial surface are likely to affect attachment (Svanborg Edén et al 1977). We noted that *E. coli* that were still viable adhered less well after treatment with subinhibitory amounts of ampicillin and amoxycillin (Svanborg Edén et al 1978). A decrease in the mannose-binding capacity of *E. coli* after streptomycin, tetracycline and trimethoprim treatment has been described (Eisenstein et al 1979). Bacteria treated with several bacteriostatic compounds showed no decrease in adherence in our system, but a decrease has been described by others (Vosbeck et al 1979). Combined treatment with 1/4 of the minimum inhibitory concentration (MIC) of ampicillin and antibodies decreased adhesion more than either treatment alone (Sandberg et al 1979). Antibodies to purified pili inhibited attachment of ampicillin-treated bacteria more efficiently than that of untreated bacteria. This suggests that the decreased attachment after ampicillin treatment is due either to a decreased synthesis of pili or to a less efficient insertion in the cell wall.

Antibacterial treatment is mostly used to kill bacteria which have already given rise to an infection. Subinhibitory amounts may then prove inefficient. Small amounts of antibacterial agents that decrease attachment may however be sufficient to prevent colonization and may be sufficient for prophylaxis against UTI. A synergy between antibodies and small amounts of antibacterial agents may also explain why the urine is sterile after treatment of UTI with single doses of such agents. The possibility that small amounts of antibacterial agents present in food may affect bacterial ecology by altering the adhesive properties of indigenous bacteria should be considered.

Conclusions

(1) The ability of *E. coli* to attach to epithelial cells of the human urinary sediment *in vitro* is related to the severity of the urinary tract infection produced *in vivo*.

(2) Whole piliated or fimbriated bacteria and their isolated fimbriae have identical haemagglutination patterns. Isolated ^{125}I-labelled fimbriae bind specifically to human uroepithelial cells. Thus, fimbriae are likely to participate in the

binding reactions of these strains. Other strains carry adhesins of a non-fimbriate nature. Although several types of adhesins can be found on *E. coli* isolated from patients with UTI, those causing mannose-resistant agglutination of human erythrocytes are the most important for attachment to human uroepithelial cells. Type 1 fimbriae may bind to mannose-containing receptors in urinary slime and be important for colonization.

(3) Globoseries glycolipids, especially globotetraosylceramide, act as receptors for *E. coli* when they attach to human uroepithelial cells and when they agglutinate human erythrocytes.

(4) Attachment is inhibited by the total antibody fraction in the urine of the infected patients. Anti-pili antibodies raised in rabbits inhibit attachment of the homologous strain and of strains carrying cross-reacting pili.

(5) Antibacterial agents such as ampicillin and trimethoprim decrease the uroepithelial attachment of *E. coli* even in doses that do not affect the viability of the bacteria.

Acknowledgements

We greatly appreciate the technical assistance of Lise-Lotte Johansson and the typing aid of Diana Bruning. This study was supported by grants from the Swedish Medical Research Council (Nos. 215 and 3967), the Faculty of Medicine, University of Göteborg, the Ellen, Walter and Lennart Hesselman Foundation for Scientific Research, and the Swedish Board for Technical Development (STU). The biochemical work was done in collaboration with M. E. Breimer, G. C. Hansson, K. A. Karlsson, G. Larsson, I. Pascher and B. E. Samuelsson at the Department of Medical Biochemistry.

REFERENCES

Allweiss B, Dostal J, Carey KE, Edwards TF, Freter R 1977 The role of chemotaxis in the ecology of bacterial pathogens of mucosal surfaces. Nature (Lond) 266:448-450

Bollgren I, Winberg J 1976 The periurethral aerobic flora in girls highly susceptible to urinary infections. Acta Paediatr Scand 65:81-87

Botta GA 1979 Hormonal and type dependent adhesion of group B streptococci to human vaginal cells. Infect Immun 25: 1084-1086

Brinton C 1965 The structure, function, synthesis and genetic control of bacterial pili, and a molecular model for DNA and RNA transport in gram negative bacteria. Trans N Y Acad Sci 27:1003-1054

Duguid JP 1968 The function of bacterial fimbriae. Arch Immunol Ther Exp 16:173-188

Eisenstein BI, Ofek I, Beachey EH 1979 Interference with the mannose binding and epithelial cell adherence of *Escherichia coli* by sublethal concentrations of streptomycin. J Clin Invest 63:1219-1228

Eshdat Y, Ofek I, Yashouv-Gan Y, Sharon N, Mirelman D 1978 Isolation of a mannose-specific lectin from *Escherichia coli* and its role in the adherence of the bacteria to epithelial cells. Biochem Biophys Res Commun 85:1551-1559
Fowler JE, Stamey TA 1977 Studies of introital colonization in women with recurrent urinary infections VII. The role of bacterial adherence. J Urol 117:472-476
Fowler JE, Stamey TA 1978 Studies of introital colonization in women with recurrent urinary infections X. Adhesive properties of *Escherichia coli* and *Proteus mirabilis:* Lack of correlation with urinary pathogenicity. J Urol 120:315-318
Freter R 1969 Studies on the mechanism of action of intestinal antibody in experimental cholera. Texas Rep Biol Med 27 (suppl 1):299-316
Freter R 1978 Possible effects of foreign DNA on pathogenic potential and intestinal proliferation of *Escherichia coli.* J Infect Dis 137:624-629
Gibbons RJ 1973 Bacterial adherence in infection and immunity. Rev Microbiol 4:49-60
Grüneberg RN 1969 Relationship of infecting urinary organisms to the faecal flora in patients with symptomatic urinary infection. Lancet 2:766-768
Hanson LÅ, Ahlstedt S, Fasth A, Jodal U, Kaijser B, Sohl Åkerlund A, Svanborg Edén C 1977 Antigens of *E. coli*, human immune response and the pathogenesis of urinary tract infection. J Infect Dis (suppl) 136:144-149
Hanson LÅ, Carlsson B, Dahlgren U, Mellander L, Svanborg Edén C 1980 The secretory IgA system in the neonatal period. In: Perinatal infections. Excerpta Medica, Amsterdam (Ciba Found Symp 77) p 187-196
Kaijser B, Larsson P, Olling S 1978 Protection against ascending *E. coli* pyelonephritis in rats and significance of local immunity. Infect Immun 20:78-81
Källenius G, Möllby R 1979 Adhesion of *Escherichia coli* to periurethral cells correlated to mannose-resistant agglutination of human erythrocytes. FEMS (Fed Eur Microbiol Soc) Lett 5:295-299
Källenius G, Winberg J 1978 Bacterial adherence to periurethral epithelial cells in girls prone to urinary tract infections. Lancet 2:540
Kallenius G, Möllby R, Svenson SB, Winberg J, Lindblad A, Svensson S, Cedergren B 1980 The P^k antigen as receptor for the haemagglutinin of pyelonephritic *Escherichia coli.* FEMS (Fed Eur Microbiol Soc) Lett 7:297-302
Korhonen TK 1979 Yeast cell agglutination by purified enterobacterial pili. FEMS (Fed Eur Microbiol Soc) Lett 6:421-425
Korhonen TK, Eden S, Svanborg Edén C 1980a Binding of purified *Escherichia coli* pili to human urinary tract epithelial cells. FEMS (Fed Eur Microbiol Soc) Lett 7:237-241
Korhonen TK, Nurmiaho EL, Ranta H, Svanborg Edén C 1980b A new method for the purification of *Escherichia coli* pili. Infect Immun 27:569-575
Leffler H, Svanborg Edén C 1980 Chemical identification of a glycosphingolipid receptor for *Escherichia coli* attaching to human urinary tract epithelial cells and agglutinating human erythrocytes. FEMS (Fed Eur Microbiol Soc) Lett 8:127-134
Lehmann JD, Smith JD, Miller TE, Barnett JA, Sanford JP 1968 Local immune response in experimental pyelonephritis. J Clin Invest 47:2541-2550
Marcus DM, Naiki M, Kundu SK 1976 Abnormalities in the glycosphingolipid content of human P^k and p erythrocytes. Proc Natl Acad Sci USA 73:3263-3267
Mårdh PA Weström L 1976 Adherence of bacteria to vaginal epithelial cells. Infect Immun 13:661-666
Ofek I, Mirelman D, Sharon N 1977 Adherence of *Escherichia coli* to human mucosal cells mediated by mannose receptors. Nature (Lond) 265:623-625
Ørskov I, Ørskov F, Birch-Andersen A 1980 A fimbria *Escherichia coli* antigen, F7, determining uroepithelial adherence. Comparison with type 1 fimbriae which attach to urinary slime. Infect Immun 27:657-666
Parsons LC, Hahibulla A, Stauffer C, Schmidt JD 1979 In vitro adherence of radioactively labelled *Escherichia coli* in normal and cystitis-prone females. Infect Immun 26:453-457
Rauvala H, Finne J 1978 Structural similarity of the terminal carbohydrate sequences of glycoproteins and glycolipids. FEBS (Fed Eur Biochem Soc) Lett 97:1-8

Rutter JM, Burrows MR, Sellwood R, Gibbons RA 1975 A genetic basis for resistance to enteric disease caused by *E. coli*. Nature (Lond) 257:135-136
Sandberg T, Stenqvist K, Svanborg Edén C 1979 Effects of subinhibitory amounts of ampicillin, chloramphenicol and nitrofurantoin on the adhesion of *Escherichia coli* to human uroepithelial cells *in vitro*. Rev Infect Dis 1:838-844
Schaeffer AJ, Amundsen SK, Schimidt LN 1979 Adherence of *Escherichia coli* to human urinary tract epithelial cells. Infect Immun 24:753-759
Stamey TA, Howell JJ 1976 Studies of introital colonization in women with recurrent urinary infections. IV. The role of local vaginal antibodies. J Urol 115:413-415
Svanborg Edén C, Hansson HA 1978 *Escherichia coli* pili as mediators of attachment to human urinary tract epithelial cells. Infect Immun 21:229-237
Svanborg Edén C, Jodal U 1979 Attachment of *Escherichia coli* to sediment epithelial cells from UTI prone and healthy children. Infect Immun 26:837-870
Svanborg Edén C, Leffler H 1980 Glycosphingolipids as possible receptors for *Escherichia coli* attaching to human urinary sediment epithelial cells. Scand J Infect Dis (suppl) in press
Svanborg Edén C, Svennerholm AM 1978 Secretory IgA and IgG prevent adhesion of *Escherichia coli* to human urinary tract epithelial cells. Infect Immun 22:790-797
Svanborg Edén C, Hanson LÅ, Jodal U, Lindberg U, Sohl Åkerlund A 1976 Variable adhesion to normal human urinary tract epithelial cells of *Escherichia coli* strains associated with various forms of urinary tract infection. Lancet 11:490-492
Svanborg Edén C, Eriksson B, Hanson LÅ 1977 Adhesion of *Escherichia coli* to human uroepithelial cells *in vitro*. Infect Immun 18:767-774
Svanborg Edén C, Sandberg T, Stenqvist K, Ahlstedt S 1978 Decrease in adhesion of *Escherichia coli* to human urinary tract epithelial cells *in vitro* by subinhibitory concentrations of ampicillin. A preliminary study. Infection 6 (suppl 1):121-124
Svanborg Edén C, Carlsson B, Hanson LÅ, Jann B, Jann K, Korhonen T, Wadström T 1979a Antipili antibodies in breast milk. Lancet 2:1235
Svanborg Edén C, Lidin-Janson G, Lindberg U 1979b Adhesion to human uroepithelial cells of faecal and urinary *E. coli* isolates from patients with symptomatic urinary tract infections or asymptomatic bacteriuria of varying duration. J Urol 122:185-188
Svanborg Eden C, Larsson P, Lomberg H 1980 Attachment of *Proteus mirabilis* bacteria to human urinary tract epithelial cells is different from that of *Escherichia coli*. Infect Immun 27:804-807
Vosbeck K, Handschin H, Menge EB, Zak O 1979 Effects of antibiotics at concentrations below the minimum inhibitory concentration on the adhesiveness of *E. coli* in a new, improved, *in vitro* assay. Rev Infect Dis 1:845-851
Uehling DT, Mizutani K, Balish E 1978 Effect of immunization on bacterial adherence to urothelium. Invest Urol 16:145-147

DISCUSSION

Sharon: If human erythrocytes are treated with trypsin this allows better access to the globoside. Are trypsinized cells much better agglutinated by the mannose-resistant strains than untrypsinized cells?

Svanborg Edén: We have not looked at that.

Sharon: Do anti-globoside antibodies inhibit attachment?

Svanborg Edén: We have not tested that.

Hughes: How many of the MR type of bacteria are inhibited by globoside?

Svanborg Edén: We have examined about 50 of our MR strains. There are different patterns but about 90% of these strains also bind to globoside. The few strains not recognizing globoside still bind to most erythrocytes, including guinea-pig red cells, in a mannose-resistant way. On the other hand, if you had asked how efficiently the attachment of different strains recognizing globoside is inhibited by globoside, the answer would have been different. The amount of globoside needed to inhibit attachment and the completeness of the inhibition obtained varies between the strains.

Hughes: So the MR character is not simply recognition of globoside; there are other characters there as well?

Svanborg Edén: Yes.

Razin: Is it possible to find *E. coli* strains which have two types of pili?

Svanborg Edén: All but about 15 of the strains with mannose-resistant agglutination of human erythrocytes also cause mannose-sensitive guinea-pig red cell agglutination. This clearly demonstrates the presence of at least two types of pili on the strains. With or without other pili, strains causing mannose-resistant haemagglutination had good attachment. The strains with only mannose-sensitive haemagglutination attached less. The non-agglutinated strains attach poorly.

Razin: Can you in practice differentiate between the strains with two haemagglutinins and those with one?

Svanborg Edén: Yes, by haemagglutination. Both types may be present at the same time. The Ørskovs suggested in a recent publication (Ørskov et al 1980) that the role of the type 1 pilus is to attach bacteria to the urinary slime. That is, all these strains, regardless of what type of disease they produce, have to have type 1 pili to enable them to colonize the urinary tract. The ones that cause disease seem to have in addition large numbers of the tissue-binding pili.

Wallach: Is there any evidence that glycoproteins can act as receptors? Have isolated glycoproteins or glycopeptides been found to act as inhibitors?

Svanborg Edén: We have not tried that. The only variable in the experiment is the glycolipid. Everything else is constant. A unique property of the glycolipids of the globoseries is that the sugar sequence has been found anchored only to lipids, not to proteins. Otherwise sugar sequences are often common for both carrier molecules. We thus have no reason to believe that glycoproteins could replace globoside in this system, but certainly it can in others. Salivary glycoproteins, for example, agglutinate oral streptococci (Ericson et al 1979) and fibroblast glycoproteins bind to mycoplasmas (Gabridge & Taylor-Robinson 1979).

Bredt: Do your strains lose their specific properties after a few passages?

Svanborg Edén: We have tried to get some kind of order into this but it is very difficult. Many of our strains are old. When I started to work with them we did a yearly survey of the adhesive capacity. We found very little change from year to year but they had already been stored. More recently we have looked at strains that we get within a week of isolation. There is a higher incidence of the mannose-resistant property in strains from symptomatic patients than in the old material. Dr Levine, did you test your challenge strain for haemagglutination when it came out of the volunteers? Did it show the same property as when you gave it to them?

Levine: We don't know that.

Taylor-Robinson: With gonococci there has been a suggestion that pili are not present in material taken directly from patients and that they occur when the organisms are cultured (Novotny et al 1975). Are the pili you talked about present on bacteria in the patient or in culture only?

Svanborg Edén: We negatively stained bacteria directly from the urine of the patients and saw different variants of bacteria. Some looked as if they had lost all their outer membranes, some seemed to be encapsulated, and some had pili. We have not done anything more systematic.

Taylor-Robinson: And you are able to say that the adhesiveness of the original isolates from the urine is the same as seen on subculture?

Svanborg Edén: No, I can't say that.

Taylor-Robinson: Nevertheless, you think it is more relevant to look at primary material than at subcultured bacteria?

Svanborg Edén: Yes, if it is available.

Richmond: John Costerton has stressed that *E. coli* in urine or exudates have enormous polysaccharide capsules which may extend far out from the surface. He says he has never seen that when he subcultures the organisms.

Svanborg Edén: We have to distinguish several phases of this process. What we measure now is possibly the first recognition phase. This may depend on pili and it can be inhibited with receptor analogues. After the initial recognition something else probably happens to the binding, such as the bacteria producing a capsule. We know that they get more nutrients when they stick to a surface, and when bacteria are fed a rich medium they may form a capsule. This second stage cannot be tested in our system because when we grow bacteria in rich media so that they form a capsule they stop adhering. We have not been able to inhibit attachment with isolated capsule polysaccharide up to 40 mg/ml. Bacteria that are resident in an individual are likely to have reached this second stage. It is very difficult to look at the initial binding on preparations obtained *in vivo*.

Sharon: Dr Silverblatt, you examined several of our strains in the electron microscope. Were they encapsulated?

Silverblatt: Those were fresh urine isolates that had been fixed in glutaraldehyde. I negatively stained them and didn't see any pili or capsule in any of the four strains that we tested. However, I don't really know what a capsule looks like by negative staining.

Sharon: Can you distinguish between mannose-sensitive and mannose-resistant pili in a mixture, Dr Svanborg Edén? Could you bind the mannose-sensitive ones, let's say to yeast cells, and see whether what is not bound is different from the whole mixture or is different in different strains?

Svanborg Edén: To purify pili we used strains that had only mannose-resistant or -sensitive pili. Since most strains carry both at the same time a method like you suggested would be very useful. The two types of pili are certainly antigenically different. We have not found any cross-reactions between type 1 pili and the mannose-resistant pili.

Vosbeck: Does globoside inhibit adhesion in strains with known adhesion factors such as CFA/I?

Svanborg Edén: We have one strain which has CFA/I and it does not bind to globoside-coated erythrocytes.

Elbein: Do you isolate the globoside and then add it back?

Svanborg Edén: We disperse it in buffer by sonication. It forms micelles that are not standardized in size or globoside content.

Elbein: Some glycolipids in other systems have a tendency to stick and be taken up by cells. Could you try adding a globoside to a mammalian cell culture which doesn't have it and see whether it would be taken up? For example, P. Cuatrescas and others added GM_1 ganglioside to transformed mouse fibroblasts. These cells were unresponsive to cholera toxin and were deficient in GM_1 as well as other gangliosides. But after taking up GM_1 the cells became responsive to cholera toxin (Fishman & Brady 1976).

Svanborg Edén: This is what we did with erythrocytes. When globoside was added, erythrocytes not containing globoside and not agglutinated by our bacteria became agglutinable. We tried coating them with other glycolipids but only the globoside worked. We tried adding globoside to rat epithelial cells, since the strains with mannose-resistant haemagglutinins did not attach to those. The results are uncertain. I think that the membrane of the urinary tract epithelial cell would less easily accept molecules from the outside than an erythrocyte would.

Feizi: Earlier we heard that the mannose-sensitive adhesive property was in many instances acquired by bacteria after culture in brain-heart broth. Do the bacteria you are studying develop mannose-susceptible adhesion? Your material provides a good opportunity for investigating whether the mannose-susceptible adhesive property can be acquired as a result of culture in this broth.

Svanborg Edén: I haven't any evidence, only some indications. In our old strains mannose-sensitive pili are common: between 60 and 90% of the strains possess type 1 pili. In the fresh isolates from newborns with symptoms of urinary tract infection the mannose-resistant property has more often been found alone. It remains to be seen whether these isolates will get more mannose-sensitive pili when they have been stored and subcultured.

Taylor-Robinson: It must be comforting to know that you have used probably the most relevant model for studying adherence, namely the patient. In relation to this and to our previous discussion, I now recall that Mårdh et al (1979) showed that adhesion of bacteria to cells was much greater when the organisms came directly from the patient than when they had been cultured *in vitro.*

Tramont: Organisms are dynamic and the antigens they express on their cell walls may change, depending on the culture medium. So when the organism comes directly from the patient it may be mannose-resistant but it can become mannose-sensitive if grown under the right conditions. Furthermore, when an organism is growing inside or on a person it is subjected to a lot of environmental pressures, such as shearing forces in urine or mucus and competition from other organisms for scarce foodstuffs, host enzymes and perhaps antibiotics that are being excreted by competing organisms. All have an effect on what antigens are ultimately expressed. What we are doing is definitely artificial in many respects. We have to try to fit what we find to what is really happening in nature. I don't have any problems with the organisms changing, because that is what happens all the time in nature.

Helenius: From the pictures we have seen it looks as if the bacteria would depend on several contacts with the cell surface for binding. Each individual contact may be relatively weak, but when multiplied over a large number of contacts the binding may become very strong. When self-associating molecules such as globosides are used as inhibitors, their physical state is therefore very important. You have shown that the globosides must be aggregated to be effective as inhibitors. Probably single globoside molecules cannot inhibit binding very well. It might be interesting to test liposomes which would contain in addition to globosides variable amounts of other lipids which would dilute out the globosides over a larger surface.

Svanborg Edén: The problem is that, depending on what you mix with, you may get shielding of the receptor. I showed you an example of lactosylceramide as a spacer increasing the efficiency with which globoside is recognized. There are other glycolipids which can take away the effect when mixed with globoside.

Helenius: Do phospholipids take away the effect?

Svanborg Edén: Sphingomyelin does. We also fractionated the urinary sediment glycolipids. The activity in the different fractions correlated with the presence of the sequence GalNAc β1-3 Gal α1-4 Gal β1-4Glc as indicated by the bar. But the inhibitory effect was greater for the intact urinary sediment glycolipid fraction than in the globoside, per estimated microgram of substance. As you say, it is certainly a complex thing. Probably it is a question of presentation.

Helenius: If we compare the binding of individual virus glycoproteins and intact viruses to cell surfaces we find that the virus binds with an apparent binding constant which is 10^6 higher. The tightness of binding adds up very rapidly with increasing multivalency of binding.

Svanborg Edén: How strong is binding between antibody and antigen?

Helenius: The binding constant between a Fab fragment and its antigen may be something like 10^3 or 10^4 M^{-1}. However, the binding of the divalent antibody to a multivalent antigen will be closer to 10^8 M^{-1}, that is the square of the individual binding constants of binding sites.

Feizi: I agree with the cooperative effect of multivalence but Dr Svanborg Edén uses purified glycosphingolipids without any auxiliary lipids. It is really surprising that she is able to detect any activity. If the purified glycosphingolipids were incorporated into cholesterol lecithin micelles the oligosaccharide moieties might be better oriented and their inhibitory activities might be substantially increased.

Choppin: The MDBK epithelial line of bovine kidney cells, which has a very high content of globoside, might be useful in this problem. It would be interesting to see whether the bacteria will bind to a glycolipid molecule that occurs naturally in a renal epithelial cell.

Candy: I take it that you are not proposing that this glycolipid is a general receptor for mannose-resistant adhesion in tissues outside the urinary tract?

Svanborg Edén: That is right.

Candy: Have you been able to compare the amount of globoside in the urinary cells from your control with that in the various groups of patients with pyelonephritis?

Svanborg Edén: No. There are reports that epithelial cells from patients bind *E. coli* better than control cells. Differences in these studies are small but consistent.

Sussman: We have compiled a list of properties of different pili, partly based on published work (Table 1). The type 1 pili, as mentioned earlier, are best expressed after about 48 h at 37 °C, in other words at the maximum stationary phase in nutrient broth, whereas MR pili are far better expressed selectively in phosphate-buffered agar, pH 7, though they may come up in the broth as well. Growth at 16 °C is also a differentiating feature, and the effect of the age of the culture is different in the two.

The mannose-sensitive (type 1) pili are antigenically homogeneous but we now have at least seven different antigenic determinants combined in at least 16 different serotypes on MR pili. Usually two determinants are present on an organism at the same time, and occasionally three. Whether these determinants are on the same pilus or whether there are two antigenically distinct pili on an organism, we don't yet know. The other difference is in relation to the stability of pilus properties after treatment with formalin.

We collected strains of *E. coli* from a large number of patients with significant bacteriuria without looking at the relationship to symptoms and divided them into five groups according to their haemagglutination and adhesion properties (Table 2). Less than 30% of such strains are totally non-adhesive.

Changes occur in the course of culture and it is sometimes difficult to find the MS pili unless the organisms are cultured for up to five passages in nutrient broth. If they are grown in broth the MR pili may be lost and the MS pili are selectively expressed.

If mass cultures are picked from cultures of any kind one may very well miss the changes in the individual organisms. There is a tendency for properties to be lost and expressed spontaneously, so if you pick individual colonies in cloning you may lose certain things. For example with some MR strains which we cultured and where we picked 100 colonies each day for several days from a broth culture, there was a loss of MR haemagglutination in five or six days (I. M. Feavers, unpublished work). Thus, if you clone under these conditions in long-term cultures you may be lucky and have the original strain, or you may not. Why this happens we don't know but

TABLE 1 (Sussman) Characteristics of mannose-sensitive (MS) type 1 pili and mannose-resistant (MR) pili associated with strains of E. coli isolated from patients with urinary tract infection[a]

Property	*MS pili*	*MR pili*
Haemagglutination	Guinea-pig erythrocytes[b,c]	Human erythrocytes[b,d]
Adhesion to human uroepithelial cells	Yes[c]	Yes[d]
Adhesion to human buccal cells	Yes[e]	Yes
Adhesion to rat and pig uroepithelial and buccal cells	Yes	No
Cultural conditions for best expression	Nutrient broth 48 h at 37°C	Phosphate-buffered (pH 7.0) nutrient agar, 24 h at 37 °C
Expression on primary isolation	Not always expressed, may require up to five broth passages	Always expressed if present; may be lost on subculture
Expression at 16 °C	Yes	No
Effect of age of culture on adhesive properties	Retained after 10 days in culture	Often lost after 7–10 days in culture
Antigenic characteristics	Common antigen	Considerable antigenic heterogeneity
Formalin (1%) 4 h at 37 °C	Retain functional and antigenic properties	Lose functional and most of antigenic properties
Heating at 80 °C for 10 min	Lose functional and antigenic properties	Lose functional and antigenic properties

[a]Based on S. N. Abraham, S. H. Parry and M. Sussman (unpublished work) unless otherwise stated
[b]Duguid et al (1979)
[c]Svanborg Edén & Hansson (1978)
[d]Källenius & Möllby (1979)
[e]Ofek et al (1977)

TABLE 2 (Sussman) The adhesive properties of E. coli isolated from urinary tract infection

		Haemagglutination				
Group	%[a]	*MR*[b]	*MS*[c]	*Adhesion to human uroepithelial cells*	*Presence*[d] *of pili*	*Suggested piliation*
A	28.9	–	–	–	–	None
B	28.9	–	+	+	+	Type 1 (MS)
C	13.5	+	–	+	+	MR colonization pili
D	24.4	+	+	+	+	Type 1 (MS) and MR colonization pili
E	4.3	–	–	Variable	?	Unknown

[a]Total of 184 *E. coli* isolated from cases of significant bacteriuria (28.9% non-adhesive, 71.1% adhesive)
[b]Mannose-resistant agglutination of human erythrocytes
[c]Mannose-sensitive agglutination of guinea-pig erythrocytes
[d]Determined by electron microscopy

we have some preliminary evidence. We have been able to mobilize plasmids coding for MR pili and we have shown transfer of MR pilus production with the mobilizing plasmid R119, but the property is lost after the first pass. We haven't been able to check the antigenic specificity of the pilus involved.

Certainly this whole thing is very fluid. Depending on how you culture your organisms and characterize them, you may gain or lose properties and you may not be working with the organism that was originally in the urinary tract, or you may have a quite different organism because of selection or spontaneous cure or loss of properties.

Sharon: In your experiments was epithelial cell adhesion mannose-sensitive or mannose-resistant?

Sussman: When it is related to type 1 pili it is mannose-sensitive; when it is type 2 it is mannose-resistant. The haemagglutination properties correlate exactly.

Taylor-Robinson: Earlier we talked about using pili as vaccines for preventing gastrointestinal disease. What are the possibilities for the urinary tract?

Svanborg Edén: Our group has been working towards the development of a vaccine for the urinary tract, possibly containing a combination of pilus antigen and capsular polysaccharide which we already know is protective against pyelonephritis in rats (Kaijser et al 1978). The problem is that there is no animal model which mimics the specificity and the early colonization phase of urinary tract infection. Once such a model is available it will be possible to study the role of adhesion and of these various antigens. I was encouraged to hear that the use of type 1 pili did not interfere with the normal flora of the intestine. Otherwise the use of mannose-resistant pilus antigens or other antigens specifically mediating attachment to the urinary tract may be a better alternative than the use of type 1 pili or a vaccine against urinary tract infection.

Taylor-Robinson: Have you thought of using primates as models?

Svanborg Edén: Yes, humans if our ethical authorities permit us to do so.

Silverblatt: We used a model of urinary tract infection in rats to evaluate the protective effect of anti-pili antibodies. This does not really mimic the human situation because we put the bugs directly into the bladder, by-passing the phase of initial colonization around the urethral meatus. However, the organisms do have to interact with the renal pelvic epithelium before they can invade the renal parenchyma. We prepared type 1 pili and immunized a group of animals both actively and passively; then we challenged them with the homologous strain as well as with another strain which had similar pilus antigens but different O, K and H antigens. We got very good protection of our immunized animals, both with the active immunization and with the passive immunization. There was cross-protection against the heterologous strain, as well (Silverblatt & Cohen 1979).

We have also measured anti-pili antibodies in a number of patients with pyelonephritis with the ELISA technique, using as an antigen type 1 pili from a single strain. We found the ELISA technique to be a good general screen for most patients. We had a prompt rise in antibody activity with all three antibody classes, which interestingly decreased rather quickly, suggesting that natural immunity has a short half-life in treated patients. However, with the ELISA technique not all patients appeared to develop anti-pili antibodies. In contrast, with indirect immunoelectron microscopy, which uses the patient's own organism as antigen, a rise in anti-pili antibodies was detected in all patients (Rene & Silverblatt 1980).

Svanborg Edén: I am not encouraged by the rat model but it is nice to know that pili are immunogenic.

REFERENCES

Duguid JP, Clegg S, Wilson MI 1979 The fimbrial and non-fimbrial haemagglutinins of *Escherichia coli.* J Med Microbiol 12:213-227

Ericson D, Bratthall D, Björck L, Myhre E, Kronvall G 1979 Interactions between human serum proteins and oral streptococci reveal occurrence of receptors for aggregated β_2-microglobulin. Infect Immun 25:279-283

Fishman PH, Brady RO 1976 Biosynthesis and function of gangliosides. Science (Wash DC) 194:906-915

Gabridge M, Taylor-Robinson D 1979 Interaction of *Mycoplasma pneumoniae* with human lung fibroblasts: role of receptor sites. Infect Immun 25:455-459

Kaijser B, Larsson P, Olling S 1978 Protection against ascending *E. coli* pyelonephritis in rats and significance of local immunity. Infect Immun 20:78-81

Källenius G, Möllby R 1979 Adhesion of *Escherichia coli* to human periurethral cells correlated to mannose-resistant agglutination of human erythrocytes. FEMS (Fed Eur Microbiol Soc) Lett 5:295-299

Mårdh PA, Colleen S, Hovelius B 1979 Attachment of bacteria to exfoliated cells from the urogenital tract. Invest Urol 16:322-326

Novotny P, Short JA, Walker PD 1975 An electron microscope study of naturally occurring and cultured cells of *Neisseria gonorrhoeae.* J Med Microbiol 8:413-427

Ofek I, Mirelman D, Sharon N 1977 Adherence of *Escherichia coli* to human mucosal cells mediated by mannose receptors. Nature (Lond) 265:623-625

Ørskov I, Ørskov F, Birch-Andersen A 1980 Comparison between F7 and type I *E.coli* fimbriae. Infect Immun 27: 657-666

Rene P, Silverblatt FJ 1980 Immune response to pilus antigen in pyelonephritis. In: Current chemotherapy and infectious disease. American Society for Microbiology, Washington, in press

Silverblatt FJ, Cohen LS 1979 Antipili antibody affords protection against experimental ascending pyelonephritis. J Clin Invest 64:333-336

Svanborg Edén C, Hansson HA 1978 *Escherichia coli* pili as possible mediators of attachment to human urinary tract epithelial cells. Infect Immun 21:229-237

Adhesion of *Neisseria gonorrhoeae* and disease

EDMUND C. TRAMONT

Uniformed Services University of the Health Sciences Medical School and Walter Reed Army Medical Center, Washington DC 20012, USA

Abstract Adhesion of gonococci to mucosal epithelial cells appears to be a critical step in the pathogenesis of gonococcal infection. Parameters affecting adhesion, the antigens involved in adhesion and the human antibody response that blocks adhesion have been studied. Gonococci adhere to a variety of human cells grown either in tissue culture or organ culture or scraped from mucosal surfaces. They adhere in greatest number to mucosal cells derived from sites of natural infection but buccal mucosal cells were used for most of these studies because of the ease of obtaining them, the homogeneous population obtained, and the similarity of the results of antibody studies with all cell types. The ability of a given strain of gonococci to attach to buccal cells (receptors) varied with the individual from whom the cells were obtained, and the capacity to support adhesion of a given strain fluctuated from day to day. The principal antigens mediating attachment are pili (filamentous protein appendages extending from the bacterial cell wall) which are antigenically heterogeneous. Lipopolysaccharide also blocks attachment but it is less efficient than pili and is not species-specific. Other antigens, probably proteins, also block attachment.

Human antibody (both local and humoral) blocked the adhesion of gonococci. The antibody was directed against pili and other antigens in naturally occurring disease and could be induced by immunizing subjects with purified pili. The results of these studies suggest two new approaches to the prevention of gonorrhoea: (1) (competitive) inhibition of adhesion with specific antigens, i.e. pili; and (2) immunization with purified attachment antigens to produce shielding antibodies.

Neisseria gonorrhoeae is a human pathogen which primarily affects the mucosal surfaces of the genital tract, rectum, oropharynx and occasionally the conjunctiva. One cannot experimentally colonize or infect any other animal species, with the exception of an occasional chimpanzee. This indicates that the organism is highly adapted to human epithelial cells. Since many human commensal bacteria also reside in these areas, adherence to these mucosal cells is not unique to the gonococcus and

1981 Adhesion and microorganism pathogenicity, Pitman Medical, Tunbridge Wells (Ciba Foundation symposium 80) p 188-201

it must possess other virulence factors that lead to an inflammatory response. However, to resist mucosal defence systems, especially the shearing forces of mucus and urine flow, the gonococcus must be well adapted and quite efficient at adhering to mucosal cells. Any discussion of this interaction must take into account the attachment or adhesion factors (antigens) of the gonococcus, the host cell tropisms or surface properties, and the kinetics of the reaction.

One difficulty in studying the attachment of gonococci to mucosal cells is the tendency of these organisms to aggregate into clumps. Careful attention must be paid to the total number of organisms in the test assay, particularly when the effect of antibody is being studied, as the large aggregates, which vary in size from day to day, can interfere with the interpretation and reproducibility of the results. The only way to alleviate this problem is to limit the number of organisms to the minimum that allows sufficient attachment for differences to be distinguished. However, the price paid for this adjustment is that the preparations have to be individually examined through a microscope, which makes the assay tedious and time-consuming.

Attachment antigens

We have studied the antigens responsible for mediating attachment of gonococci both by direct or 'competitive' inhibition and by determining the specificity of blocking antibody. Freshly isolated gonococci form distinct colony types on artificial media and these colonies quickly transform on subculture. Organisms from the freshly isolated colonies known as colony types T1 and T2 are of increased virulence (Kellogg et al 1963) and attach (adhere) more readily to eukaryotic cells than do organisms from colony types T3, T4 or T5. The principal difference between these colony types lies in the presence and/or quantity of pilus antigens (Swanson 1973).

Gonococcal pili (Brinton et al 1978) are filamentous protein appendages which extend from the bacterial cell wall and mediate attachment to host cells. Purified preparations of pili directly inhibit this attachment. The inhibition appears to be competitive, with higher concentrations of pili inhibiting attachment of gonococci more strongly (Fig. 1). This is presumably due to the covering or masking of specific receptors for the pili. Purified pili inhibit both heterologous and homologous strains but are most efficient in blocking the attachment of the homologous strain (Fig. 2). In a study of recurrent infections the only antigenic difference between two strains infecting the same hosts four months apart was found to be pili (Tramont et al 1979). This inhibition also appears to be species-specific: pili isolated from *E. coli* do not inhibit the attachment of gonococci, and those isolated from gonococci do not inhibit attachment of *E. coli* (Buchanan et al 1978, Tramont et al 1980).

Genital and humoral antibodies developing after natural infection are also capable of blocking the attachment of gonococci (Table 1) (Tramont 1977) and the

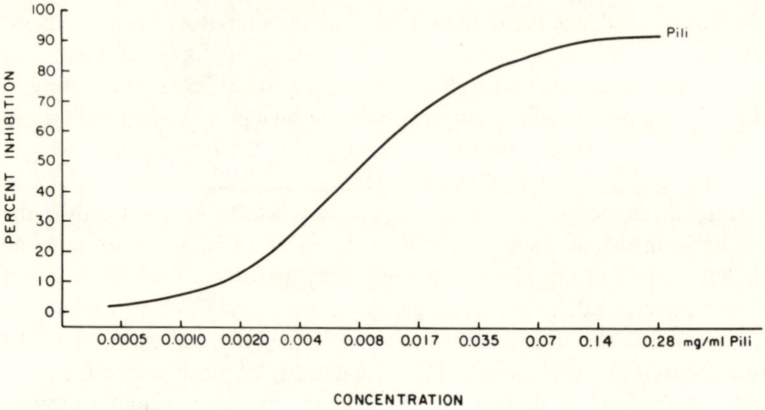

FIG. 1. Inhibition of attachment of gonococci with purified pili. Gonococci of colony type T1 or T2 or both were suspended in Medium 199 supplemented with 2% (w/v) bovine serum albumin. These were gently vortexed to break up large clumps of organisms and adjusted to contain 1 x 10^7 bacteria/ml. Human buccal epithelial cells were scraped loose with a wooden applicator, suspended in 0.067 M-phosphate-buffered saline (PBS), pH 4.5, washed twice in PBS, counted and adjusted in Medium 199 to 2 x 10^5 cells/ml. Equal volumes (0.25 ml) of buccal cells, gonococci, antibody or inhibiting antigen (pili) and PBS were incubated at 37 °C for 30 min on a tumbling shaker apparatus. A 50:1 ratio of organisms to epithelial cells was used. The cells were then washed in PBS to remove unattached bacteria, dried onto glass slides and stained by Gram's method. The number of organisms attached per 50 cells was counted. Cells incubated without gonococci were used as controls. Since the number of organisms attaching to the control cells varied each day from 450 to 1500 organisms the amount of inhibition was expressed as a percentage, taking the average number of organisms attached to three control slides as 100% attachment. An unabsorbed vaginal secretion was included with each absorption study (Tramont et al 1980).

principal antigen to which this antibody is directed is pili (Fig. 3) (Tramont et al 1980). In addition, immunization of human volunteers with a vaccine prepared from purified gonococcal pili raised antibody capable of blocking the attachment of homologous and heterologous gonocci (Table 2). Finally, studies of the attachment (adhesion) of pili to cells have, in general, produced the same results as studies of intact organisms (Buchanan et al 1978, Pearce & Buchanan 1978).

Other antigens also block the attachment of gonococci. Lipopolysaccharide can 'competitively' block attachment but this is not an entirely species-specific reaction. Lipopolysaccharide purified from several species of bacteria blocked the attachment of gonococci (Table 3). This was probably due to the receptor sites for gonococci on the buccal cells being covered or masked in a random non-specific manner by this very sticky hydrophobic substance. Nevertheless, lipopolysaccharide purified from the parent gonococcal species was two to four times more efficient in blocking attachment than lipopolysaccharide purified from other bacteria (Table 3).

Isolated outer membrane complexes (bacterial cell walls devoid of cytoplasm)

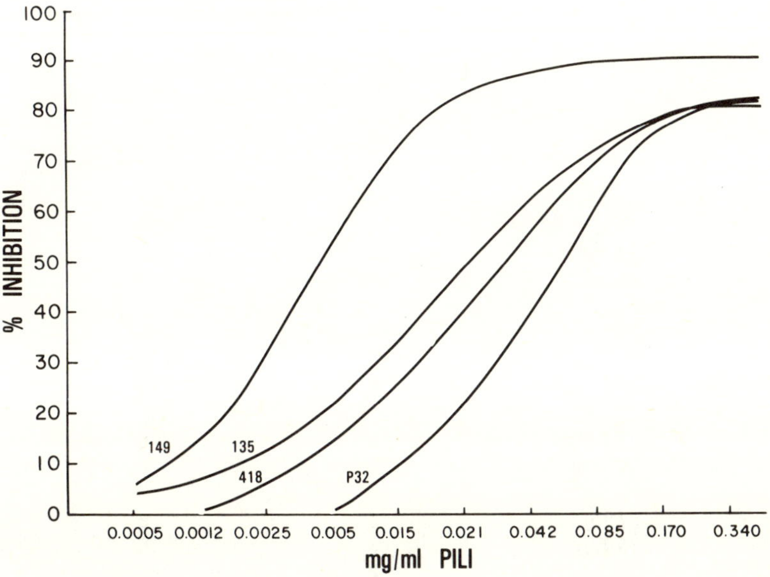

FIG. 2. Inhibition of attachment of strain 149 of *N. gonorrhoeae* with pili of different strains (149, 135, 418, P32). The pili were incubated with the buccal cells first. See Fig. 1 for other experimental details.

also competitively blocked attachment. Outer membrane complexes purified from piliated colony types were 10–20 times more efficient at blocking attachment than outer membrane complexes from non-piliated colony types. The inhibition of attachment by these complexes from non-piliated gonococci could be attributed at least in part to the lipopolysaccharide content of the complexes.

There is good evidence that other antigens are involved in gonococcal attachment (Lambden et al 1980) and that the human antibody response can be directed primarily against them. When vaginal secretions or sera were absorbed with pili purified from either the heterologous or homologous infecting strain, not all of the inhibitory activity against all the strains was removed (Table 4). Furthermore, Swanson (1977) has demonstrated a leucocyte association factor that mediates adherence to polymorphonuclear cells or ingestion by them. Finally, non-piliated colony type organisms (T4) adhere to eukaryotic cells, albeit to a lesser degree than piliated gonococci. These other antigens could be antigenically different pili.

Host cell properties (tropisms)

Neisseria gonorrhoeae organisms attach to a variety of human cells, including cervical mucosal cells (Mårdh & Weström 1976, Tramont 1977), vaginal mucosal

TABLE 1 Inhibition of epithelial cell attachment of the homologous strain of N. gonorrhoeae by serum and vaginal secretions

		Inhibition of epithelial cell attachment by:	
Patient	*Date*	*Serum*[a]	*Vaginal secretions*
134	9 June	32	64
	17 June	32	128
	23 June	32	128
	30 June	ND[b]	8
136	17 June	4	128
	22 June	16	128
	30 June	8	128
143	7 July	8	8
	14 July	8	8
149	7 July	16	8
	25 July	16	8
	8 August	32	8
418	29 January	ND	64
	19 February	ND	32
	4 March	ND	64
	23 April	ND	32
	19 May	ND	64

[a]Reciprocal titre, 50% inhibition of the total number of organisms attached to 50 buccal cells.
[b]ND: not done

cells (Mårdh & Weström 1976), Fallopian tube mucosal cells (Johnson et al 1977), sperm (James et al 1976) and buccal cells (Tubbett et al 1976, Tramont 1977). *N. gonorrhoeae* also attaches to human polymorphonuclear leucocytes and to a variety of other eukaryotic cells grown in tissue culture (Swanson 1977). In general, the cell-specificity of attachment suggests the greatest affinity (tropism) for cells obtained from the natural sites of infection, including human buccal cells (which is probably related to the ability of gonococci to colonize the oropharynx). Because of this avidity for buccal cells, the ease of obtaining them, the homogeneous cell population, and the concordance of the attachment and the antibody studies no matter whether buccal, cervical or vaginal mucosal cells were used (Tramont 1977), we used human buccal cells in most of our studies.

Buccal cells obtained from different donors were found to differ in their capacity to support attachment of any given strain of gonococcus (Table 5) (Tramont 1976). However, the ability of any one source of buccal cells to support attachment of a given strain appeared also to be in a state of flux: buccal cells obtained from

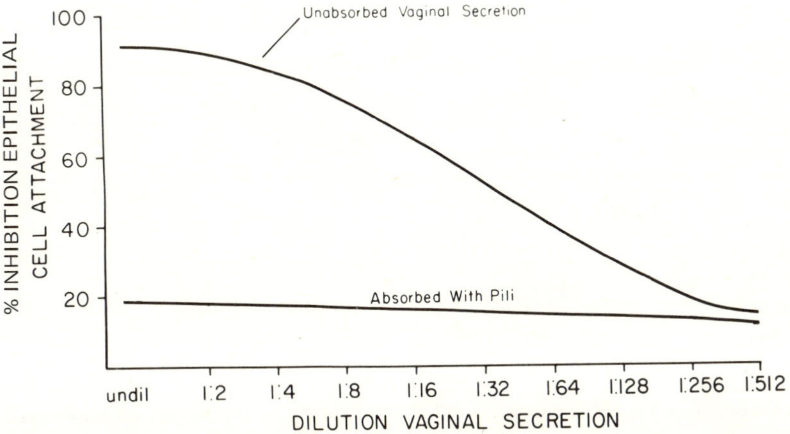

FIG. 3. Inhibition of attachment of gonococci to epithelial cells by vaginal secretion. The secretion inhibited attachment of the homologous strain. Absorption of the secretion with purified pili removed all the inhibiting ability. Removal of anti-pilus antibody from the secretion was confirmed using a solid-phase radioimmunoassay (Tramont et al 1980).

TABLE 2 Inhibition of epithelial cell attachment of four strains of gonococci in a volunteer[a] immunized with purified gonococcal pilus vaccine P32

Strain	*Preimmunization*	*Postimmunization*[b]
P32	1:1	1:8
135	< 1:1	1:4
1174	< 1:1	1:2
418	< 1:1	1:4

[a]Volunteer vaccinated with two doses of 1 mg pili purified from strain P32.
[b]Reciprocal titre, 50% inhibition of the total number of organisms attached to 50 buccal cells.

TABLE 3 Concentrations of lipopolysaccharide necessary to inhibit epithelial cell attachment by 50%

LPS from	*Dry wt. LPS mg/ml*	*Protein concentration mg/ml*
Pseudomonas aeruginosa	4	< 0.01
Salmonella typhimurium	2	< 0.01
Escherichia coli	2	< 0.01
Neisseria gonorrhoeae	1	< 0.01

Lipopolysaccharides (LPS) were freeze-dried and suspended in distilled water to the desired concentration. Protein concentrations were determined by the method of Lowry.

TABLE 4 Inhibition of attachment of gonococci to epithelial cells by vaginal secretion absorbed with homologous pili

Vaginal secretion from patient infected with strain 418	*Strain*			
	418	*135*	*222*	*P32*
Unabsorbed	1:32	1:4	1:8	1:32
Absorbed with pili isolated from strain 418	0	0	1:4	1:16
Absorbed with pili isolated from strain P32	1:16	1:2	1:4	1:16

TABLE 5 Variation in the ability of buccal cells from different donors to support adhesion of different gonococcal strains

	Strain[a]					
Donor	*104*	*128*	*129*	*142*	*156*	*186*
A	1099	0	33	475	0	246
B	1042	0	396	675	126	164
C	1061	347	594	678	216	369
D	852	368	198	375	144	82

[a]Each strain was tested against buccal cells obtained from four different donors. Values are numbers of organisms attached per 50 cells.

the same donor on different days varied in their capacity to support attachment of that strain (Table 6). The sex of the buccal cell donor did not appear to affect these tropisms, but Forslin et al (1979) found that more gonococci adhered to vaginal mucosal cells obtained from donors in the postmenstrual phase than adhered to cells from donors in the premenstrual phase. Acute salpingitis and disseminated gonococcal infection are more likely to occur during or shortly after menstruation than before. Although it is tempting to speculate that hormonal fluctuations influence these changes in tropisms this seems an unlikely explanation for the changes observed in male buccal cells. Nevertheless, these changes could influence the susceptibility of a host to becoming infected on any given day, or the incubation period or the probability of dissemination of the infection.

The receptor sites for the gonococcus are not known. Lectins that block fucose, *N*-acetylgalactosamine and *N*-acetylglucosamine did not block attachment (Watt et al 1978). Extrapolation from studies on pili reveals a confusing picture. Buchanan et al (1978) found that sialidase (neuraminidase) and low concentrations of gangliosides inhibited attachment to buccal cells but that sialic acid (specifically) had no

TABLE 6 Variation in the ability of buccal cells from a single donor to support attachment of a single gonococcal strain

	Day of collection of buccal cells							
Strain	*1*	*5*	*7*	*12*	*14*	*19*	*21*	*26*
104	1072	918	126	147	137	971	897	1002
418	617	701	71	88	141	431	421	781
135	912	895	254	225	397	921	1088	1115

Values refer to number of organisms attached per 50 buccal cells. The number of organisms of each strain attached decreased between days 7 and 14 but strains 104 and 418 were affected to a greater extent. These changes were not due to alterations in the attachment antigens since attachment was not affected with buccal cells obtained from other donors. The buccal cells were collected fresh every day.

effect. This suggested that the receptor most probably resembled a ganglioside. Lambden et al (1980) have presented evidence that on some strains two distinct types of pili (alpha and beta) may exist, one of which (alpha) can be inhibited by sialic acid. The alpha type of pili were also not inhibited by galactose, fucose, mannose, glucose, *N*-acetylglucosamine and *N*-acetylgalactosamine, or by the mucopolysaccharides hyaluronic acid and chondroitin sulphate A, B and C (Buchanan et al 1978).

Increasing the number of organisms inoculated increased both the total number of organisms that attached and the number of cells to which they attached.

Kinetics of attachment of gonococci to buccal cells

The factors determining attachment of gonococci to mucosal cells are complex, involving the relative balance of inhibitory forces such as electrostatic repulsion, steric interference and the shearing effects of mucus and urine flow, and the forces that promote adherence, such as van der Waals forces or hydrophobic interaction.

The net charge on epithelial cells, the gonococcus and gonococcal pili is negative, which would be a strong repulsive force reducing adhesion. How this barrier is overcome is not known, but pili undoubtedly play a critical role (Heckels et al 1976). Attachment of most strains of gonococci is favoured at a pH level (4.5) that would result in a less negative charge. However, some strains attach better at pH 6.5–8.0, and recent evidence has corroborated these results with pili (Lambden et al 1980). These differences may be related to alterations in the cell wall induced by the ecology of the sites from which the organisms were isolated – namely the male urethra (pH 6.8–7.4) or the cervix (pH 4.5).

Iron salts dramatically enhance the attachment of gonococci (James et al 1976) as well as pili (Buchanan et al 1978). Other highly positive charged molecules such as protamine sulphate also favour the attachment of pili, possibly through the formation of ligands between pili and the cell surface.

Treatment with trypsin, α-chymotrypsin, tryptophan or lysozyme reduces attachment, but this is probably related to conformational changes of the pili or cell wall, or both.

The role of steric hindrance is still speculative. Physical factors of this kind are known to influence functions such as phagocytosis and the avidity of the antibody–antigen reaction.

The external shearing forces applied to the gonococcus are obviously formidable but of a non-specific nature.

As mentioned already, the ability of cells to support attachment appears to be in a state of flux. Treatment of buccal cells with exoglycosidases interferes with attachment. The adhesion of gonococci is also temperature-dependent, being greatest at 35–37 °C. These observations imply that factors other than charge and hydrophobic effects influence the attachment of gonococci to mucosal cells.

In summary, it appears that gonococci have well-developed means of attaching to mucosal cells of their natural host. It also appears that there is a human immune response that results in antibodies which block this attachment. A major antigen to which this antibody response is directed is pili, although other antigens are also involved. The ability of eukaryotic cells to support attachment appears not only to be in a state of flux but also to vary with each donor. Neuraminidase and gangliosides inhibit the attachment of gonococcal pili, which suggests that the receptors for pili resemble gangliosides. Different strains attach better at different pH levels, although most attach best at pH 4.5. Iron enhances attachment.

The results of these studies offer two new approaches to the prevention of gonorrhoea: (1) a gonococcal vaccine that will induce blocking antibody; and (2) blocking substances that can be applied directly to the mucosal area at risk so that attachment is directly (competitively) inhibited.

Acknowledgements

The experiments described here were done with Jennie Ciak and Daniel McChesney. I am indebted to them for their invaluable assistance and to Mrs Helen Carafelli for her help in preparing this manuscript.

REFERENCES

Brinton CC, Bryan J, Dillon JA, Guerina N, Jacobson LJ, Labik A et al 1978 Uses of pili in gonorrhea control: role of bacterial pili in disease, purification and properties of gonococcal pili, and progress in the development of gonococcal pilus vaccine for gonorrhea. In: Brooks GF et al (eds) Immunobiology of *Neisseria gonorrhoeae.* The American Society for Microbiology, Washington DC, p 155-178

Buchanan T, Pearce WA, Chen KS 1978 Attachment of *Neisseria gonorrhoeae* pili to human cells, and investigation of the chemical nature of the receptor for gonococcal pili. In: Brooks GF et al (eds) Immunobiology of *Neisseria gonorrhoeae.* American Society for Microbiology, Washington DC, p 242-249

Forslin L, Danielson D, Falk V 1979 Variation in attachment of *Neisseria gonorrhoeae* to vaginal epithelial cells during the menstrual cycle and early pregnancy. Med Microbiol Immunol 167:231-238

Heckels JE, Blackett B, Everson JS, Ward ME 1976 The influence of surface charge in the attachment of *Neisseria gonorrhoeae* to human cells. J Gen Microbiol 96:359-364

James AN, Knox JM, Williams RP 1976 Attachment of gonococci to sperm. Influence of physical and chemical factors. Br J Vener Dis 52:128-135

Johnson AP, Taylor-Robinson D, McGee ZA 1977 Species specificity of attachment and damage to oviduct mucosa by *Neisseria gonorrhoeae.* Infect Immun 18:833-839

Kellogg DS Jr, Peacock WL, Deacon WE, Brown L, Prikle CI 1963 *Neisseria gonorrhoeae.* I. Virulence genetically linked to clonal variation. J Bacteriol 85:1274-1279

Lambden PP, Robertson JN, Watt PJ 1980 Biological properties of two distinct pilus types produced by isogenic variants of *Neisseria gonorrhoeae* P 9. J Bacteriol 141:393-396

Mårdh PA, Weström L 1976 Adherence of bacteria to vaginal epithelial cells. Infect Immun 13:661-666

Pearce WA, Buchanan TA 1978 Attachment role of gonococcal pili. J Clin Invest 61:931-943

Swanson J 1973 Studies on gonococcus infection. IV. Pili: Their role in attachment of gonococci to tissue culture cells. J Exp Med 137:571-589

Swanson J 1977 Surface components associated with gonococcal-cell interactions. In: Roberts R (ed) The gonococcus. Wiley, New York, p 369-399

Tramont EC 1976 Specificity of inhibition of epithelial cell adhesion of *Neisseria gonorrhoeae.* Infect Immun 14:593-595

Tramont EC 1977 Inhibition of adherence of *Neisseria gonorrhoeae* by human genital secretions. J Clin Invest 59:117-124

Tramont EC, Hodge WC, Gilbreath MJ, Ciak J 1979 Differences in attachment antigens of gonococci in reinfection. J Clin Lab Med 93:730-735

Tramont EC, Ciak J, Boslego J, McChesney DG, Brinton CC, Zollinger W 1980 Antigenic specificity of a genital antibody response to the infecting strain of *Neisseria gonorrhoeae.* J Infect Dis 142:23-31

Tubbett GM, Veale DR, Hutchison JGP, Smith H 1976 The adherence of pilate and non-pilate strains of *Neisseria gonorrhoeae* to human and guinea-pig epithelial tissues. J Med Microbiol 9:263-273

Watt PJ, Ward ME, Heckels JE, Trust TJ 1978 Surface properties of *Neisseria gonorrhoeae:* attachment to and invasion of mucosal surfaces. In: Brook GF et al (eds) Immunobiology of *Neisseria gonorrhoeae.* American Society for Microbiology, Washington DC, p 253-257

DISCUSSION

Taylor-Robinson: An effective gonococcal vaccine is not going to be easy to develop, because of the antigenic heterogeneity of the pili and because non-piliated strains stick also.

Tramont: The heterogenicity of pili may be a problem in gonococcal pilus vaccine development. However, there is evidence that the serological differences we observe between pili may not be as important in terms of attachment. Indeed, I have presented evidence that antisera to one pilus blocks the attachment of heterologous pili. Non-piliated strains do stick. This may be related to other surface antigens, such as lipopolysaccharide, or it may be a quantitative problem. We have never been able to culture gonococcal organisms in which we cannot detect pilus antigens by immunological techniques. We cannot demonstrate pili usually, but they still are capable of binding, albeit less, to specific anti-pilus antibody. Theoretically, pilus antibody will block attachment of so-called 'non-piliated organisms' as well as piliated organisms. However, I feel that other cell wall attachment antigens are also present, and that a vaccine designed to raise antibodies to block attachment may have to include other attachment antigens besides pili.

Elbein: I was curious about the big variation in binding with different buccal cells. One possibility is that there are things like mucins on the surface of these cells which could be different from one group of people to another. The other possibility is that protease or protease inhibitors in saliva and other secretions may change in different conditions.

Tramont: There is good evidence that the change in the buccal cells is a loss of fibronectin.

Elbein: But the point is that the fibronectin could change as a result of protease activity or some other change which could remove it.

Tramont: What influences those changes? You can artificially change them but they also change in a normal healthy individual. That may have some importance for who is susceptible to infection and when. Dr Sussman earlier mentioned his student who mapped attachment to her cells and found that attachment increased when she was symptomatic.

Sussman: It wasn't really the attachment that was increased but the MS pilus state that normally appeared on her bacteria was not found. Normally there were Gram-positive organisms in the urethral flora and sometimes MS *E. coli*. When she became symptomatic Gram-negative organisms appeared which were MR rather than MS.

Richmond: If the number of bacteria adhering to buccal cells is plotted, certain strains from a given buccal smear give two peaks, implying that there are two categories of cells present. With other strains those differences can't be distinguished. Also, in a buccal cell preparation from any one person the proportion of living cells varies from day to day, giving another source of differences.

Beachey: We also have seen variations in adherence to epithelial cells from donor to donor and from day to day in our studies with streptococci and *E. coli.* To overcome this problem we now pool cells from three or more individuals for each adherence test. Have you tried that as a solution to the problem of epithelial cell variability?

Tramont: That is why I began by referring to the ratio of organisms to buccal cells that we work with. This is particularly critical for demonstrating antibody differences. Even working with our very sensitive test system we still have low titres. Because we are limited in the total number of organisms that we can put into our test system, we use only one donor. But we screen the donors for their ability to support attachment of the host strain. For example, my cells might support attachment of bug 120 while my technician's cells would support bug 180.

Beachey: Every time we have looked at the viability of oral epithelial cells from adults, at least 96 to 99% have been dead by the trypan blue exclusion test. In newborn infants within two hours after birth we often see viability of up to 50%. In some experiments, we counted the number of organisms stuck to trypan blue-excluding cells as compared to the number stuck to stained cells and found no real difference in adhering abilities between the viable and the dead cells, with either *E. coli* or streptococci. The problem with the newborn cells is that their adhering ability is considerably lower than that of adult cells for the first 48 h after birth (Ofek et al 1977), and, therefore, we were really comparing low adhering abilities of dead and live cells.

Richmond: Can you plot the distributions of cells per nucleus, for example?

Beachey: We haven't done that.

Sharon: You sounded quite pessimistic about the prospects of getting purified pili uncontaminated with other materials, especially if the pili are very sticky. But I understand that *Neisseria gonorrhoeae* pili also consist of subunits of about 2 000 molecular mass. Has anybody tried dissociating these pili into their subunits, so that they are not denatured? This of course gives a protein of relatively low molecular mass which is much easier to purify and then to reassemble. Such an approach may eventually give you highly purified pili.

Tramont: That has been done. The critical question is whether they stimulate effective antibodies. If you look at these pili with a microscope you will see that they no longer aggregate as pure crystals. This may be important from the functional standpoint.

Sharon: Maybe you have not found the right conditions yet.

Tramont: Purification of pili that can be given as a vaccine is a major technical problem. Vast quantities can be lyophilized but when this preparation was reassembled it had no pilus crystals. This makes a difference in antibody titres but whether it makes any functional difference is unknown at this time.

Sharon: Has this been published?

Tramont: No.

Taylor-Robinson: Why do the pili to be used as the vaccine have to be so pure? Why can't there be a little impurity in the vaccine?

Tramont: It probably is good if it is impure but if you give a vaccine which has lots of pili and a little bit of LPS, the question comes up of which antigen is giving the protection. If it is LPS why bother with the pili? Making lots of LPS is a lot easier than making pili.

Levine: But that is a big difference. If you put LPS into someone's arm it is a red-hot arm. If you put in pili with trace doses of LPS then it is acceptable.

Tramont: If small amounts of LPS with contaminated pili are protective you can give small amounts of LPS as a vaccine without pili too.

Taylor-Robinson: So from a scientific point of view you need pure pili so that you can find out what is stimulating protection, but from a practical point of view you may not mind if there is some impurity?

Levine: That is right. Our studies, for example, have shown that with a lower dosage of the pilus protein you don't get the O antibody response; it is only when you put in enormous protein doses that each antigen apparently acts as an adjuvant for the other. There may be nothing wrong with having some impurity.

Sharon: Must it be pilus protein or can you put in LPS with some other protein?

Levine: If you think pilus is the important antigen, you just have to live with a little bit of contaminating O.

Sharon: From the practical point of view, can you immunize effectively with small amounts of LPS together with some pilus protein?

Tramont: I doubt it. If we absorb vaccine sera with purified LPS, it still blocks attachment at the very same level. But if we absorb out all of the pilus antibody then it doesn't block attachment any longer.

Mirelman: How about producing monoclonal antibodies with an antigen that is not very pure?

Tramont: That is now being done.

Levine: One of the questions we tend to ask about new pathogens and disease is whether there is disease-induced immunity. If there is, it is nice to know that mother nature can stimulate protection. Kearns et al (1973) looked at acute gonococcal and multiple gonococcal infections and found that people who had repeated bouts had the highest levels of local and circulating antibody, suggesting that multiple previous infections did not protect. If that is true, what do you expect the vaccine to do that mother nature doesn't do? Is it quantity of antibody, or quality of antibody, or the site of action?

Tramont: The critical question to ask is the level of antibody at the time the patient is exposed. That study hasn't been done. Furthermore, it is hard to do it because it is difficult to get antibody levels before someone gets the disease. It is also difficult to examine infected individuals since local antibody usually falls to low levels within a month of infection. Also, the efficiency of that antibody is not very great. For example, if it takes 1000 molecules to block the attachment of 100 gonococci and you only have 500 because your titre is waning, you are going to be in trouble. Whereas if you had that many antibody molecules against a meningococcus in your circulation, you would probably be able to kill a million injected organisms since the efficiency is much greater because of the amplifying effects of the complement system. Thus the critical questions of quantity of antibody and of how long it lasts must be addressed. A possible solution would be a local vaccine put on a tampon, with reimmunization of women every month at the time of menses. An advantage of dealing with sexually transmitted diseases is that protecting half of the population will protect the other half!

Choppin: In terms of the adherence of bacteria with and without pili, you alluded to the possibility that the protein is there even though you don't see the pilus.

Will those organisms grow a pilus in an hour in the presence of an inhibitor of protein synthesis? If so, is it just a polymerization problem?

Tramont: No, they won't grow in the presence of, say, sublethal doses of chloramphenicol.

Vosbeck: You screened a large number of gonococcal strains for adhesion to tissue culture cells. Do they all stick?

Tramont: Some strains stick better to certain tissue culture lines but the variation is not as great as with buccal cells. Of all the tissue culture cells the Flow-1000 cells support attachment best.

Vosbeck: So you have an ideal tissue culture system and you could abandon buccal cells.

Svanborg Edén: We have compared the adhesion of *Proteus mirabilis* to urinary tract epithelial cells with the adhesion of *E. coli.* We see a very marked variation with the menstrual cycle in the binding of *Proteus* strains which we do not see with *E. coli* strains. The *Proteus* strains bind only to the squamous epithelium. It is known that glycolipids mature in the epithelium, depending on the hormonal status.

Feizi: I have no information on changes in cell surface glycolipids during the menstrual cycle, but it is well known that changes do occur in the expression of blood group A and B antigens in epithelial cells in the genital tract. For example, the expression of blood group antigens in epithelial cells of the endometrium reach their height at the late proliferative phase and fade out rapidly during the early secretory phase (Szulman 1960). These antigens were visualized by immunofluorescence in cryostat sections of tissues and were presumably on glycoproteins. It is possible that similar changes occur on glycosphingolipids on the surface of these cells.

Mirelman: Are the *P. mirabilis* strains inhibited by globosides?

Svanborg Edén: None of those tested are.

Choppin: The cells that I mentioned before, the MDBK line of bovine kidney cells, have a very high concentration of globoside. As they become confluent and age, the concentration of globoside goes up; after cell division it goes down. In fact the relative amount of ceramide to globoside decreases with age. In other words the older the cell gets the longer its glycolipid oligosaccharide chains grow. This is a rather striking phenomenon.

REFERENCES

Kearns DH, Steibert GB, O'Reilly R, Lee L, Logan L 1973 Paradox of the immune response to uncomplicated gonococcal urethritis. N Engl J Med 289:1170-1174

Ofek I, Beachey EH, Eyal F, Morrison JC 1977 The postnatal development of binding of streptococci and lipoteichoic acid by oral mucosal cells of man. J Infect Dis 135:267-274

Szulman AE 1960 The histological distribution of blood group substances A and B in man. J Exp Med 111:785-807

Invasion of erythrocytes by malaria merozoites: evidence for specific receptors involved in attachment and entry

RUSSELL J. HOWARD and LOUIS H. MILLER

The Malaria Section, Laboratory of Parasitic Diseases, National Institute of Allergy and Infectious Diseases, The National Institutes of Health, Bethesda, Maryland 20205, USA

Abstract Invasion of erythrocytes by malaria merozoites involves attachment of the merozoite membrane at the point of collision with the erythrocyte, then reorientation of the merozoite such that its apex is opposed to the erythrocyte membrane, followed by invagination of the erythrocyte membrane and interiorization of the parasite. Specific recognition of erythrocyte surface components by the merozoite has been shown by studies on the specificity of merozoites of different malaria species for a limited range of host-species erythrocytes, for erythrocytes of particular maturity, and for erythrocytes possessing particular blood-group determinants. Different malaria species capable of invading erythrocytes of the same host also exhibit differences in specificity for components on enzyme-treated erythrocytes. The attachment phase of merozoite invasion has been isolated from subsequent steps by treatment of merozoites with cytochalasin B – they then attach to but do not invade susceptible erythrocytes. There is now evidence for other differences between initial attachment steps and subsequent invasion steps from studies on merozoite treatments *in vitro* which affect invasion but not attachment. It has also been shown that addition of *N*-acetyl-D-glucosamine to cultures of *Plasmodium falciparum* inhibits merozoite invasion. Elucidation of the sequence and nature of molecular interactions of merozoite and erythrocyte membrane molecules during invasion will be based on the fundamental ultrastructural observations and on the specificity of attachment and invasion steps already described.

This review will describe our current understanding of the invasion of erythrocytes by the malaria parasite or merozoite. Ultimately we wish to understand the molecular mechanism of the invasion process, but studies at a molecular level are in their infancy. Ultrastructural studies on the attachment and invasion of erythrocytes by malaria merozoites have revealed several stages in the merozoite–erythrocyte interaction. Other experiments, performed largely *in vitro*, have convincingly demonstrated that merozoite attachment and invasion involves the specific recognition of

---

1981 Adhesion and microorganism pathogenicity. Pitman Medical, Tunbridge Wells (Ciba Foundation symposium 80) p 202-219

molecules on the erythrocyte membrane. Considered together, these approaches suggest some of the limiting properties of the receptor molecules involved. These properties will be useful for identifying putative receptor molecules in future biochemical studies.

A description of the invasion process

The clinical symptoms of malaria, such as anaemia, fever and capillary obstruction, are related to the phase of asexual parasite multiplication within erythrocytes. Attachment of the invasive form of the malaria parasite, the merozoite, to erythrocytes and the subsequent invasion of these cells are essential steps in the proliferation of the blood parasites. After invasion the merozoite rapidly differentiates, grows in size, and eventually undergoes repeated division (schizogony) to produce up to 20 daughter parasites within the infected cell. The infected cell (schizont) is ruptured and the released merozoites infect additional red cells. The lifetime of extracellular merozoites is less than 30 minutes (Dennis et al 1975, Johnson et al 1980).

Merozoite morphology

Comparative studies of the ultrastructure of merozoites and exclusively intra-erythrocytic forms of the malaria parasite (reviewed by Aikawa 1971) have illuminated several unique morphological features of the merozoite. These can be assumed to relate to the unique properties of merozoites, namely their release from the infected cell, their ability to survive brief extracellular exposure, and their ability to attach to and invade host erythrocytes (Dennis et al 1975). At the merozoite apex is a complex of organelles called the apical complex which includes a pair of electron-dense organelles (the rhoptries), smaller organelles of similar density (the micronemes), and three polar rings which circumscribe the apex. A complex multi-lamellar series of membranes (pellicle membranes) underneath the outer merozoite membrane is also unique to merozoites, as is a diffuse layer of material on the merozoite surface (the cell coat) which has been shown to be associated with parasite antigens (Miller et al 1975a). All invasive forms of members of the subphylum Apicomplexa possess similar organelles at their apex and invade host cells with the apex against the host cell membrane, suggesting the importance of these organelles in attachment to host cells and invasion.

The sequential steps in merozoite invasion: microscopic studies

The sequence of events in merozoite attachment and invasion has been studied by light microscopy and video photography (Dvorak et al 1975) and at the ultrastructural level by electron microscopy (Ladda 1969, Ladda et al 1969, Bannister et al 1975, Aikawa et al 1978) and can be summarized as follows (1) The merozoite attaches to the host cell surface at the point of collision on the parasite's exterior ('initial attachment'). (2) Within seconds the parasite reorients such that its apical complex is directly opposed to the red cell surface ('reorientation and apical attachment'). The parasite's external membrane appears to remain in contact with the erythrocyte membrane during reorientation. (3) After 'apical attachment' there is a brief (5–10 seconds) but extensive deformation of the erythrocyte involving distortion to a spherical shape, then return to the original biconcave disc morphology ('erythrocyte deformation'). (4) An intimate junction of the merozoite membrane at the apical complex and the erythrocyte membrane has been described by electron microscopy (Aikawa et al 1978). It is not known whether this step ('apical complex–erythrocyte junction formation') precedes or follows erythrocyte deformation. On the cytoplasmic side of the erythrocyte membrane opposite the merozoite's apex a thickening of electron-dense material is evident. The erythrocyte membrane is slightly deformed at this point to form a concave depression shaped around the merozoite apex. At this time, several elongated membrane-bound vacuoles are seen within the erythrocyte's cytoplasm in the region of the erythrocyte–merozoite junction. (5) At the point of cell contact, an extensive invagination of the erythrocyte membrane develops. The merozoite remains associated with the erythrocyte membrane during this invagination, resulting in sheathing of the merozoite in the deepening cleft of the host cell ('invagination of the host cell'). This process takes 10–20 seconds (Dvorak et al 1975). Ultrastructural studies showed that the initial zone of intimate membrane contact at the merozoite's apex appears to move as a band at the orifice of the invagination (Aikawa et al 1978). Behind the band of intimate merozoite–erythrocyte membrane interaction (i.e. within the invagination) the two membranes are not so tightly joined and the electron-dense thickening under the erythrocyte membrane is absent. Freeze–fracture studies have shown that the erythrocyte membrane is devoid of intramembranous particles within the invagination containing the merozoite (McLaren et al 1979). There is a sharp demarcation between the intramembranous particles on the erythrocyte membrane outside the invagination and that within (M. Aikawa & L. H. Miller, unpublished data). The band of intimate membrane contact is possibly located at the sharp demarcation of intramembranous particles. During internalization the outer diffuse cell coat on the merozoite is stripped from its surface (Ladda et al 1969, Bannister et al 1975, Aikawa et al 1978). (6) Finally, the parasite is completely enclosed within membrane contiguous with the erythrocyte membrane by fusion of the invaginated membrane at the posterior end of the parasite.

The ultrastructural studies showed that there are distinct stages in the merozoite–erythrocyte interaction. Different types of molecular interaction are suggested by this sequence. The complexity and rapidity of merozoite invasion highlight the importance of several experimental means (described below) which allow only partial completion of invasion and thereby provide systems for study of the sequence of molecular interactions.

Specificity in invasion

The most illuminating property of the merozoite–erythrocyte interaction with respect to the likely molecular mechanisms involved is its specificity. This specificity can be described at several levels. The first level is *specificity for the host cell.* Merozoites of *Plasmodium falciparum* invade human erythrocytes but cannot invade erythrocytes of the rhesus monkey *Macaca mulatta* (Haynes et al 1976). The *in vivo* infectivity of the simian malaria *P. knowlesi* for various host species is related to the ability of *P. knowlesi* merozoites to invade erythrocytes of these species *in vitro* (Butcher et al 1973, Miller et al 1973). These results indicate that differences in the surface properties of erythrocytes from different species form the basis for specific recognition and invasion by the merozoite.

Differential susceptibility of erythrocytes to invasion by the merozoite is not necessarily an all-or-none phenomenon, such as in the case referred to above. It has been shown that a particular species of malaria parasite may under identical conditions quantitatively invade more erythrocytes of one susceptible host species than of another susceptible host species. McGhee (1953) demonstrated that merozoites of *P. lophurae* invaded more duck erythrocytes than chick erythrocytes when exposed to both in chick embryo. Rhesus monkey erythrocytes are more susceptible than human erythrocytes to invasion by *P. knowlesi* merozoites *in vitro* (Butcher et al 1973, Miller et al 1973). The chance of a collision between merozoites and erythrocytes leading to invasion clearly differs for different species of host erythrocyte. For *P. knowlesi* merozoites with human erythrocytes *in vitro*, it was shown that some merozoites even attach to and deform erythrocytes before detaching and going on to invade another erythrocyte successfully (L. H. Miller, unpublished observations). The reasons for this quantitative difference in invasion of different susceptible host cells by a particular species of merozoite are not known.

An additional level of specificity of merozoites for the host cell concerns the predilection of parasites of different malaria species for host erythrocytes of a particular maturity – *specificity for either immature reticulocytes or mature erythrocytes.* There are examples of absolute specificity and relative specificity at this level, just as there are for the invasion of erythrocytes from different hosts described above. *P. vivax* displays a very strong preference for human reticulocytes, almost never being seen within a more mature erythrocyte. *P. falciparum,* on the other hand, also preferentially invades reticulocytes but invades as well a significant proportion of mature cells. During the first one or two days of infection of mice with virulent or avirulent strains of *P. yoelii* the parasites show a marked preference for invasion of reticulocytes. After four days, however, the virulent strain is distinguished by its capacity to invade mature erythrocytes as well as reticulocytes, whereas the avirulent strain is restricted to reticulocytes (R. Carter, unpublished observations). It appears that the avirulent strain of *P. yoelii* is restricted absolutely to invasion of reticulocytes. The virulent strain can invade immature and mature erythrocytes but probably requires many more collisions with mature cells for successful invasion.

Another level of merozoite specificity for host erythrocytes is shown by differences in the invasion of erythrocytes from the same host species but *differing in blood group phenotype.* Differences in blood group antigens on human erythrocytes of the ABH, MN, Ss series, plus other major antigens, have no effect on the invasion of *P. knowlesi* merozoites (Miller et al 1975b). However, the Duffy blood group system has been shown to be associated with the capacity of merozoites of *P. knowlesi* and *P. vivax* to attach to and invade human erythrocytes. Before these results are described a brief introduction to the Duffy blood group system is appropriate.

The Duffy antigen and merozoite invasion

There are three codominant allelic forms of the gene coding for Duffy antigen, called Fy^a, Fy^b and *Fy*. The presence of the gene products $\mathrm{Fy^a}$ and $\mathrm{Fy^b}$ on erythrocytes is indicated by agglutination with specific antiserum (anti-$\mathrm{Fy^a}$ and anti-$\mathrm{Fy^b}$, respectively). Erythrocytes that fail to be agglutinated by either antiserum are homozygous for a third allele, *Fy,* and are called Duffy-negative. Another antiserum, anti-$\mathrm{Fy^3}$, reacts with erythrocytes carrying $\mathrm{Fy^a}$ or $\mathrm{Fy^b}$ but not Duffy-negative erythrocytes (FyFy).

It has long been known that a high percentage of African and American blacks are completely resistant to mosquito-induced or blood-induced *P. vivax* infection (Boyd & Stratman-Thomas 1933, Bray 1958, Young et al 1955). For 20 years it has also been known that erythrocytes from the majority of African and American blacks are Duffy-negative (Sanger et al 1955). The Duffy-negative genotype is extremely rare in other racial groups without black admixture (Mourant et al 1976). A direct correlation between refractoriness to *P. vivax* and lack of the Duffy determinants $\mathrm{Fy^a}$ and $\mathrm{Fy^b}$ was established by blood typing black and white volunteers who had been exposed to the bites of *P. vivax*-infected mosquitoes (Miller et al 1976). The blacks with FyFy genotype were resistant to infection, whereas the remaining blacks and all whites with Duffy-positive determinants had contracted malaria.

In vitro experiments with the simian parasite *P. knowlesi,* which also infects human erythrocytes, showed a similar requirement for $\mathrm{Fy^a}$ or $\mathrm{Fy^b}$ in order for these merozoites to invade (Miller et al 1975b). *P. knowlesi* merozoites do, however, attach to Duffy-negative human erythrocytes before detaching (Miller et al 1975b). They can briefly attach to several Duffy-negative erythrocytes but fail to continue with subsequent steps in the invasion sequence. Pretreatment of susceptible Fy(a+b–) human erythrocytes with anti-Fy^a markedly reduced invasion by *P. knowlesi* (Miller et al 1975b). Inhibition of invasion of susceptible Fy(a–b+) human erythrocytes by *P. knowlesi* with anti-Fy^b serum (54–61% inhibition in three experiments) has also been demonstrated (L. H. Miller, unpublished work). Because of the lower titre of anti-Fy^b serum than of anti-Fy^a serum it was necessary to pretreat the Fy(a–b+) erythrocytes with undiluted anti-Fy^b and maintain 10% anti-Fy^b in the incubation medium during the invasion assay in order to demonstrate this inhibition. Parallel treatments of Fy(a+b–) erythrocytes with anti-Fy^b did not reduce invasion. These results indicate that a receptor for *P. knowlesi* on human erythrocytes is closely associated physically with the Duffy antigen on the red cell membrane. Evidence for a close genetic association of a receptor with the Duffy antigen came from studies with two Duffy-negative individuals from non-black populations (one a white Australian, the other an American Indian) whose erythrocytes were also refractory to invasion by *P. knowlesi* (Mason et al 1977). Despite this compelling evidence for a requirement for Fy^a or Fy^b for invasion of human erythrocytes by *P. knowlesi* or *P. vivax* it is premature to state that the Duffy antigenic determinant is a receptor on the erythrocyte surface, in view of the demonstration that Duffy-negative human erythrocytes were rendered susceptible to *P. knowlesi* by treatment with trypsin or neuraminidase (Mason et al 1977). These treatments did not generate a serologically detectable Fy^a or Fy^b determinant. There is therefore an urgent need for molecular identification of the Fy^a, Fy^b and Fy determinants and isolation of the molecules which carry these determinants.

All the above studies describe the specificity of merozoites for different host erythrocytes: erythrocytes from different host species; erythrocytes of different maturity; and erythrocytes with different blood group phenotypes. They suggest the presence of specific receptor molecules on the parasite for components of the host erythrocyte. Another level of specificity in merozoite–erythrocyte interaction is evident from comparison of the *specificities of different species of malaria parasite for erythrocytes of one type.* These studies suggest differences in the merozoite receptors of closely related malaria species.

P. vivax and *P. knowlesi* cannot invade Duffy-negative human erythrocytes. In contrast, *P. falciparum* appears to invade Duffy-negative erythrocytes and cells with Fy^a or Fy^b determinants equally (Miller et al 1977).

Experiments in which susceptible erythrocytes were treated with various enzymes before incubation *in vitro* with parasites also indicated differences in the requirements for invasion by different malaria species. Chymotrypsin treatment of Duffy-positive human erythrocytes renders them resistant to invasion by *P. knowlesi,* while the invasion of *P. falciparum* merozoites was unaffected by chymotrypsin treatment. Conversely, trypsin or trypsin plus neuraminidase markedly reduced infection of human erythrocytes by *P. falciparum* but caused little or no change in susceptibility to *P. knowlesi* (Miller et al 1977).

Experimental isolation of the attachment phase of invasion

Isolation of particular phases of the invasion sequence will be a prerequisite for understanding the temporal sequence of molecular interactions. So far only one phase has been isolated. The attachment phase has been isolated from the subsequent interactions in studies with cytochalasin B-treated *P. knowlesi* merozoites which attach to, but do not invade, rhesus monkey erythrocytes (Miller et al 1979). The detailed mechanism of action of cytochalasin B on merozoites (possibly through multiple effects) is not of interest when it is used operationally to study factors which affect merozoite attachment.

Cytochalasin-treated merozoites only attach to erythrocytes from species that are susceptible to invasion. Assay of the relative numbers of treated *P. knowlesi* merozoites which attached to rhesus monkey and human erythrocytes revealed a quantitative difference in attachment affinity. Attachment to rhesus monkey erythrocytes was greater than to human erythrocytes, corresponding to the relative invasion rates of these erythrocytes. It was also shown that several manipulations of merozoites *in vitro* which markedly reduced invasion had no effect on attachment (for example, incubation of merozoites at 4 °C or at pH 7.9) (Johnson et al 1980). Electron microscope studies showed that the apical end of cytochalasin-treated merozoites was usually against the erythrocyte membrane. A thickening of electron-dense material under the erythrocyte membrane at the point of attachment was observed with cytochalasin-treated merozoites and rhesus monkey erythrocytes, just as has been described for one phase of the normal invasion sequence ('apical complex–red cell junction formation'). Vacuoles bounded by a unit membrane were also seen in the erythrocyte cytoplasm near the attachment site (Aikawa et al 1978). Cytochalasin treatment causes a block in subsequent steps as invagination of the erythrocyte membrane does not proceed beyond formation of a small concavity in the membrane opposite the merozoite apex (Miller et al 1979).

Attachment of cytochalasin B-treated *P. knowlesi* merozoites has been shown to be identical to both Duffy-positive and Duffy-negative erythrocytes, despite the inability of *P. knowlesi* merozoites to invade Duffy-negative erythrocytes (Miller et al 1979). However, the mode of attachment was quite distinct. Cytochalasin B-treated merozoites formed a junction with Duffy-positive cells, just as described with rhesus monkey cells, whereas no junction was observed with Duffy-negative cells. Instead, the merozoites were attached by a cylinder of filaments between the apical complex and the erythrocyte (Miller et al 1979).

Inhibition of merozoite invasion

It may prove possible in the future to identify which molecules are involved in merozoite invasion if soluble fragments of erythrocyte or merozoite surface components can be shown to compete with the receptor molecules on the interacting cells and reduce the level of attachment or invasion, or both. Monoclonal antibodies against merozoite or erythrocyte surface components may also prove useful probes of the interacting molecules. There are already encouraging results indicating that these approaches may be successful.

Synchronous cultures of *P. falciparum* in human erythrocytes have been used to assay the reinvasion of merozoites in the presence of a variety of substances (Weiss et al 1980). Several monosaccharides showed no effect on merozoite invasion at 100 mM concentration. *N*-Acetyl-D-glucosamine almost completely blocked reinvasion at levels above 25 mM and did not appear to exert this effect through toxicity to intracellular parasite development. This study suggests that carbohydrates containing *N*-acetyl-D-glucosamine may be involved in the invasion of human erythrocytes by *P. falciparum.* In contrast, studies on the invasion of purified *P. knowlesi* merozoites into rhesus monkey erythrocytes failed to show any inhibitory action of a range of sugars including *N*-acetyl-D-glucosamine (Johnson et al 1980). This difference may reflect the different specificities of *P. falciparum* and *P. knowlesi* merozoites for human and rhesus erythrocytes, respectively, or the differences in merozoite invasion assay used.

Preliminary experiments on the effects of added protease inhibitors on the invasion of rhesus erythrocytes by *P. knowlesi* merozoites (J. G. Johnson & L. H. Miller, unpublished results) indicated that 50 μM concentrations of TPCK and TLCK markedly reduced invasion. Other protease inhibitors tested had no effect (Table 1). Although these results suggest that proteases may be involved in merozoite invasion, we still cannot exclude a non-specific inhibitory effect of TPCK and TLCK on invasion or on some process internal to the parasite and unrelated to the mechanism of invasion.

The potential of monoclonal antibodies of highly specific inhibitors of merozoite invasion, and thereby as probes of antigenic determinants important in the invasion mechanism, has recently been demonstrated. Passive transfer of monoclonal antibodies specific for antigens exclusive to *P. yoelii* merozoites protected mice against this rodent malaria (Freeman et al 1980). It was suggested that these monoclonal antibodies directed against merozoite surface components might block merozoite receptor determinants for erythrocytes, and thereby block infection. It will be of great interest to learn the nature of the merozoite surface antigens recognized by these monoclonal antibodies.

Invasion of rhesus erythrocytes by purified *P. knowlesi* merozoites has been shown to be inhibited by treating the merozoites with trypsin (1 μg/ml) for 10 minutes at 23 °C (Johnson et al 1980). Extension of these studies may allow the

TABLE 1 Effect of protease inhibitors on invasion of rhesus erythrocytes by Plasmodium knowlesi merozoites

Enzyme inhibitor[a]	*Solvent*	*Amount added (μl)*	*Final inhibitor concentration*	*% invasion*[b]
STI	Water	20	0.1 mg/ml	23.8
None	Water	20	0	35.0
PMSF	Acetone	5	50 μM	30.0
TPCK	Acetone	5	50 μM	0
TPCK	Acetone	5	5 μM	18.0
None	Acetone	5	0	3.8
TLCK	50 mM-Potassium phosphate, pH 5.0	5	50 μM	< 0.1
TLCK	50 mM-Potassium phosphate, pH 5.0	5	5 μM	26.0
None	50 mM-Potassium phosphate, pH 5.0	5	0	24.8

[a]STI, soybean trypsin inhibitor; PMSF, phenylmethylsulphonyl fluoride; TPCK, L-1-tosylamide-2-phenylethyl chloromythyl ketone; TLCK, *N*-α-*p*-tosyl-L-lysine chloromethyl ketone HCl.

[b]Viable merozoites (1 ml) were incubated with various protease inhibitors or controls containing the same solvent for 2 min at 23 °C before adding the test rhesus erythrocytes for invasion at 37 °C, as described in detail elsewhere (Johnson et al 1980).

preparation of relatively large proteolytic fragments of merozoite surface molecules which retain receptor specificity for erythrocyte components.

The mechanism of merozoite invasion

None of the molecules on the merozoite and red cell surface which interact during attachment and invasion have been identified. It is therefore only possible for us to point out some of the main questions which require a molecular explanation and, on the basis of the results summarized above, suggest some of the properties of the molecules involved.

Attachment and reorientation

As the merozoite appears to be able to attach specifically to susceptible erythrocytes by any part of its surface, the receptor(s) for initial attachment must be distributed over the entire merozoite. How does reorientation occur such that the apical complex becomes attached to the red cell membrane? Do the same parasite and red cell molecules remain associated as the parasite moves during reorientation, or is there continuous association and disassociation of interacting molecules which keeps the merozoite and host cell together yet allows for movement of one membrane over another? Are submembrane contractile elements of the merozoite involved in reorientation? What directs the merozoite to attach by its apex after an initial attachment by another part of its surface has been made? Perhaps the number or affinity of receptors at the apical attachment site is different from (greater than?) that over the rest of the merozoite's surface. We would predict that at least one specific molecular recognition of erythrocyte surface molecules by a merozoite surface component occurs during this phase.

Apical attachment and junction formation

When cytochalasin B-treated *P. knowlesi* merozoites attach to Duffy-negative human erythrocytes, fibrous filaments are seen by electron microscopy between the merozoite apex and the erythrocyte. An intimate junction of the two membranes is not formed. Cytochalasin B-treated merozoites do form an intimate membrane junction with Duffy-positive human erythrocytes, suggesting that the Duffy determinants are involved in this step.

Comparative studies of the molecular events which accompany attachment of cytochalasin B-treated merozoites with Duffy-positive versus Duffy-negative erythrocytes will give us insight into the mechanism of junction formation. This phase of merozoite invasion also appears to involve specific molecular recognition of erythrocyte components by the merozoite.

Invagination of the red cell membrane

The invagination step is thought to involve the insertion of merozoite rhoptry material into the erythrocyte membrane. Some electron micrographs show decreased electron density in this organelle during invasion (Aikawa et al 1978), and there is an apparent continuity of electron-dense material between the rhoptry interior of *P. knowlesi* merozoites and the erythrocyte membrane at the site of invagination (Miller et al 1979).

The nature of rhoptry material is unknown. It has been suggested that this material includes lipid which can form membranes when released from the rhoptry into the erythrocyte membrane, thereby creating an invagination (Bannister et al 1977). It has also been suggested that a parasite protein of unusually high histidine content which was isolated from cytoplasmic granules in *P. lophurae* (Kilejian et al 1975) may also occur within rhoptry organelles and be related to erythrocyte membrane invagination (Kilejian 1976). Identification of the chemical nature of rhoptry material is urgently required.

Interiorization

Although there is good evidence for specificity in the initial attachment and apical attachment phases of invasion, there is no information on the question of molecular recognition events specific for a particular parasite–host cell interaction during the later phases of invagination and interiorization. It is possible that the mechanism of invagination of the host cell membrane and internalization is very similar if not identical for several malaria species. The invasion of erythrocytes of different host species would have required the evolution of receptors for specific recognition of the host cell in the early phases of erythrocyte–merozoite interaction during invasion. After the invasion sequence has passed a particular stage, the subsequent stages (which are presumably activated by preceding specific molecular interactions) may share common molecular mechanisms in a variety of merozoite-susceptible host cell combinations.

Summary

The invasion of erythrocytes by malaria merozoites is a rapid process that involves several stages that have been described at the ultrastructural level. It is suggested that the molecular interactions between merozoite and erythrocyte surfaces will alter during the different stages of invasion. At least some of the molecular interactions must involve specific recognition of components on the host cell surface by the merozoite. Merozoites of a particular malaria species can be shown to display specificity for invasion at several levels: specificity for a limited range of host species erythrocytes, specificity for host erythrocytes of a particular maturity, and specificity for erythrocytes which possess particular blood group determinants. Examples of absolute specificity and relative differences in erythrocyte specificity are known for the first two levels of specificity described above. Closely related species of malaria parasite capable of infecting erythrocytes of the same host also display differences in specificity. These differences are revealed after enzyme treatments of otherwise susceptible erythrocytes which affect the invasion of one species but not another. With one malaria parasite (*P. knowlesi*) it is possible to isolate the attachment stage from later stages of merozoite invasion by pretreating the merozoites with cytochalasin B. Systems of this type will be required in order to relate particular molecular interactions of the merozoite and erythrocyte to a particular stage in the invasion sequence. So far, none of the molecules involved either in specific recognition of the erythrocyte by the merozoite or in the later events of erythrocyte invagination and interiorization, which we suggest may be common to several malaria species, have been identified. The accumulated observations on merozoite invasion provide a framework of information from which we must derive important questions that can be tested at a molecular level.

REFERENCES

Aikawa M 1971 Fine structure of malaria parasites. Exp Parasitol 30:284-320

Aikawa M, Miller LH, Johnson JG, Rabbege J 1978 Erythrocyte entry by malarial parasites: a moving junction between erythrocyte and parasite. J Cell Biol 77:72-82

Bannister LH, Butcher GA, Dennis ED, Mitchell GH 1975 Structure and invasive behavior of *Plasmodium knowlesi* merozoites *in vitro*. Parasitology 71:483-491

Bannister LH, Butcher GA, Mitchell GH 1977 Recent advances in understanding the invasion of red cells by merozoites of *Plasmodium knowlesi*. Bull WHO 55:163-169

Boyd MF, Stratman-Thomas WK 1933 Studies on benign tertian malaria. 4. On the refractoriness of Negroes to inoculation with *Plasmodium vivax*. Am J Hyg 18:485-489

Bray RS 1958 The susceptibility of Liberians to the Madagascar strain of *Plasmodium vivax*. J Parasitol 44:371-373

Butcher GA, Mitchell GH, Cohen S 1973 Mechanism of host specificity in malarial infection. Nature (Lond) 244:40-42

Dennis ED, Mitchell GH, Butcher GA, Cohen S 1975 *In vitro* isolation of *Plasmodium knowlesi* merozoites using polycarbonate sieves. Parasitology 71:475-481

Dvorak JA, Miller LH, Whitehouse WC, Shiroishi T 1975 Invasion of erythrocytes by malaria merozoites. Science (Wash DC) 187:748-749

Freeman RR, Trejdosiewicz AJ, Cross GAM 1980 Protective monoclonal antibodies recognizing stage-specific merozoite antigens of a rodent malaria parasite. Nature (Lond) 284:366-368

Haynes JD, Diggs CL, Hines FA, Desjardins RE 1976 Culture of human malaria parasites, *Plasmodium falciparum*. Nature (Lond) 263:767-769

Johnson JG, Epstein NN, Shiroishi T, Miller LH 1980 Factors affecting the ability of isolated *P. knowlesi* merozoites to attach to and invade erythrocytes. Parasitology, in press

Kilejian A 1976 Studies on a histidine-rich protein from *Plasmodium lophurae*. In: van den Bossche H (ed) Biochemistry of parasites and host–parasite relationships. Elsevier/North-Holland, Amsterdam, p 441-448

Kilejian A, Liao T-H, Trager W 1975 Studies on the primary structure and biosynthesis of a histidine-rich polypeptide from the malaria parasite, *Plasmodium lophurae*. Proc Natl Acad Sci USA 72:3057-3059

Ladda RL 1969 New insights into the fine structure of rodent malaria parasites. Mil Med 134:825-865

Ladda RL, Aikawa M, Sprinz H 1969 Penetration of erythrocytes by merozoites of mammalian and avian malarial parasites. J Parasitol 55:633-644

Mason SJ, Miller LH, Shiroishi T, Dvorak JA, McGinniss MH 1977 The Duffy blood group determinants: their role in the susceptibility of human and animal erythrocytes to *Plasmodium knowlesi* malaria. Br J Haematol 36:327-335

McGhee RB 1953 The infection by *Plasmodium lophurae* of duck erythrocytes in the chicken embryo. J Exp Med 97:773-782

McLaren DJ, Bannister LH, Trigg PI, Butcher GA 1979 Freeze-fracture studies on the interaction between the malaria parasite and the host erythrocyte in *Plasmodium knowlesi* infections. Parasitology 79:125-139

Miller LH, Dvorak JA, Shiroishi T, Durocher JR 1973 Influence of erythrocyte membrane components on malaria merozoite invasion. J Exp Med 138:1597-1601

Miller LH, Aikawa M, Dvorak JA 1975a Malaria (*Plasmodium knowlesi*) merozoites: immunity and the surface coat. J Immunol 114:1237-1242

Miller LH, Mason SJ, Dvorak JA, McGinnis MH, Rothman IK 1975b Erythrocyte receptors for (*Plasmodium knowlesi*) malaria: the Duffy blood group determinants. Science (Wash DC) 189:561-563

Miller LH, Mason SJ, Clyde DF, McGinnis MH 1976 The resistance factor to *Plasmodium vivax* in blacks. The Duffy blood-group phenotype, *FYFY*. N Engl J Med 295:302-305

Miller LH, Haynes JD, McAuliffe FM, Shiroishi T, Durocher JR, McGinnis MH 1977 Evidence for differences in erythrocyte surface receptors for the malarial parasites, *Plasmodium falciparum* and *Plasmodium knowlesi*. J Exp Med 146:277-281

Miller LH, Aikawa M, Johnson JG, Shiroishi T 1979 Interaction between cytochalasin B-treated malarial parasites and erythrocytes. J Exp Med 149:172-184
Mourant AE, Kopec AC, Domaniewska-Sobczak K 1976 The distribution of the human blood groups and other polymorphisms, 2nd edn. Oxford University Press, London
Sanger R, Race RR, Jack J 1955 The Duffy blood groups of New York Negroes: the phenotype Fy (a–b–). Br J Haematol 1:370-374
Weiss MM, Oppenheim JD, Vanderberg JP 1980 *Plasmodium falciparum: in vitro* assay for substances that inhibit penetration of erythrocytes by merozoites. Exp Parasitol, in press
Young MD, Eyles DE, Burgess RW, Jeffrey GM 1955 Experimental testing of Negroes to *Plasmodium vivax.* J Parasitol 41:315-318

DISCUSSION

Bredt: Can you remove your attached merozoites by trypsin?

Howard: Trypsin-treated merozoites of *P. knowlesi* don't attach to rhesus monkey erythrocytes. Johnson et al (1980) showed that trypsin treatment of merozoites appears to cleave only three proteins from the set of merozoite proteins labelled by lactoperoxidase-catalysed radioiodination. It is possible that one of these proteins is involved in specific merozoite attachment, because after this trypsin treatment (1 μg/ml, 10 min at 23 °C) merozoite attachment to erythrocytes and invasive capacity are also lost.

Hughes: The junctional complex that you described is reminiscent of the type of adherence junction seen when mammalian cells adhere to one another and also when they stick to a solid substratum (for example, see Heath & Dunn 1978, Heaysman & Pegrum 1973). It is clear that this is a stabilization of the initial contact and involves a functional cytoskeleton, especially microfilaments. I suspect that this is also going to be important in the stable adhesion of bacteria and protozoa to mammalian cells. We have talked a lot about the initial recognition of bacteria and protozoa when they stick to mammalian cells. This interaction is probably sticking the cells together under conditions in which the adhesion assays are being done, that is in conditions of low stress. But *in vivo* it may be that the adhesion sites have to be stabilized first by multiplication of the interactions between an adherent microorganism and the host cell, followed by a re-ordering of the cytoskeleton of the latter to stabilize the contact. So your finding of a cytoplasmic specialization within the erythrocyte at the site of contact could be important quite generally.

Howard: The intimate membrane junction between merozoite and erythrocyte membranes and the electron-dense material under the erythrocyte membrane associated with the junction represent a dynamic interaction between these cells – the junction and electron-dense material move to the orifice of the erythrocyte membrane invagination as the merozoite enters. This dynamic aspect, and the events such as release of rhoptry material from the merozoite, swelling and contraction of the erythrocyte and erythrocyte membrane invagination which parallel junction formation, all suggest intercellular communciation and the sequential switching of different processes. The mechanisms of these changes which parallel junction formation are unknown – enzymic reactions may possibly be involved.

Hughes: There is a parallel here also in migrating fibroblasts, where adhesive plaques which are stabilized junctions make and break rapidly as the fibroblast moves over the cell surface (Abercrombie 1961). Even in mammalian cells it is a very dynamic structure and I think the parallels are really very close.

Taylor-Robinson: Is there a parallel with the manner of attachment of *Mycoplasma pneumoniae?*

Razin: There may be a parallel from the point of view of the parasites but from the point of view of the eukaryotic cells nothing like this is seen as there is no internalization of the parasite in the case of the attached mycoplasmas.

Hughes: Professor Richmond, you mentioned that epithelial cells send processes around bacteria adhering to their surface. That seems to indicate some active reorganization of microfilamentous contractile elements by the epithelial cells around the bacterium. Similar things may be happening in the interactions triggered by binding of malarial merozoites to the erythrocyte surface, but with spectrin possibly being substituted for actin filaments.

Richmond: Dr Svanborg Edén's photographs showed finger-like processes that had been sectioned through. In many of the interactions between merozoites and erythrocytes this same phenomenon may be taking place.

Hughes: Does anybody know the effects of cytochalasin B on adhesion in other systems, for example bacteria and epithelial cells?

Helenius: In macrophages cytochalasin B inhibits phagocytosis of bacteria and the membrane reorganization that goes with it.

Howard: The concentration of cytochalasin B used to block merozoite invasion was relatively high (10 μg/ml; Miller et al 1979), i.e. above the minimum level used generally to block contractile systems in other cells. This concentration was chosen operationally in a method designed to block merozoite invasion at the step of merozoite apical attachment to the erythrocyte and thereby provide an assay of attachment independently of invasion. In this assay the merozoites are briefly treated with cytochalasin B before the addition of erythrocytes.

Wallach: Processes known to produce endocytosis in erythrocytes involve segregation of the endoskeletal components, including spectrin, from their generally uniform distribution in the membrane plane. Spectrin and cytochalasin B do not interact. The principal interaction of cytochalasin B and red cell surface components is with the sugar transport system and we are not talking about that. What we appear to see in the endocytosis of the parasites is a lateral segregation, from the tip of the merozoite, of endoskeletal components. These may constitute the bands of the moving junction, i.e. the thickening on the inner surface of the invaginating surface membrane. This is feasible since endoskeletal components can associate and dissociate quite rapidly, depending not only on the spectrin molecule but also on associated proteins and their phosphorylation state. Among the associated proteins are ones that bind spectrin to the membrane, and these depend on the phosphorylation status of the entire cell. Perhaps what the merozoite does is to 'inject' an agent that interferes with the metabolic state of the red cell and allows depolymerization of the endoskeletal network.

Choppin: Does the band become part of the membrane of the merozoite rather than remaining on the membrane of the cell?

Howard: The electron-dense band is below the erythrocyte membrane. We have no definitive information yet on possible transfer of components from the erythrocyte membrane to the merozoite as you suggest, or on transfer from the merozoite to the erythrocyte membrane. There is speculation on the transfer of contents of the merozoite rhoptry organelles into the erythrocyte membrane during invagination of this membrane.

Elbein: Have you looked for reversible inhibitors of either protein synthesis or glycosylation and used these to see whether you could stop the process at some point and then reverse it by washing out?

Howard: Unfortunately there are some technical limitations which at present restrict experiments on the invasion of erythrocytes by biosynthetically labelled merozoites. Although *P. knowlesi* can be cultured in rhesus monkey erythrocytes *in vitro* sufficiently long for good incorporation of radiolabelled amino acids and sugars into protein/glycoprotein, merozoites derived from these cultures are not invasive when purified by the standard technique. The standard method for purification of *invasive* parasites involves a brief (less than 90 min) incubation *in vitro* of mature infected cells freshly derived from an infected monkey. During this time amino acid precursors are incorporated into the parasites but sugar incorporation into glycoprotein is very poor. Furthermore, J. G. Johnson, N. Epstein and L. H. Miller (unpublished work) showed that trypsin treatment of *P. knowlesi* merozoites which had been labelled with $[^3H]$isoleucine for 60 min within mature infected cells failed to cleave any of the radiolabelled merozoite proteins. In contrast, the same trypsin treatment cleaves some proteins labelled by lactoperoxidase-catalysed radioiodination of merozoites. It appears that proteins labelled by $[^3H]$isoleucine late in parasite maturation do not get incorporated into the merozoite surface membrane. The approach you have suggested, using inhibitors of protein synthesis to identify proteins important in invasion, could be attempted with synchronous cultures of the parasite. However the assay for reinvasion would then be less specific for invasion *per se:* namely, the appearance of parasites in new erythrocytes after rupture of merozoites from infected cells.

Feizi: Plapp et al (1980) have reported that Rh-negative cells contain Rh_0(D) antigen but it is in the inner surface of the erythrocyte membrane. You describe a system where trypsin treatment or neuraminidase treatment of erythrocytes renders them susceptible to malaria although they are Duffy-negative. Could it be that the so-called Duffy-negative cells have this antigen in cryptic form?

Howard: Genetically the Duffy-negative phenotype appears to be the product of a separate allele.

Feizi: Genetically the Rh-positive and Rh-negative phenotypes behave like alleles but it would seem that the *location* of the antigen is genetically determined.

Howard: Treatment of Duffy-negative cells with trypsin or neuraminidase failed to generate a serologically detectable Duffy antigen. Obviously the sensitivity of the serological test and its ability to detect an antigen which could in some way be cryptic, due to other membrane components shielding it from the antibody probe, are critical in any attempt to answer this point.

Feizi: The membranes have to be solubilized or turned inside out for Rh_0(D) to be detected, in the Rh-negative cells.

Howard: The Duffy antigenic determinant might only be a small part of a larger molecule or complex of several gene products. The receptor for the merozoite may not be the antigenic determinant but some other part of the same molecule. Molecular identification of the Duffy antigenic determinants and the molecule(s) which carry them is obviously required to answer these questions.

Mirelman: Do erythrocyte ghosts inhibit merozoite attachment to intact erythrocytes?

Howard: That has been examined; unfortunately *P. knowlesi* merozoites do not attach to rhesus monkey erythrocyte ghosts.

Choppin: Can you biosynthetically label the material in the rhoptry body, and if so does it wind up in the cell?

Howard: That would be a good experiment but there are technical problems still to be overcome – such as the difficulty of obtaining pure biosynthetically labelled merozoites which are invasive, and then the requirement for complete purification of the erythrocyte membrane from the parasite after invasion, in order to examine what material has been incorporated from the parasite. In addition, we have no means yet of identifying the subcellular location of internal components from merozoites. We have no marker for rhoptry material. The identification of electron-dense material in rhoptry organelles and its involvement in erythrocyte membrane invagination has resulted from electron microscopy studies. Purification of the rhoptry organelles would be a major advance in this area.

Taylor-Robinson: You talked about the specificity of receptors. In relation to that, does the malaria parasite have its particular life cycle because it is the most advantageous one for it, even though there may be receptors on other cells? Or does it fail to stick, for example, to the vascular epithelium because that would be a disadvantage or because there don't happen to be receptors there for it?

Howard: A proportion of each generation of newly invaded asexual malaria parasites within circulating erythrocytes switch into a differentiation pathway that leads to sexual forms. These remain within circulating erythrocytes for extended periods, whereas asexual forms tend to be removed from the circulation in a host with sufficient immunity. The sexual stages are taken up during mosquito feeding and then continue the cycle of transmission to another vertebrate host. The possession of receptors on the merozoite for erythrocyte surface components therefore allows them to enter a cell which will be taken up during mosquito feeding and which will allow it to infect other hosts. In addition, the erythrocyte interior is a location for parasite growth that is relatively hidden from the host's immune system and it is an environment which lacks potentially destructive host organelles such as lysosomes.

Taylor-Robinson: But it doesn't attach to the vascular endothelial lining?

Howard: Merozoites don't appear to invade epithelial cells *in vivo*. However, the range of cells which have been incubated with merozoites *in vitro* in order to test for their capacity to be invaded (apart from their capacity to sustain parasite growth) is limited. It is possible that merozoites could invade transformed cell lines which express erythrocyte membrane determinants or normoblasts and other precursors of the erythrocytic lineage.

Bredt: Can you differentiate between the initial merozoite attachment and apical attachment? Can you determine what forces direct the merozoite movement from initial attachment by any part of its surface membrane to apical attachment?

Howard: The receptor(s) for initial attachment appear to be distributed over the entire merozoite surface. In order for the merozoite to move from initial attachment to apical attachment there must be either an association–dissociation of the merozoite and erythrocyte receptors, or movement of the merozoite membrane relative to the initial receptors bound to the erythrocyte, which in this case would remain associated. In the first case a gradient of receptor density or affinity over the merozoite membrane might direct the merozoite attachment from an initial point on its membrane to the merozoite apex where the receptor density or affinity might be greatest. In the second case submembrane contractile elements of the parasite might move the membrane about a fixed point (the initial contact site) until the apex of the merozoite and erythrocyte membrane are together.

Silverblatt: I was intrigued by the rejected merozoite in the film you showed. Could internalization of the first parasite have altered the erythrocyte membrane so that it was no longer susceptible to subsequent invasion?

Howard: That is an interesting question. Invasion is not necessarily an all-or-none phenomenon. In some infections the merozoites interact with many erythrocytes before invading one of them, whereas another species of merozoite with the same cells will hit an erythrocyte and enter immediately. The particular case shown in the film was with *P. knowlesi* merozoites, a simian species, interacting with human cells. In this case many collisions and reorientations are often observed before they enter. There is not thought to be an irreversible change in the membrane of the invaded cell and therefore a block to invasion of subsequent merozoites. Multiply-infected erythrocytes are not uncommon, especially when the blood parasitaemia is very high.

REFERENCES

Abercrombie M 1961 The bases of the locomotory behaviour of fibroblasts. Exp Cell Res (Suppl 8):188-198

Heath JP, Dunn GA 1978 Cell to substratum contacts of chick fibroblasts and their relation to the microfilament system. A correlated interference-reflexion and high voltage electron microscopic study. J Cell Sci 29:197-212

Heaysman JEM, Pegrum SM 1973 Early contacts between fibroblasts – an ultrastructural study. Exp Cell Res 78:71-78

Johnson JG, Epstein N, Shiroshi T, Miller LH 1980 Factors affecting the ability of isolated *P. knowlesi* merozoites to attach and invade erythrocytes. Parasitology, in press

Miller LH, Aikawa M, Johnson JG, Shiroshi T 1979 Interaction between cytochalasin B-treated malarial parasites and erythrocytes. Attachment and junction formation. J Exp Med 149:172-184

Plapp FV, Evans JP, Tilzer LL 1980 Detection of Rh_o(D) antigen on the inner surface of the Rh-negative erythrocyte membrane. Fed Proc 39:547

Plasmodial modifications of erythrocyte surfaces

DONALD F. H. WALLACH, ROSS B. MIKKELSEN and RUPERT SCHMIDT-ULLRICH

Tufts-New England Medical Center, Therapeutic Radiology Department, Radiobiology Division, 171 Harrison Avenue, Boston, Massachusetts 02111, USA

Abstract The maturation of malarial parasites in red blood cells produces major alterations in the composition and properties of the host cell surface. Existing surface-exposed proteins are modified and new antigenic glycoproteins are synthesized by the parasite and inserted into the membrane. Some of the neoproteins are associated with surface excrescences on the host cells and in some cases these foster adhesion of those cells to capillary endothelium. The erythrocyte endoskeleton is degraded and in association the infected cells become deformed and lose pliability. An increase in intracellular Ca^{2+} may contribute to the changes in host cell surfaces.

Malaria is a major problem in world health and afflicts populations of Africa, tropical Asia and South America with serious morbidity in adults and deaths in children. It is characterized by intermittent febrile paroxysms, anaemia and splenic enlargement, due primarily to the periodic invasion, destruction and reinvasion of erythrocytes by the malaria parasite. The various phases of the red cell cycle of malaria include several phenomena of intercellular adhesion which are important to the pathogenicity of the organisms causing malaria.

The life cycle of the malarial parasite

The perpetuation of malaria involves a complicated host–vector life cycle. Fig. 1 shows this for human malaria. The causative organisms are protozoans of the genus *Plasmodium*, transmitted by mosquito vectors of the genus *Anopheles.* The saliva of infected anopheles mosquitoes contains malarial forms (sporozoites). When a person is bitten by an infected mosquito, sporozoites are injected into the blood. They rapidly disappear from the blood through uptake by liver macrophages. The parasites next invade hepatic parenchymal cells where they undergo asexual replica-

1981 Adhesion and microorganism pathogenicity. Pitman Medical, Tunbridge Wells (Ciba Foundation symposium 80) p 220-233

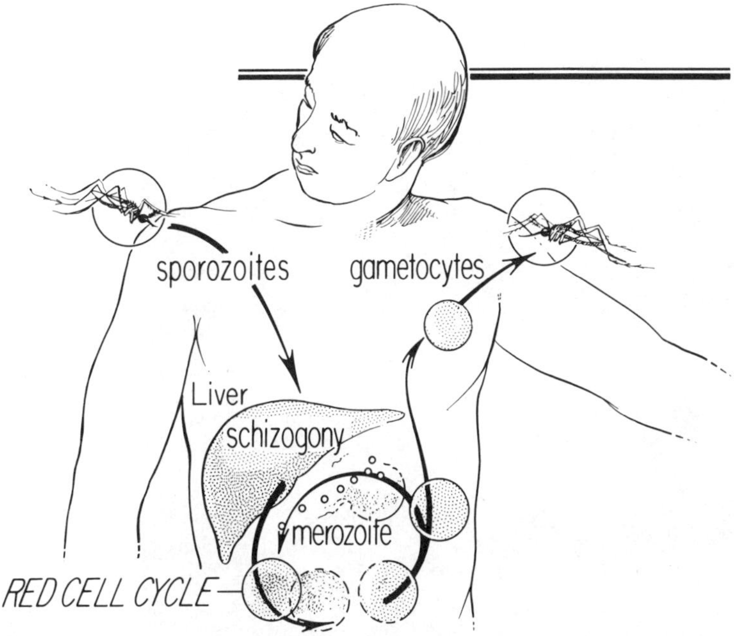

FIG. 1. Plasmodial life cycle in human malaria.

tion (schizogony). The liver stages of some plasmodial species can persist for long periods which are not associated with overt disease. However, liver-stage reservoirs account for recrudescences of malaria long after apparent cures.

Maturation in hepatocytes leads to release of merozoites from ruptured host cells. These are infectious only for erythrocytes. After invasion of an erythrocyte a second cycle of asexual replication occurs and the internalized merozoite develops first into a trophozoite and then into a multinucleate schizont. Rupture of the infected red cell releases 6–12 new merozoites. The duration of the cycle (48–72 h) depends on the species of *Plasmodium.* The released merozoites bind to and invade new host cells (Fig. 2), giving rise to additional cycles of erythrocyte schizogony.

During erythrocyte schizogony some parasites develop into male and female gametocytes, which are infective to anopheles mosquitoes. Within the mosquito stomach these develop into gametes which initiate the vector phase of parasite development.

Changes in host cell surfaces and vascular sequestration

Plasmodium-specific erythrocyte surface antigens have been detected by immune fluorescence (Collins et al 1967) and complement fixation (Eaton & Coggeshall

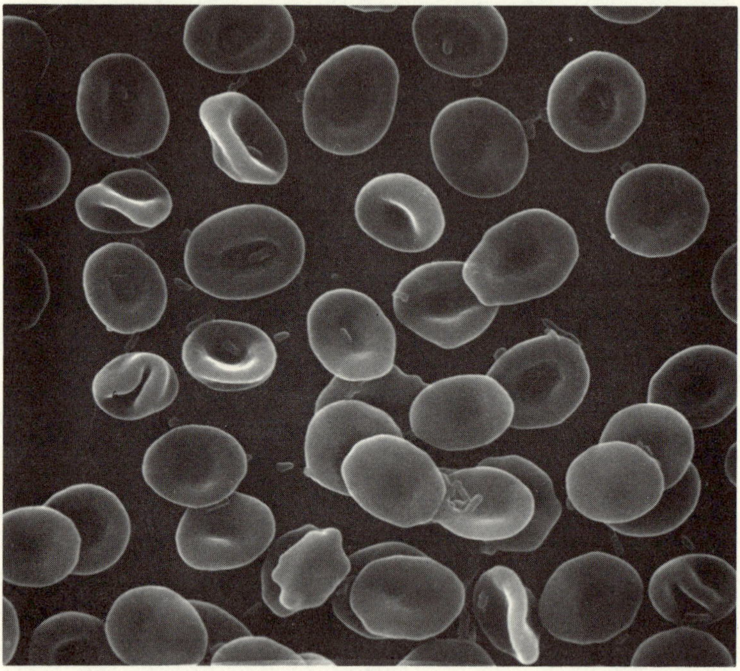

FIG. 2. *P. falciparum* malaria. Scanning electron micrograph shows attachment of merozoites to human erythrocytes. × 2000. (Reproduced by courtesy of Dr P.-S. Lin.)

1938). Also, as *P. knowlesi* matures in infected rhesus monkey erythrocytes, these blood cells become agglutinable by sera from animals chronically infected with or immune to infection by the same strain of *P. knowlesi* (Brown et al 1968). This is due to the appearance of strain-specific antigens on the host cell surfaces.

Transmission electron microscopy of *P. knowlesi* (Miller et al 1971a, Kilejian et al 1977), as well as of some other plasmodial species (Miller 1969, Desowitz et al 1969, Luse & Miller 1971), has revealed electron-dense 'knobs' or plaques lying just below (or apposed to) the membranes of infected cells (Fig. 3). For *P. falciparum,* immunoelectron microscopy has shown that these 'knobs' contain parasite-specific antigens that are exposed at the exterior of the host cell surface (Kilejian et al 1977).

The membrane excrescences of red cells parasitized by *P. coatneyi* or *P. falciparum,* form what seem to be junctions with the endothelial membranes of small vessels in tissues such as the spleen. These parasites mature preferentially in such locations, a process known as 'deep vascular schizogony'. It has been proposed (cf. Luse & Miller 1971) that plasmodially induced excrescences mediate adhesion of infected cells to endothelial walls, thereby producing deep vascular sequestration. However,

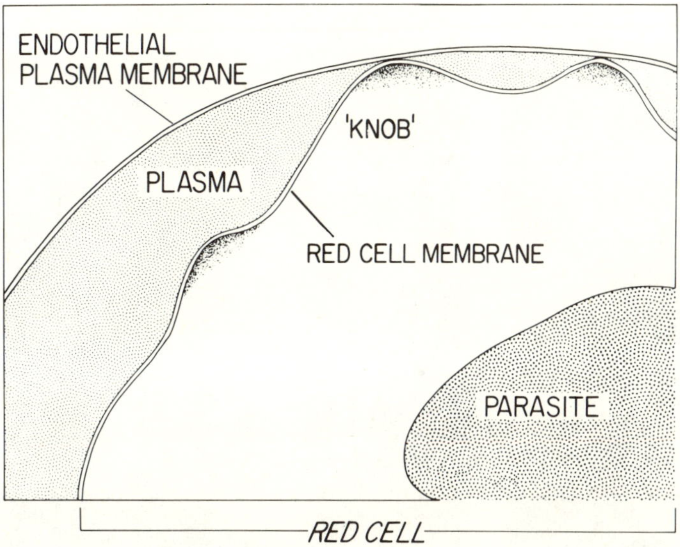

FIG. 3. Scheme of adhesion between endothelial plasma membrane of small vessels and host red cell membrane 'knobs'.

some plasmodia (e.g. *P. knowlesi*) induce excrescences on host cell surfaces but exhibit only moderate deep vascular maturation, and some (e.g. *P. brasilianum* and *P. malariae*) seem to mature in free circulation.

Where deep vascular sequestration occurs the role of surface excrescences has not been established. The surface charge and the charge distribution on host erythrocyte surfaces appear normal in cells infected with *P. coatneyi* or *P. knowlesi*, although the small radii of curvature of the excrescences might provide an electrostatic basis for adhesion (Miller et al 1972). It is also possible that deep vascular sequestration has a mechanical basis: red cells infected with *P. knowlesi* are less deformable than uninfected cells and have anomalously high viscosities. These abnormalities could favour capillary retention (Miller et al 1971b). Poor deformability and associated high blood viscosity characterize other red cell abnormalities involving endoskeletal defects (cf. Wallach 1979), including plasmodially infected erythrocytes (Weidekamm et al 1973, Wallach & Conley 1977, Schmidt-Ullrich & Wallach 1978).

Changes in host cell membrane proteins

Biochemical experiments have shown that erythrocyte schizogony produces major changes in host cell membrane proteins. Electrophoretic studies by Weidekamm et

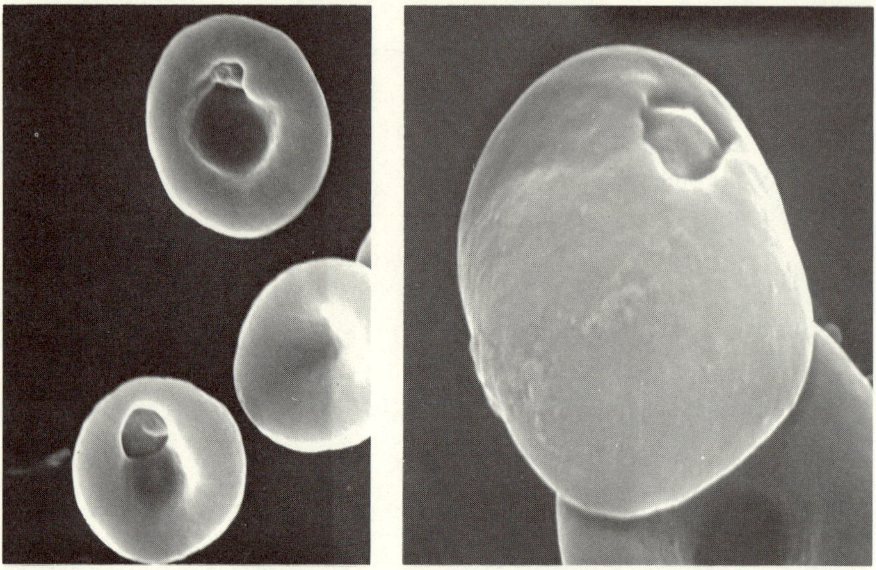

FIG. 4. *P. falciparum* malaria. Scanning electron micrographs of parasitized human erythrocytes. *Left:* Early infection. The host cells are still biconcave. × 6000. (*Reduced to 75%.*) *Right:* Host cell near rupture. × 10 000. (*Reduced to 75%.*) (Reproduced by courtesy of Dr P.-S. Lin.)

al (1973) on membrane preparations from mouse red cells infected with *P. berghei* revealed degradation of spectrin, the endoskeletal protein of high molecular weight on the internal membrane surface of normal erythrocytes. This process appeared to be closely correlated with host cell deformation, as revealed by scanning electron microscopy. Related host cell deformations occur as *P. falciparum* matures in human erythrocytes (Fig. 4). Spectrin is also depleted during the maturation of *P. knowlesi* in rhesus monkey erythrocytes (Wallach & Conley 1977, Schmidt-Ullrich & Wallach 1978) and this depletion has been attributed to the action of parasite proteases (Weidekamm et al 1973, Yuthavong et al 1978).

Parasite-free host cell membranes, as well as intact parasites, can be isolated from red cells infected with *P. knowlesi* (Wallach & Conley 1977). Erythrocytes, both normal and infected, are freed of leucocytes and their external surfaces are labelled with ^{125}I by lactoperoxidase-catalysed iodination. The erythrocyte membranes are vesiculated in the cold and at physiological ionic strength by nitrogen decompression. Host cell membranes and intact released parasites are isolated in good yield and purity by differential and density gradient centrifugation. Dodecyl sulphate–polyacrylamide gel electrophoresis of the membranes from uninfected and parasitized cells show major differences between the proteins that are exposed to the cell surfaces. These differences include (a) the appearance of seemingly new glycoproteins with a relative molecular mass (M_r) of 125 000–130 000 in the membranes of parasitized cells; (b) the deletion or possible cleavage of a normal glycoprotein of M_r about 98 000; and (c) the apparent loss in the parasitized membranes of a component of M_r about 52 000. The results also suggest that there is both parasite-induced modification of existing surface proteins and insertion of new proteins during schizont maturation.

Trigg et al (1977) used galactose-oxidase/NaB^3H_4 to compare the labelling of surface sugars of normal rhesus monkey erythrocytes with that of cells infected by *P. knowlesi*, after neuraminidase treatment. Dodecyl sulphate–polyacrylamide gel electrophoresis of membranes from uninfected cells revealed peaks of radioactivity at M_r values of about 170 000, 120 000, 90 000, 50 000 and 35 000, all of which were missing in schizont-infected cells. The changes at about M_r 120 000, 90 000 and 50 000 may correspond to the membrane modifications that we reported (Wallach & Conley 1977).

These studies have been followed up by isoelectric focusing and by immunochemical and metabolic techniques (Schmidt-Ullrich & Wallach 1978, Schmidt-Ullrich et al 1979, 1980a, b, Miller et al 1980). The first approach revealed that host cell membranes from parasitized cells contain components that do not occur in the membranes of normal erythrocytes but are found in the parasite. These components focus at pH$<$ 5.2. Isoelectric focusing also showed depletion of a major component in normal host cell membranes and focusing above pI 6.5. Isoelectric focusing, followed by dodecyl sulphate–polyacrylamide gel electrophoresis at right angles, showed that membrane proteins from normal and parasitized erythrocytes and purified parasites can be resolved into at least 50 individual protein spots each. Three parasite-induced proteins (pI 4.8–5.2; M_r 65 000–95 000) consistently appeared in host cell membranes. No components were common to normal erythrocyte membranes and parasites (Schmidt-Ullrich & Wallach 1978).

New components in plasma membranes have also been studied immunochemically. Rhesus monkeys were immunized with purified *P. knowlesi* schizonts (Wallach & Conley 1977) in complete Freund's adjuvant. Crossed immune electrophoresis of host cell membranes solubilized with Triton X-100, using the monkey immune serum, revealed seven major immunoprecipitates, none of which were seen with membranes from normal erythrocytes. The responsible antibodies could be absorbed out with infected erythrocytes, showing that antigens were exposed at the external surfaces of the host cells (Schmidt-Ullrich et al 1979).

Two of the plasmodium-specific host cell surface antigens, 1 and 13, were precipitated by sera of rhesus monkeys that became naturally immune against *P. knowlesi* by repeated infection and cure. These antigens are therefore *in vivo* immunogens. Moreover, analyses of membranes from erythrocytes parasitized by two different strains of *P. knowlesi* (Malaysian; Philippine) showed that both strains produced components 1 and 13. Electrophoretic analyses of immunoprecipitates indicate that component 13 is the 65 000 M_r neoprotein.

Schmidt-Ullrich et al (1980a) have tested the possibility that the immunogenic antigens might also occur on membranes of erythrocytes infected with species of plasmodia that cause malaria in humans. Sera of Gambian patients particularly immune to infection with *Plasmodium falciparum* and of rhesus monkeys immune to infection with *Plasmodium knowlesi* were reacted with Triton X-100-solubilized, ^{125}I-labelled membranes from *P. knowlesi*-infected erythrocytes. This was followed by indirect immune precipitation with *Staphylococcus aureus*, Cowan strain I, followed by dodecyl sulphate–polyacrylamide gel electrophoresis. Both types of sera precipitated plasmodium-specific antigens with M_r values of about 125 000, 90 000 and 65 000–50 000 from membranes of parasitized erythrocytes but not from membranes of normal cells. Sera of individuals transiently exposed to infections with *P. falciparum* and *P. vivax* reacted similarly. This indicates that *P. knowlesi* antigens share determinants with antigens induced by *P. falciparum* and *P. vivax*.

Metabolic labelling of parasite-synthesized components of host cell membranes

Kilejian (1979) has compared the proteins in membrane-enriched fractions from human red cells parasitized *in vitro* with either a 'knob'-producing strain of *P. falciparum*, K^+, or a strain, K^-, that does not produce knobs. Protein separation by dodecyl sulphate–polyacrylamide gel electrophoresis and analysis of gels by protein staining or distribution of metabolically incorporated [^{3}H]proline revealed a major protein of M_r roughly 80 000 that is produced by K^+ trophozoites or schizonts but not by any stage of K^- parasites or by early K^+ forms. The synthesis of this protein by the parasites correlates with the appearance of knobs on host cell membranes.

Host cell membrane proteins produced by *P. knowlesi* have been metabolically labelled with radioactive amino acids and sugar precursors (Schmidt-Ullrich et al 1980b). Infected rhesus monkey erythrocytes, collected at the late trophozoite or early schizont stages (85–90% synchrony; 30–45% parasitaemia) and freed of leucocytes and thrombocytes, were purified by Ficoll-hypaque density gradient centrifugation, and the parasitized erythrocytes, enriched to 85–90% (less than 0.001% leucocytes), were transferred to RPMI 1640 medium, 10% in calf serum, in flat sealed FEB Teflon bags and kept in humidified 8% CO_2, 80% N_2 at 37 °C. Incubation was for 6 or 12 h in 0.1 mCi [^{14}C]glucosamine or 0.1 mCi ^{14}C-labelled amino acids/20 ml medium (5×10^8 cells/ml). After culture, host cell membranes were purified from the washed infected erythrocytes for analysis.

The infected cells actively incorporated labelled amino acids and sugar precursors for 6 h with little additional uptake thereafter. Dodecyl sulphate–polyacrylamide gel electrophoresis of host cell membranes showed that both isotopes were incorporated into the 60 000–90 000 M_r glycoproteins focusing between pH 4.5 and 5.2. [^{14}C]Glucosamine was also incorporated into glycolipids. Clearly, the parasite-synthesized proteins, glycoproteins and glycolipids can be exported into the host cell membrane within 6 h. The principal proteins identified by metabolic labelling have the same relative molecular masses and pI values as the parasite-specific neoproteins defined by other techniques (Schmidt-Ullrich & Wallach 1978).

Calcium modifications in plasmodially parasitized erythrocytes

Increased concentrations of Ca^{2+} within erythrocytes can severely impair the structure and function of the red cell cytoskeleton. Spectrin tends to aggregate and segregate in the plane of the membrane. Also, the peptide chains of spectrin may be cleaved by Ca^{2+}-requiring proteases and cross-linked by a Ca^{2+}-activated *trans*-glutaminase. With its cytoskeleton impaired the cell becomes distorted, poorly deformable and possibly anomalously adhesive. It is thus vulnerable to vascular sequestration.

Intraerythrocytic $[Ca^{2+}]$ is normally maintained at $\leqslant 10^{-6}$ M because of slow influx, plasma membrane binding, e.g. by polyphosphoinositides, and active extrusion by the Ca^{2+} pump/ATPase, activated by a Ca^{2+}-binding protein (calmodulin). When normal erythrocytes are depleted of ATP they cannot extrude Ca^{2+}. They then accumulate Ca^{2+}, contract, become distorted, lose their normal pliability and pinch off spectrin-depleted, lipid-enriched vesicles (Lutz et al 1977). These consequences of ATP depletion can be simulated by artificially enriching the cytoplasmic space with Ca^{2+}. Weidekamm et al (1977), who lysed normal erythrocytes in a hypotonic high Ca^{2+} (4.5 mM) medium and resealed the cells in 160 mM-NaCl, obtained small spectrin-depleted vesicles and ghosts showing aggregation of spectrin instead of the usual ghosts. Allan et al (1976) raised cytoplasmic $[Ca^{2+}]$ by exposing red cells to the Ca^{2+}-ionophore A23187 in the presence of Ca^{2+}. They also observed the release of spectrin-depleted vesicles. Anderson et al (1977), also using A23187 in Ca^{2+}-containing media, observed anomalies in spectrin peptides apparently related to Ca^{2+}-activated proteases and *trans*-glutaminases. Protease activation appeared at intracellular Ca^{2+} levels $\geqslant 5 \times 10^{-6}$ M but *trans*-glutaminase activation required 500×10^{-5} M-Ca^{2+}. The deterioration of spectrin reported for plasmodium-infected red cells (Weidekamm et al 1973, Wallach & Conley 1977, Schmidt-Ullrich & Wallach 1978, Yuthavong et al 1978) may thus relate to the compartmentation of Ca^{2+} in infected red cells. However, information about erythrocytic Ca^{2+} levels and fluxes during plasmodial infection is not available.

We have initiated an investigation of Ca^{2+} metabolism in *P. chabaudi*-infected rat erythrocytes. Heparinized blood was freed of leucocytes and thrombocytes by cellulose powder percolation and Ficoll-hypaque gradients. Erythrocytes were suspended in RPMI 1640 medium (containing 1 mM-Ca^{2+}) and analysed for Ca^{2+} directly or after initiation of efflux by a 50-fold dilution into Ca^{2+}-free phosphate-buffered saline. For Ca^{2+} analysis cell suspensions were layered over silicon oil and centrifuged for 30 s at 10 000 *g* to separate cells from the medium. We estimated the trapped extracellular-space Ca^{2+}, using ^3H_2O and $[^{14}C]$polyethylene glycol

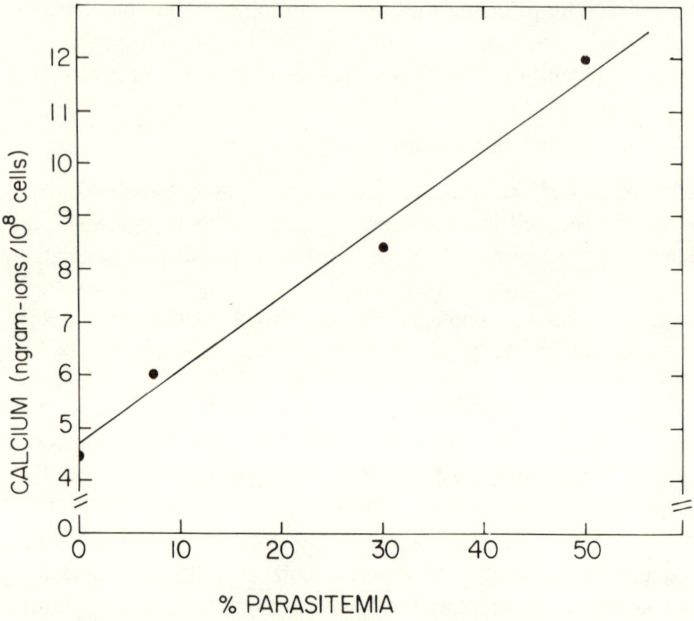

FIG. 5. Calcium content of rat erythrocytes infected with *P. chabaudi.* Each data point represents the mean of triplicate determinations for one rat of stated parasitaemia. The standard deviation within triplicate samples was less than 10%.

(M_r 4000) as a marker for calculating extracellular space. The silicon oil and medium were removed and the cells were suspended in pure 0.5M-HCl plus 0.1% $LaCl_3$; insoluble material was pelleted at 10 000 *g*-min and the supernatant fluid was analysed by atomic absorption spectroscopy.

We find an approximately linear relationship between parasitaemia and total cell Ca^{2+} content (Fig. 5). At 50% parasitaemia the Ca^{2+} content of *P. chabaudi*-infected cells is about three times that of normal cells. Parallel studies of *P. knowlesi*-infected monkey erythrocytes also demonstrated a 25-fold increase over normal in total cell Ca^{2+}. *P. chabaudi*-infected erythrocytes also differ from normal in their Ca^{2+} efflux. However, the early phase (< 1 min) of Ca^{2+} efflux from infected cells is parallel to that from normal erythrocytes, suggesting that a large part of the increase in intracellular Ca^{2+} is localized in the erythrocytic rather than the parasitic compartment. Also, after efflux ($t > 4$ min) there is an apparently enhanced uptake of Ca^{2+} by infected but not by control erythrocytes. This may be due to parasite metabolism.

The raised levels of intracellular Ca^{2+} in plasmodially infected erythrocytes may indicate conditions that would allow Ca^{2+}-induced endoskeletal modifications, with expected consequences for the adhesion, filtrability and stability of infected red cells. Not resolved are questions of how the Ca^{2+} economy of host cells may be affected by the parasite. It is interesting, however, to contemplate the possibility that neoproteins introduced into host cell membranes by erythrocytic plasmodial stages have specific membrane functions, and that one of these is to influence host cell Ca^{2+}.

Acknowledgements

This work was supported by US Army Contract DAMD 17-74-C-4118 (D.F.H.W.), National Institutes of Health grant A116087 (to R.B.M.) and WHO grant MAL T16/181/MZ/9 (to R.S-U.).

REFERENCES

Allan DM, Billah MM, Finean JB, Michell RH 1976 Release of diacylglycerol-enriched vesicles from erythrocytes with increased intracellular [Ca^{2+}]. Nature (Lond) 261:58-60

Anderson DF, Davis JL, Carraway KL 1977 Calcium-promoted changes of the human erythrocyte membrane. J Biol Chem 252:6617-6623

Brown IN, Brown KN, Hills, LA 1968 Immunity to malaria: the antibody response to antigenic variation by *Plasmodium knowlesi.* Immunology 14:127-138

Collins WE, Skinner JC, Coifman RE 1967 Fluorescent antibody studies in human malaria. V. Response of sera from Nigerians to five *Plasmodium* antigens. Am J Trop Med Hyg 16:568-571

Eaton MD, Coggeshall LT 1938 Complement fixation in human malaria with an antigen prepared from the monkey parasite *Plasmodium knowlesi.* J Exp Med 69:379-398

Desowitz RS, Miller LH, Buchanan RD, Pernipanich B 1969 The sites of deep vascular schizogony in *Plasmodium coatneyi* malaria. Trans R Soc Trop Med Hyg 63:198-202

Kilejian A 1979 Characterization of a protein correlated with the production of knob-like protrusions on membranes of erythrocytes infected with *Plasmodium falciparum.* Proc Natl Acad Sci USA 76:4650-4653

Kilejian A, Abati A, Trager W 1977 *Plasmodium falciparum* and *Plasmodium coatneyi* immunogenicity of 'knob like' protrusions on infected erythrocyte membranes. Exp Parasitol 42:157-164

Luse SA, Miller LH 1971 *Plasmodium falciparum* malaria: ultrastructure of parasitized erythrocytes in cardiac vessels. Am J Trop Med Hyg 30:655-660

Lutz HU, Shih-Chun L, Palek J 1977 Release of spectrin-free vesicles from human erythrocytes during ATP depletion. J Cell Biol 73:548-560

Miller LH 1969 Distribution of mature trophozoites and schizonts of *Plasmodium falciparum* in the organs of *Aotus trivirgatus,* the night monkey. Am J Trop Med Hyg 18:860-865

Miller LH, Fremount NH, Luse SA 1971a Deep vascular schizogony in *Plasmodium knowlesi* infected *Macacca mulatta.* Am J Trop Med Hug 20:816-824

Miller LH, Vsami S, Chien S 1971b Alterations of the rheological properties of *P. knowlesi*-infected red cells: a possible mechanism for capillary obstruction. J Clin Invest 50: 1451-1455

Miller LH, Cooper GW, Chien S, Freemount HN 1972 Surface charge on *Plasmodium knowlesi* and *P. coatneyi*-infected red cells of *Macacca mulatta.* Exp Parasitol 32:86-95

Miller LH, Johnson JG, Schmidt-Ullrich R, Haynes D, Wallach DFH, Carter R 1980 Determinants on surface proteins of *Plasmodium knowlesi* merozoites common to *Plasmodium falciparum.* J Exp Med 151:790-798

Schmidt-Ullrich R, Wallach DFH 1978 *Plasmodium knowlesi*-induced antigens in membranes of parasitized rhesus monkey erythrocytes. Proc Natl Acad Sci USA 75:4946-4953

Schmidt-Ullrich R, Wallach DFH, Lightholder J 1979 Two *Plasmodium knowlesi* specific antigens on the surface of schizont-infected rhesus monkey erythrocytes induce antibody production in immune hosts. J Exp Med 150:86-99

Schmidt-Ullrich R, Miller LH, Wallach DFH, Lightholder J 1980a Antigens common to *P. knowlesi* and human plasmodia. J Immunol, in press

Schmidt-Ullrich R, Wallach DFH, Lightholder J 1980b Metabolic labeling of *P. knowlesi*-specific glycoproteins in membranes of parasitized rhesus monkey erythrocytes. Cell Biol Int Rep 4:55-561

Trigg PI, Hirst S, Shakespeare P, Tappenden L 1977 Labelling of membrane glycoprotein in erythrocytes infected with *Plasmodium knowlesi.* Bull WHO 55:205–210

Wallach DFH 1979 Membrane pathobiology of malaria. Cell Biol Int Rep 3:395-408

Wallach DFH, Conley M 1977 Altered membrane proteins of monkey erythrocytes infected with simian malaria. J Mol Med 2:119-136

Weidekamm E, Wallach DFH, Lin PS, Hendricks J 1973 Erythrocyte membrane alterations due to infection with *Plasmodium berghei.* Biochim Biophys Acta 323:539-546

Weidekamm E, Brdiczka D, Ci Pauli G, Wildermuth M 1977 Biochemical characterization of segregated membrane vesicles from human erythrocytes with increased intracellular Ca^{2+}. Arch Biochem Biophys 179:486-494

Yuthavong Y, Wilairat P, Panijpan B, Potiwan C, Beale G 1978 Alterations in membrane proteins of mouse erythrocytes of malaria parasites. Comp Biochem Physiol 63B:83-85

DISCUSSION

Howard: Dr Richard Carter and I have discussed why the malaria parasite would go to the trouble of putting highly immunogenic molecules on the surface of an asexual infected cell. He proposed that it was because the parasite only wishes to induce transient asexual parasitaemia and the associated host-threatening clinical symptoms. In terms of perpetuation of the parasite population in other hosts, the only role of asexual parasites is to provide a pool of parasites for generation of sufficient sexual forms to infect mosquitoes. Sexual forms remain in the circulation long after clearance of asexual parasites. The asexual parasites, according to this hypothesis, have therefore evolved a mechanism for their own removal which allows host survival, survival of the sexual forms transmitted to mosquitoes, and consequently growth of the parasite in other hosts.

Feizi: Is the 90 000 M_r glycoprotein that is precipitated by immune serum, of parasite origin, or is it perhaps band 3 protein?

Wallach: No, the antigens have plasmodial and strain specificities and appear unrelated to any normal membrane component.

Feizi: The parasites may somehow render band 3 protein immunogenic; for example, a glycolipid of parasite origin may complex with the protein and act as an adjuvant. The glycolipid could confer strain specificity.

Wallach: I think you are saying that glycolipids might migrate with proteins in the molecular weight range of band three. We made a whole series of antisera against normal monkey red cell components, including band three, but found no cross-reactivity with the neoantigens. It is still possible, however, that these are glycolipids associated with protein, but the isoelectric focusing phases of the two-dimensional separations are expected to break down such associations.

Feizi: Do you detect glycolipids in the electrophoresed immune precipitates?

Wallach: Gangliosides would show up; the neutral glycolipids would not.

Helenius: In eukaryotic cells plasma membrane glycoproteins are synthesized in the endoplasmic reticulum. They are then carried out to the plasma membrane by a sequence of membrane fusion and transport steps. Here, I understand, you are proposing that proteins synthesized by the parasitic ribosomes are transported through the membrane of the parasite, through the membrane of the vacuole which surrounds the parasite, and finally insert themselves into the plasma membrane of the host cell.

Wallach: Possible mechanisms involved in the transfer of parasite-synthesized proteins to the host cell membrane have concerned us. Significantly, perhaps, the parasite isn't sitting passively. In some species it undergoes rather violent contortions within the cell. The question of whether continuities sometimes occur between parasite and host cell membranes has been raised on a number of occasions, but has not been sufficiently examined. It may simply be that we haven't had a real look at the transfer problem.

Helenius: There are examples in the eukaryotic cells where proteins synthesized in the cytoplasm pass through one membrane and finally attach to another. In mitochondria and chloroplasts some of these proteins become integral membrane components.

Howard: In several malaria-host systems, deep crypts and invaginations of the outer erythrocyte membrane have been described. Perhaps the parasite-derived proteins destined for the erythrocyte surface membrane are inserted into the vacuolar membrane surrounding the parasite. Fusion of this membrane with the deep invaginations of the outer erythrocyte membrane, or fusion of vesicles from the vacuolar membrane, would effectively transfer parasite proteins to membrane contiguous with the erythrocyte surface.

Hughes: I am impressed by the number of glycosylation reactions going on in what I thought was a metabolically uninteresting cell, the red blood cell. Are the parasites contributing these enzymes or is some vestigial enzyme system of the red blood cell activated by the infection?

Wallach: We don't know enough about the glycosylation reactions. We were surprised to find that they existed at all and have not had the chance to explore whether vestigial red cell enzymes are involved. We were very worried about contamination by white cells and about a number of other possible artifacts. I think we have ruled out the obvious possibilities but have not yet found a way to exclude the one you propose.

Razin: Can you inhibit glycosylation, for example with tunicamycin?

Wallach: We are testing this.

Choppin: Does the vacuolar membrane with the parasite inside have these proteins in it also, and if so which way are they oriented?

Wallach: I don't know whether the vacuolar proteins are glycosylated or not, but some glycoproteins are detectable in the parasite itself. There is a labelling gradient between the parasite and the surface membrane.

Howard: The surface antigens you have identified do not appear to be specific for a particular strain or variant of the parasite, yet it is known with *P. knowlesi* infections of rhesus monkeys that there is an agglutinogen on the surface of mature infected cells which show antigenic variation and apparent switching during sequential relapse parasitaemias of a single animal. Are other methods required to identify the variant antigens?

Wallach: Neither of the antigenic components defined by M_r and pI is a single species of molecule. There is strain-specific microheterogeneity. We are focusing on the interspecies component because of its possible relevance to vaccine production. I discussed several sources of antigens for potential vaccines: blood stage components, including host cell membrane and merozoite antigens, and sporozoite antigens. There is evidence that antibodies to all of these can give protective immunity in the proper system. Recent work (Freeman et al 1980) on *P. yoelii* using monoclonal antibodies (which are not ideal if one is looking for cross-reacting interspecies antigen) indicates that one can get protective immunity with an anti-merozoite antibody. Work on sporozoite antigens shows that one can get stage-specific protective immunity (Yoshida et al 1980). Our own work indicates an association of protective immunity with host cell membrane antigens but some of these also appear on the merozoites. Common to all antigenic sources are very serious problems of production. We lack the culture techniques to generate sufficient amounts of whatever antigen we choose to use. I believe we have to look to DNA recombinant techniques for production of sufficient amounts of antigen. In this context it is important to reach decisions as to what proteins to clone.

Feizi: I am worried about the effects of immunization against an antigen closely associated with host red cell membranes. When I was studying immune haemolytic disorders some years ago, I encountered a most dramatic case of acute intravascular haemolysis associated with a positive antiglobulin (Coombs test). This was due to complement-coating of the patient's red cells. The patient was a young woman with *P. falciparum* malaria and in her serum she had antibodies which gave a high agglutination titre with her own red cells at the height of her illness. When she recovered, these antibodies no longer agglutinated her red cells to the same extent and they gave only low titre agglutination comparable to that with normal cells. This woman almost died of acute haemolysis. So if you hyperimmunize people against the parasite you may have to combat severe haemolytic anaemia if they were likely to become parasitaemic.

Wallach: I am aware of similar cases but it is not, I understand, a frequent occurrence. Moreover, in a hyperimmune or an adequately vaccinated individual infected cells are expected to be eliminated at a very low level of parasitaemia (i.e. shortly after initiation of the red cell cycle). That is different from what happens in individuals who develop immunity through exposure to *P. falciparum.*

Newell: I understand that the 90 000 M_r protein shows species-specificity, but could this be the protein that shows immune modulation?

Wallach: It is possible. There is a correlation between that and a test which is related to antigenic variation. We are not dealing with one antigen but with a group of antigens of which one category is variant-specific.

REFERENCES

Freeman RR, Trejdosiewicz AJ, Cross GAM 1980 Protective monoclonal antibodies recognizing stage-specific merozoite antigens of a rodent malaria parasite. Nature (Lond) 284:366-368

Yoshida N, Nussenzweig RS, Potochjak P, Nussenzweig V, Aikawa M 1980 Hybridoma produces protective antibodies against the sporozoite stage of malaria parasite. Science (Wash DC) 204:71-73

Interaction of chlamydiae with host cells and mucous surfaces

J. H. PEARCE, I. ALLAN and S. AINSWORTH

Department of Microbiology, University of Birmingham, P.O. Box 363, Edgbaston, Birmingham B15 2TT, UK

Abstract For chlamydiae, as obligate intracellular parasites, attachment to and ingestion by host cells are essential steps in reproduction. Their attachment site appears to be heat-sensitive; it has not been correlated with any morphological entity. Antibody blocks chlamydial attachment to cells and, for certain *Chlamydia psittaci* and *Chlamydia trachomatis* strains which are highly infective for cell cultures, *N*-acetylglucosamine appears to contribute to cell receptor specificity. Sialic acid residues have been suggested as receptors for other *C. trachomatis* strains.

The guinea-pig inclusion conjunctivitis strain of *C. psittaci* becomes associated with the conjunctiva during incubation of inoculated tissue fragments *in vitro*. However, although antibody from tears neutralizes infectivity of this strain *in vivo*, association of the organism with tissue fragments is not inhibited, suggesting that antibody neutralization *in vivo* is not mediated by prevention of attachment to cells.

Chlamydial infectivity for cell monolayers is greatly increased by centrifugation. The process is temperature-dependent and involves cooperative interactions between directional force and the pressure generated during centrifugation. Enhanced infectivity appears to result from changes induced in the cell surface. These changes may favour non-specific interactions in attachment, since antibody inhibition of infectivity on static cell monolayers is overcome by centrifugation.

Chlamydiae are generally considered to be host-dependent bacteria whose essential defect is a lack of energy-generating mechanisms. However, as obligate intracellular parasites they differ in one important respect from most of the microorganisms considered in this symposium. Adhesion or attachment to the host cell surface is only the first stage of interactions with the surface which lead on to ingestion and, ultimately, reproduction.

The genus comprises two species with very different host ranges. Strains of *Chlamydia trachomatis* cause ocular and urogenital infection only in humans and

1981 Adhesion and microorganism pathogenicity. Pitman Medical, Tunbridge Wells (Ciba Foundation symposium 80) p 234-251

other primates; a distinct subset, the LGV strains, are responsible for the disease lymphogranuloma venereum. *Chlamydia psittaci* strains cause varied infections in many animal species; those in domestic animals and birds are economically important. The host restriction of *C. trachomatis* strains is an obstacle to exploration of their mechanisms of pathogenicity, especially now that non-human primates are unavailable to most laboratories. There has thus been increased interest in the guinea-pig model of conjunctival and urogenital disease induced by the *C. psittaci* strain, 'guinea-pig inclusion conjunctivitis' (GP-IC).

Chlamydiae also differ from one another in their capacity to infect tissue cell cultures. This is true for strains of both species and appears to result from attachment differences. The *C. trachomatis* LGV strains and a few *C. psittaci* strains are highly infective for static cell monolayers. The ocular and urogenital *C. trachomatis* strains are, in the main, poorly infective. Their infectivity is greatly enhanced by centrifugation of inoculated monolayers, apparently as a result of the changes this induces in the cell surface. We have termed these two infection modes 'spontaneous' and 'centrifuge-assisted' infection, respectively (Allan et al 1976).

Despite the importance of chlamydiae as pathogens we are only beginning to understand chlamydial attachment and ingestion and we know almost nothing about interference with these events by the host immune response. It is convenient to summarize developments under three headings: chlamydial interactions with cell cultures in spontaneous infection; infection of the conjunctiva by the guinea-pig inclusion conjunctivitis strain and interference by antibody; and interactions in centrifuge-assisted infection. Our work has been concerned particularly with the second and third topics.

Spontaneous chlamydial infection of cell cultures

Before we consider the phenomena of attachment it may be helpful to sketch the early stages of the chlamydia developmental cycle. Entry of the infective particle into the host cell is followed by differentiation to a non-infective metabolically active form (Manire & Wyrick 1979). Ingestion occurs by a process akin to phagocytosis which is not under the control of cytoskeletal structures (Gregory et al 1979a). This process is not equivalent to viropexis since the latter appears to be an energy-independent process (Patterson et al 1979) and chlamydial ingestion is inhibited by sodium fluoride (Kuo & Grayston 1976). Lysosomal fusion with the vacuole membrane surrounding the particle is inhibited and appears to be dependent on surface properties of the particle (Friis 1972); the inhibition is presumably mediated through the vacuole membrane. The vacuole membrane must also mediate energy and nutrient supplies from the host cell and possibly its composition is 'selected' during ingestion, as is seen in endocytosis by 'professional' phagocytes (Willinger et al 1979). Conjecturally, then, attachment–ingestion events for chlamydiae may influence both intracellular survival and the early metabolic events of multiplication.

Attachment as a discrete event in chlamydial infection of cells has been demonstrated most clearly by Byrne (1978), following earlier work by Friis (1972) which cast doubt on whether attachment and ingestion were separable. Temperature-independent attachment to cell monolayers has been observed for an ocular strain of *C. trachomatis*, TW5 (Kuo & Grayston 1976), and for the GP-IC strain of *C. psittaci* (Allan & Pearce 1979a). The chlamydial attachment site may be protein in nature. For the TW5 and an LGV strain of *C. trachomatis* (Kuo & Grayston 1976) and the 6BC strain of *C. psittaci* (Byrne 1976) attachment was inactivated by incubation at 56 or 60 °C.

Matsumoto and colleagues (Matsumoto 1979) have shown the presence of a patch of projections on the chlamydial surface. The projections occur on strains of *C. psittaci* and on ocular, genital and LGV strains of *C. trachomatis*, and are probably characteristic of the genus (Gregory et al 1979b). However, they do not appear to be involved in attachment (Gregory et al 1979b), since organisms which had settled on cell monolayers (specific attachment) or on polylysine-coated glass (non-specific) showed similar proportions of exposed projection arrays.

Trypsin treatment of cells temporarily prevents chlamydial attachment, suggesting that cell receptors themselves, or influential adjacent structures, are trypsin-sensitive (Byrne 1976). However, only two studies have so far appeared which provide information on the nature of these receptors. The first developed from observations that a broad distinction could be made between ocular strains of *C. trachomatis* and the highly cell-infective LGV strains. Only the ocular strains showed enhanced infectivity when cells were pretreated with the polycation DEAE–dextran (Kuo et al 1972). Kuo et al (1973) then reported that neuraminidase treatment of cell monolayers greatly reduced infectivity of the ocular strain but not that of LGV or a *C. psittaci* strain. Rinsing of monolayers with *N*-acetylneuraminic acid before inoculation had a similar discriminating effect. Kuo et al (1973) suggested that sialic acid residues on the cell surface might act as receptors for ocular *C. trachomatis* strains. Competition studies with possible receptor-containing materials were not done and these interesting observations have not been followed up. It has been shown with radiolabelled organisms that polycation enhancement results from increased attachment (Kuo & Grayston 1976).

More recently Levy (1979) has examined attachment of the 6BC *C. psittaci* strain and an LGV strain of *C. trachomatis.* Wheat germ agglutinin, a lectin specific for *N*-acetyl-D-glucosamine, inhibited attachment of both strains completely and pretreatment of the lectin with the free amino sugar prevented the inhibition. Two other lectins with different sugar specificities caused up to 40% inhibition of attachment. More importantly, *N*-acetylglucosamine itself did not block chlamydial attachment. These effects suggest that other residues contribute, with *N*-acetylglucosamine, to cell receptor specificity.

Overall it appears that ocular and LGV strains of *C. trachomatis* can be grouped separately and that the LGV strains are similar to the *C. psittaci* 6BC strain in receptor requirements. At present we have little information about the many other *C. psittaci* strains which infect domestic animals and birds. It remains to be seen whether there are differences between the different serotypes of ocular and genital strains of *C. trachomatis* which have been described. For many of these strains infectivity is poor when the inoculum has not been centrifuged, although it is possibly measurable if there is a sufficient concentration of organisms in the inoculum. Pursuing the question of receptors for such strains will be technically difficult unless these are much further adapted to cell culture – with the attendant risks.

If definition of an attachment step in chlamydial infection is recent, so also is evidence of interference by antibody. Byrne & Moulder (1978) have shown that this important immune mechanism functions for chlamydiae by demonstrating that antibody inhibits attachment of radiolabelled *C. psittaci* strain 6BC to cell cultures. We have similarly observed inhibition for MRC4f, an LGV-like strain of *C. trachomatis* (Ainsworth et al 1979). Our procedure was developed after we had observed that MRC4f, when fully neutralized for spontaneous infection of cell monolayers, was only partly neutralized for centrifuge-assisted infection (Ainsworth et al 1976). In itself this suggested that centrifugation was altering the interaction between the organism and the cell surface. Technically it allowed us to show, by centrifuge-assisted infection, that after organism–antibody mixtures had been inoculated into static cell monolayers the supernatants contained more infectivity, as a result of inhibition of attachment, than control supernatants from inocula of organisms alone.

The divergent effect of antibody on spontaneous and centrifuge-assisted infection is even more marked for the GP-IC strain of *C. psittaci.* Full neutralization of spontaneous infection of monolayers was achieved yet there was a negligible loss of infectivity when this was measured after centrifugation of cell monolayers (Allan et al 1977, Ainsworth et al 1979). One can therefore, by measuring the infectivity of test suspensions after centrifugation, follow the fate of GP-IC–antibody complexes inoculated on mucosal surfaces and test whether antibody can inhibit attachment in a more realistic cell environment than the monolayer preparation. This is the subject of the next section.

Interaction of the GP-IC strain with the guinea-pig conjunctiva and interference by antibody

Here we describe recent work (S. Ainsworth, unpublished) on the role of local antibody in immunity to conjunctival infection. We have attempted to answer two questions. First, can immune tears neutralize GP-IC infectivity for normal eyes?

Second, given evidence of neutralization, does it correlate with inhibition by tears of the association of GP-IC with conjunctival tissue fragments?

Murray and colleagues have shown that GP-IC infection of the guinea-pig conjunctiva is self-limiting and is followed by a period of immunity. In exploring its basis they were unable to show a contribution from local antibody in tears when this was tested by its ability to neutralize GP-IC infectivity for normal eyes (Murray et al 1973). Human tears, however, reduced *C. trachomatis* infectivity for owl monkey eyes (Nichols et al 1973). Success here may have been due to the use of smaller challenge doses. Subsequently, evidence from studies of drug-modified immune responses in guinea-pigs has indicated that humoral mechanisms contribute to immunity in both conjunctival and urogenital infections (Modabber et al 1976, Rank et al 1979).

We have used a simple procedure to measure chlamydial association with tissue in guinea-pigs. Whole eyelids are dissected free of underlying connective tissue, cut in half and inoculated with 1-μl volumes of GP-IC. After the tissue has been incubated for 1 h in a humidified atmosphere and washed by repeated rinsing the associated organisms are released by sonic treatment and measured by centrifugation titration of infectivity. Association of organism–antibody complexes with tissue can be measured similarly, as indicated earlier.

Technically, infectivity is a poor marker for monitoring the fate of organisms. However, restrictions posed by low yields of the GP-IC strain from cell culture and low efficiency of radiolabelling have made it, until recently, the only one feasible. From preliminary studies, incubation for 1 h represents a compromise between increase in association with time and loss of infectivity from tissue as organisms are ingested and differentiate to the non-infective form. Despite these uncertainties, over a series of experiments we recovered about 70% (mean value) of inoculum infectivity in washings and 11% associated with tissue. Suspension of organisms in normal serum, tears or buffer has had no detectable effect on the proportion associating with tissue, nor has this varied over an 80-fold concentration range of inoculum.

We first examined neutralization of GP-IC with immune serum, as a test of the experimental procedures. After 15 min incubation with heated immune serum and inoculation onto tissue the organisms showed reduced association with tissue. As a percentage of inoculum infectivity, 4.6% (SE ± 1.28, $n = 20$) of organisms associated after incubation with immune serum compared with 11.2% (SE ± 1.33, $n = 20$) for organisms incubated with normal serum. Organism–antibody mixtures were substantially neutralized in spontaneous infection of cell monolayers. Full retention of infectivity after centrifugation indicated that no inactivation had occurred.

For tests of neutralization of infectivity for the guinea-pig eye the inocula used for tissue association must be heavily diluted. Otherwise, even with 95% neutralization, enough infective organisms remain to cause overwhelming eye infection. The 50% infective dose (ID_{50}) for the guinea-pig eye is about 10 ELD_{50} (50% egg

TABLE 1 Inhibition of GP–IC association with conjunctival tissue by tears sampled from guinea-pig donors at intervals after infection

Tear pool used in inhibition (donation day)	*% GP–IC association*	*Tear pool sIgA Ab (titre^{-1})*	*Immune status of tear donor*
0	11.1	0	Normal
18	12.6	64	ND
28	ND	256	Immune
38	2.8[a]	128	Immune
63	8.2	64	Immune
80	3.3	8	Partially immune
98	9.2	ND	ND
143	2.1[a]	4	Susceptible
Buffer	10.1,1.5[a]	–	–

Association is % inoculum infectivity recovered after manual rinsing of tissue. ND: not done.
[a]Mechanical agitation applied during rinsing.

lethal doses). Guinea-pig eyes were inoculated with GP-IC–antibody mixtures containing 10, 33, 50 and 100 ELD_{50} of GP-IC. Judged by the absence of clinical signs and infected epithelial cells in conjunctival scrapings, protection was complete at 10 and 33 ELD_{50} but was hardly discernible at the higher doses. Thus a reduction of about 2.5-fold in the association of GP-IC with conjunctival tissue correlated with an increase of about threefold in ID_{50} for the guinea-pig eye.

Immune tears were sampled from donor guinea-pigs and tear pools were prepared at intervals over 140 days after the initiation of eye infection. The rise and fall in the antibody titre of secretory IgA (measured by microimmunofluorescence titration) generally accorded with acquired immunity: peak titres were found over days 30 to 40, falling markedly as immunity waned (at about day 80) until, by day 140, antibody levels were minimal and animals had regained almost full susceptibility (see Table 1).

Neutralization of infection and inhibition of association by tears were assessed as before, using mixtures of organisms (about 2×10^8 ELD_{50} ml^{-1}) incubated with equal volumes of tears diluted 1:4 in buffer. Graded doses from 10 to 1000 ELD_{50} were inoculated (six eyes per dose). Infection was significantly reduced with all tear pools up to day 75 samples. Maximum neutralization was found with day 56 and 75 pools, with complete protection of eyes at a dose of 100 ELD_{50} (Fig. 1), equivalent to a 15- to 20-fold rise in ID_{50}. However, results so far available do not show that neutralization of tear pools significantly inhibits the association of organisms and tissue (Table 1). Certain values require confirmation (day 38 and day 143 pools). Only with the day 80 pool was there a significant reduction in association.

If these results accurately predict events on the intact conjunctiva, what other

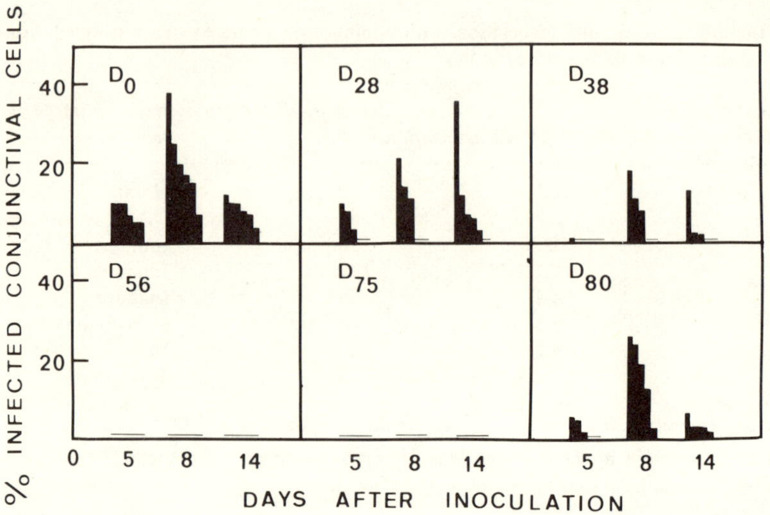

FIG. 1. Neutralization of infection in eyes of normal guinea-pigs inoculated with mixtures of GP–IC (100 ELD_{50}) and tears sampled from immune animals at intervals (D_0 – D_{80}) after infection.

humoral protection mechanism may operate? Local antibody might neutralize infectivity by sensitizing organisms to extracellular or intracellular (via lysosomal fusion) inactivation. Or antibody might inhibit ingestion by masking further sites of interaction or by clumping or by anchoring organisms to the epithelial surface. Whether we have the techniques to distinguish between these possibilities is questionable. Recently, after a major reappraisal of amino acid utilization by chlamydiae (I. Allan, unpublished), the problem of low incorporation of radiolabel has been overcome. In future work we can recheck our observations for a range of organism: tear ratios. More important, it will be possible to follow organisms on the conjunctiva and directly measure association *in vivo*.

Chlamydial interaction with the cell surface in centrifuge-assisted infection

As a basic procedure for enhancing chlamydial infection of cell monolayers, centrifugation assumes further importance in generating infected cell cultures for the study of intracellular events in multiplication. It is important therefore to assess how far centrifugation may distort the early events in spontaneous infection.

As already indicated, inhibition of attachment by antibody is overcome in centrifuge-assisted infection. Use of the technique may account for previous failures to detect significant neutralization by antisera (Graham & Layton 1971, Howard 1975). It should also be pointed out that attachment and infection are not invariable consequences of centrifugation. For the GP-IC strain of *C. psittaci,* cell surface properties can be modulated by a single growth passage of cells in an appropriate serum, switching cells from susceptibility to resistance and vice versa (Allan & Pearce 1977).

Enhancement of infection by centrifugation has previously been attributed to the increased number of collisions resulting from sedimentation. This is unlikely at the low centrifugal forces of 1000–2000 *g* that are successfully used for both chlamydiae and viruses. Moreover, for GP-IC, the infectivity titre did not alter when a constant inoculum suspended in volumes varying over a 16-fold range was centrifuged (Allan et al 1977).

In addition to the effects of antibody, three further observations with GP-IC have indicated that centrifugation induces cell surface changes. First, attachment is greatly impaired by centrifugation at 4 °C but is unaffected in spontaneous infection (Allan & Pearce 1979a). Secondly, the pressure generated during centrifuging appears to cooperate with directional force in the enhancement of infectivity. Thus, layering immiscible liquid over the inoculum and centrifuging at 69 *g* gives an infectivity similar to that in the control inoculum centrifuged at 1580 *g* (Allen et al 1977). The low centrifugal force also rules out sedimentation and the requirement for increased pressure suggests some effect of deformation of the cell surface. Thirdly, successive transient cell states that develop during centrifuging affect attachment (Allan & Pearce 1979a). Recognition of these effects came when we analysed the kinetics of two forms of attachment which we termed 'productive' and 'unproductive' binding.

In 'productive binding' organisms attach to, enter and infect cells. This is responsible for the normal infectivity detected after centrifugation. Organisms may also bind 'unproductively'; that is, they attach but are not ingested and can be stripped from cells by trypsin treatment. Remarkably, such organisms can be induced to infect by a combination of cold shock and recentrifugation of cell monolayers. The kinetics of these two types of attachment showed that they occurred sequentially and not simultaneously during centrifugation (Fig. 2). Thus, it appears that when centrifuging begins, only productive binding occurs. This is followed by a refractory state in which little or no binding takes place, and then by a third state during which only unproductive binding occurs.

By introducing inocula onto monolayers which are already being centrifuged one can show that the refractory state develops independently of the presence of organisms. Inoculation after 20 min resulted in a 10-fold reduction in titre. Correspondingly, since the normal cell state is resumed in stationary monolayers, infectivity can be boosted by repeated centrifugation for 20 min (Allan & Pearce 1979a).

In diagnostic laboratories several different cell treatments are used empirically to increase the 'catch' of infectivity from clinical material. Our colleague Jane Hill has found a much smaller proportion of non-productive to productive binding with

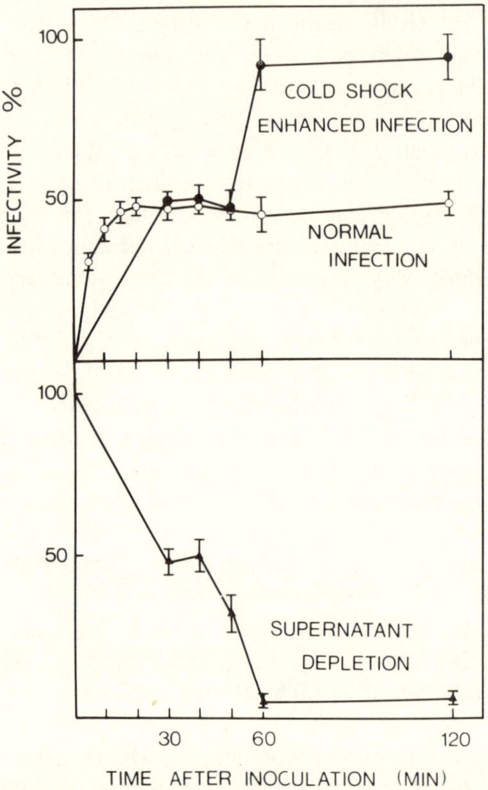

FIG. 2. Kinetics of normal infection, cold-shock-enhanced infection and supernatant depletion. Cell monolayers were centrifuged with GP–IC for varying times. For normal infection (productive binding only, ○) monolayers were incubated for development of infected cells after removal of supernatants. Replicate monolayers were examined for cold-shock-enhanced infection (productive and unproductive binding, ●) after removal of supernatants, cold shock and recentrifugation for 60 min. Supernatants were titrated for cold-shock-enhanced infectivity (▲) on fresh monolayers. All infectivities are expressed as a percentage of the inoculum infectivity, summed from that in monolayer and supernatant at 60 min, measured by cold shock enhancement; results are means ± SD of four replicates. (From Allan et al 1979a by permission of *The Journal of General Microbiology.*)

these treatments. But the same phenomenon of initial rapid attachment, levelling off to a non-binding state, was seen in all situations.

Persuasive as our observations may be, they provide no direct evidence that the rapid attachment at the beginning of centrifugation is due to changes induced on cell surfaces. Precisely how attachment is mediated remains unknown. It may result from specific interactions after moderation of weak repulsive charges in the vicinity of receptors. On the other hand, centrifugation may generate a wholly non-specific interaction. This is suggested by the fact that organisms whose spontaneous infectivity has been inhibited by antibody or treatment with phospholipase C (Allan & Pearce 1979b) retain full infectivity when they are centrifuged. Receptor-competition studies comparing attachment in the two infection modes may make it possible to decide between these alternatives. Studies along such lines are in progress.

Conclusions and prospects

Our understanding of chlamydial attachment mechanisms is primitive. Their molecular basis, role in pathogenesis, relationship with the species and tissue origin of infection, and susceptibility to interference by host defences are virtually unknown.

Progress has been hampered by lack of sensitive tracer techniques. It is hoped that this can now be resolved with improved radiolabelling. But lack of a suitable host cell model for many of the important pathogens – the ocular and genital strains of *C. trachomatis* – remains an obstacle to attachment studies, if centrifugation enhancement of infectivity is ruled out. Possibilities include buccal mucosal and exfoliated urethral cells, as described elsewhere in this symposium.

Acknowledgements

Our work has been supported by grants from the Medical Research Council.

REFERENCES

Ainsworth S, Allan I, Pearce JH 1976 Neutralisation and adherence of chlamydiae to mucous surfaces. Proc Soc Gen Microbiol 4:14-15

Ainsworth S, Allan I, Pearce JH 1979 Differential neutralisation of spontaneous and centrifuge-assisted chlamydial infectivity. J Gen Microbiol 114:61-67

Allan I, Pearce JH 1977 Serum modulation of cell susceptibility to chlamydial infection. FEMS (Fed Eur Microbiol Soc) Lett 1:211-214

Allan I, Pearce JH 1979a Modulation by centrifugation of cell susceptibility to chlamydial infection. J Gen Microbiol 111:87-92

Allan I, Pearce JH 1979b Host modification of chlamydiae:presence of an egg antigen on the surface of chlamydiae grown in the chick embryo. J Gen Microbiol 112:61-66

Allan I, Ainsworth S, Pearce JH 1976 Host-induced modification of the chlamydial surface? Proc Soc Gen Microbiol 4:14

Allan I, Spragg SP, Pearce JH 1977 Pressure and directional force components in centrifuge-assisted chlamydial infection of cell cultures. FEMS (Fed Eur Microbiol Soc) Lett 2:79-82

Byrne GI 1976 Requirements for ingestion of *Chlamydia psittaci* by mouse fibroblasts (L cells). Infect Immun 14:645-651

Byrne GI 1978 Kinetics of phagocytosis of *Chlamydia psittaci* by mouse fibroblasts (L cells): separation of the attachment and ingestion stages. Infect Immun 19:607-612

Byrne GI, Moulder JW 1978 Parasite-specified phagocytosis of *Chlamydia psittaci* and *Chlamydia trachomatis* by L and Hela cells. Infect Immun 19:598-606

Friis RR 1972 Interaction of L cells and *Chlamydia psittaci:* entry of the parasite and host responses to its development. J Bacteriol 110:706-721

Graham DM, Layton JE 1971 The induction of chlamydia group antibody in rabbits inoculated with trachoma agents and demonstration of strain-specific neutralizing antibody in sera. In: Nichols RL (ed) Trachoma and related disorders. Excerpta Medica, Amsterdam, p 145-157

Gregory WW, Byrne GI, Gardner M, Moulder JW 1979a Cytochalasin B does not inhibit ingestion of *Chlamydia psittaci* by mouse fibroblasts (L cells) and mouse peritoneal macrophages. Infect Immun 25:463-466

Gregory WW, Gardner M, Byrne GI, Moulder JW 1979b Arrays of hemispheric surface projections on *Chlamydia psittaci* and *Chlamydia trachomatis* observed by scanning electron microscopy. J Bacteriol 138:241-244

Howard LV 1975 Neutralisation of *Chlamydia trachomatis* in cell culture. Infect Immun 11:698-703

Kuo CC, Grayston JT 1976 Interaction of *Chlamydia trachomatis* organisms and HeLa 229 cells. Infect Immun 13:1103-1109

Kuo CC, Wang SP, Grayston JT 1972 Differentiation of TRIC and LGV organisms based on enhancement of infectivity by DEAE-dextran in cell culture. J Infect Dis 125:313-317

Kuo CC, Wang SP, Grayston JT 1973 Effect of polycations, polyanions and neuraminidase on the infectivity of trachoma-inclusion conjunctivitis and lymphogranuloma venereum organisms in HeLa cells: sialic acid residues as possible receptors for trachoma-inclusion conjunctivitis. Infect Immun 8:74-79

Levy NJ 1979 Wheat germ agglutinin blockage of chlamydial attachment sites: antagonism by N-acetyl-D-glucosamine. Infect Immun 25:946-953

Manire P, Wyrick PB 1979 Cell envelopes of chlamydiae: adaptation for intracellular parasitism. In: Schlessinger D (ed) Microbiology 1979. American Society of Microbiology, Washington, p 111-115

Matsumoto A 1979 Recent progress of electron microscopy in microbiology and its development in future: from a study of the obligate intracellular parasites, chlamydia organisms. J Electron Microsc 28 (suppl): S57-S64

Modabber F, Bear SE, Cerny J 1976 The effect of cyclophosphamide on the recovery from a local chlamydial infection. Immunology 30:929-933

Murray ES, Charbonnet LT, McDonald AB 1973 Immunity to chlamydial infections of the eye. 1. The role of circulatory and secretory antibodies in resistance to reinfection with guinea-pig inclusion conjunctivitis. J Immunol 110:1518-1525

Nichols RL, Oertley RE, Fraser CEO, McDonald AB, McComb DE 1973 Immunity to chlamydial infections of the eye. 6. Homologous neutralisation of trachoma infectivity for the owl monkey conjunctivae by eye secretions from humans with trachoma. J Infect Dis 127:429-432

Patterson S, Oxford JS, Dourmashkin RR 1979 Studies on the mechanism of influenza virus into cells. J Gen Virol 43:223-229

Rank RG, White HJ, Barron AL 1979 Humoral immunity in the resolution of genital infection in female guinea pigs infected with the agent of guinea pig inclusion conjunctivitis. Infect Immun 26:573-579

Willinger M, Gonatas N, Frankel FR 1979 Fate of surface proteins of rabbit polymorphonuclear leukocytes during phagocytosis. 2. Internalisation of proteins. J Cell Biol 82:45-56

DISCUSSION

Taylor-Robinson: You have shown serum modulation of attachment (Allan & Pearce 1977). Are the alterations in attachment due to membrane changes?

Pearce: I think they must reflect cell surface differences. Earlier, tissue cultures were discussed as defined systems. However, for the GP-IC strain of *Chlamydia psittaci* the serum used for growing cells can markedly affect chlamydial attachment during centrifugation. GP-IC attached well to cells grown in serum A but only poorly to cells grown in serum B. Surface differences imposed by the sera must surely have been responsible for these effects.

Hughes: Does this happen if sera A and B are from the same species?

Pearce: Yes, we observed it for fetal bovine and calf sera.

Feizi: How do you make tears flow in the guinea-pigs?

Pearce: It is important not to stimulate reflex tear flow which would dilute antibody present in tear secretions (Little et al 1969). Samples of 1 μl were taken from each eye by gentle suction with a micropipette and tear pools were prepared from 6 or 12 eyes.

Hughes: If you have one adhesion assay where everything is stationary, with no stress and no shear, and another assay system where there are quite large shear forces, as in your centrifugal assay, are you sure the adhesions are the same in each case? That is to say, is it possible that when there is no shear and you have adhesions that are weak, whatever that means, mechanistically, and adhesions that are strong, you can reverse the weak adhesions but not the strong ones with things like antibodies? When there is stress or shear you eliminate the weakly adherent organisms and obtain organisms that are well adapted to sticking to that surface. In other words is some sort of selection involved?

Pearce: There is no evidence that the organism population is heterogeneous in its ability to attach to cells. Spontaneous and centrifuge-assisted infectivity are properties of one organism (Allan & Pearce 1979).

Hughes: I am not saying that you have a heterogeneous population of cells, some of which are weakly adhesive and some of which are strong, but that you have a homogenous population of cells and it so happens that at any one time some have reached a state of stable adherence and others have not.

Pearce: Spontaneous infection increases linearly with time. However, for a given incubation period low or high proportions of cells on a monolayer can be infected by proportional changes in inoculum concentration, indicating that, over that time, all cells are capable of allowing stable association of organisms.

Hughes: What percentage of organisms sticks in stationary assays and what percentage sticks under shear?

Pearce: Measured as infectivity the proportion associating in static incubation is about 1% of that associated after centrifugation.

Watts: Could you say more about the method of centrifugation?

Pearce: A cell monolayer is formed on a glass coverslip contained in a specimen vial. Organism suspension is present as a shallow layer above the monolayer and the whole is centrifuged at 1600 *g*. The mechanism of the enhancement is not sedimentation because varying the inoculum volume over an eight-fold range does not affect infectivity in static incubation, indicating that the latter is not restricted by collision rate (Allan et al 1977).

Bredt: How can 85% of the chlamydiae be removed from a given volume if they don't reach the cells by centrifugation? They certainly have to come down and to contact the surface.

Pearce: I mean that 85% have become attached to that monolayer.

Bredt: But first they have to come down to the monolayer. Is there convection during centrifugation, or something like that?

Pearce: There are probably convection currents and also Brownian motion (Valentine & Allison 1959).

Bredt: But there must be some force which brings the chlamydiae down to the cells before they can attach.

Wallach: Is the centrifugal field applied exactly perpendicular to the monolayer?

Pearce: The vials containing the monolayers are held in swinging buckets.

Wallach: How thick is the layer?

Pearce: The volume is 0.2 ml, giving a depth of about 1 mm.

Candy: If it were just convection currents you should get the same effect by agitating the culture. The centrifugal force must change the cells in some way.

Wallach: With such a thin layer strong convection currents are unlikely. However, you may be forcing nuclei down onto the bottom of the layer. I don't know how that affects attachment.

Beachey: What is the mechanism of the internalization of *Chlamydia* once the organisms attach to the surface of the tissue cells?

Pearce: Ingestion of chlamydiae is apparently by phagocytosis but is insensitive to cytochalasin B or colchicine (Moulder 1979), drugs which inhibit function of cytoskeletal structures in phagocytosis. Entry might be by viropexis, as defined for influenza virus (Patterson et al 1979), a process which seems akin to receptor-mediated endocytosis (Goldstein et al 1979).

Beachey: Do the chlamydia have a specific enzyme that either turns on an ingestive process or allows the organisms to burrow through the cell membrane?

Pearce: Nothing like that is known.

Howard: Why is it so difficult to radiolabel chlamydiae and isolate them?

Pearce: Most workers have found very low incorporation of labelled amino acids into chlamydiae. From recent work (I. Allan, unpublished) it appears that certain amino acids have a regulatory effect in that they are required in large excess over the amount incorporated if multiplication is to occur. Others may be present at low level without inhibition of multiplication. Labelling with a 'regulator' amino acid results in poor incorporation, whereas with a 'non-regulator' amino acid incorporation is considerable.

Howard: Have you tried using labelled precursors of DNA or RNA?

Pearce: Incorporation appears low with these also.

Tramont: Did you try centrifugation with just one tissue culture line or other cell lines.

Pearce: We examined detailed kinetics of association for only McCoy cells but centrifugation enhancement of infection has been found for many other cell lines.

Tramont: Have you used an organ culture system?

Pearce: Yes, some years ago. GP-IC is highly infective for guinea pig conjunctival tissue (Moore et al 1974). In what context were you thinking of use of organ cultures?

Tramont: For measuring attachment or adhesion in the same way as has been done for gonococci.

Pearce: Availability of human or other primate tissue is a problem. Exfoliated or scraped cells could be easily obtained.

Taylor-Robinson: As you mentioned, we have used the Fallopian tube organ culture model for studying the effects of chlamydiae (Hutchinson et al 1979). Although centrifugation is required to enhance infection in McCoy cells, I thought that this might not be needed in an organ culture system with human cells. However, the same phenomenon was seen because we could enhance infection 5 to 20 times by centrifuging the inocula onto the organ cultures. Despite their similarities, there is obviously something different about the organ culture and *in vivo* situations because centrifugation isn't needed to initiate a human infection!

Tramont: Centrifuging produces a non-specific effect on all cells. There is another question as to whether something is going on in the organism. If you centrifuge an organ culture or a McCoy cell or any other tissue culture model then centrifugation must change the organism as well as the cells themselves.

Pearce: Cell surface changes do occur during centrifuging, as is shown by the development of the refractory and non-productive binding states that I mentioned. Evidence that changes occur at the onset of centrifuging in such a way as to enhance infection is wholly circumstantial. It has seemed to us that the requirement for increased pressure suggests deformation of a flexible surface; this is more likely for the host cell plasma membrane. The dependence of centrifuge-assisted infection of GP–IC on raised incubation temperature also argues for cell rather than organism-mediated surface events, bearing in mind that temperature has little influence at the inoculation step on GP–IC spontaneous infection.

Tramont: Certainly gonococci are sensitive to high temperature.

If you leave the culture for a long time do you eventually get attachment of 90%?

Pearce: Infectivity and therefore presumably attachment increase linearly with time in spontaneous infection (Allan & Pearce 1979). However, thermal inactivation limits the increase in infectivity with prolonged incubation.

Taylor-Robinson: In the organ culture system we got a self-limiting infection. Multiplication occurred but then the infection petered out after maybe one or two cycles. It seems to me that the only way of keeping it going would be to centrifuge the organ cultures repeatedly.

Pearce: In organ culture systems there are several reasons why successive cycles of infection may not occur. Perhaps with *C. trachomatis* strains the restriction is that only a small proportion of ingested organisms give rise to productive infection rather than that the efficiency of attachment of organisms to cells is low.

Freter: How does this compare to the natural infection of conjunctivitis? If the same inoculum went into an *in vivo* model would you get an infectivity of 1% or 90%?

Pearce: With GP–IC, infection of the guinea-pig conjunctiva appears to be highly efficient, in accord with the low ID_{50}. Five minutes after inoculation of organisms on to the conjunctiva in the intact animal, up to 10% of the inoculum has become associated with conjunctival tissue (S. Ainsworth, unpublished). It appears to be a very rapid attachment process.

Freter: The rate of attachment is equivalent to centrifugation. What do you mean by 100% infectivity?

Pearce: This is the infectivity of a suspension after centrifugation with a cell monolayer totalled with that for the supernatant titrated on fresh monolayers. For some GP–IC preparations we get close to one infective unit per two particles.

Freter: So 100% would be 50% of the particles?

Pearce: Yes.

Freter: If only 1 in 10 particles were infective normally and if this became one in two after centrifugation, then, even after diluting, you would see that effect and you wouldn't have to postulate convection or anything like that.

Taylor-Robinson: You say that infection occurs after centrifuging the immune complex, but what do you think happens? Does centrifugation break up the complex so that the organism can enter the cell?

Pearce: It seems unlikely that complexes would dissociate during centrifugation. The simplest explanation we can offer, for organisms like GP–IC, which attach poorly to cell monolayers, is that presence of a few antibody molecules per organism is sufficient to impair the weak association that leads to spontaneous infection (Ainsworth et al 1979). Centrifugation promotes a strong association which antibody fails to impede, and attachment, ingestion and multiplication follow, given that the presence of antibody on the chlamydial surface does not interfere with inhibition of lysosomal fusion with the phagocytic vacuole.

Taylor-Robinson: You said that *C. trachomatis* was specific for humans and subhuman primates but I recall that Doris Graham infected the respiratory tract of mice with strains of *C. trachomatis* (Graham 1967).

Pearce: Yes, but that was to produce pneumonitis. Production of ocular and genital disease has generally been considered a highly host-specific effect. However, this view will have to be revised because of recent observations that ocular disease can be induced in cats and guinea-pigs (R. M. Woodland, M. A. Monnickendam & S. Darougar, personal communication 1980).

Taylor-Robinson: Based on what we know about attachment, are there any tricks up your sleeve or other peoples' sleeves whereby the isolation of chlamydiae could be further improved?

Pearce: We need information for more *C. trachomatis* strains to be sure that problems in isolating these are really due to poor attachment. Beyond that, improvements seem likely to depend on what happens after ingestion. Variations in medium and cell cultures affect the extent of productive infective infection in ways that we don't understand. This is highlighted by the problem of chlamydial dormancy where, in cell culture, apparently non-multiplying organisms are triggered into growth when an appropriate amino acid is added to spent medium (Hatch 1975).

Taylor-Robinson: Is the cell population homogeneously sensitive? If not, and if some cells in a tissue culture system are very sensitive, it should be possible to clone them out.

Pearce: Most workers find that infectivity increases linearly with inoculum dose, which suggests that cells are similar in sensitivity. But there must be differences, probably phenotypic, to account for the fact that treating a cell population with cycloheximide can increase the proportion of cells showing productive infection (Ripa & Mårdh 1977).

REFERENCES

Ainsworth S, Allan I, Pearce JH 1979 Differential neutralisation of spontaneous and centrifuge-assisted infectivity. J Gen Microbiol 114:61-67

Allan I, Pearce JH 1977 Serum modulation of cell susceptibility to chlamydial infection. FEMS (Fed Eur Microbiol Soc) Lett 1:211-214

Allan I, Pearce JH 1979 Host modification of chlamydiae: differential infectivity for cell monolayers of chlamydiae grown in eggs and monolayers. J Gen Microbiol 112:53-59

Allan I, Spragg SP, Pearce JH 1977 Pressure and directional force components in centrifuge-assisted chlamydial infection of cell cultures. FEMS (Fed Eur Microbiol Soc) Lett 2:79-82

Goldstein JL, Anderson RGW, Brown MS 1979 Coated pits, coated vesicles, and receptor-mediated endocytosis. Nature (Lond) 279:679-685

Graham DM 1967 Growth and immunogenicity of TRIC agents in mice. Am J Ophthalmol 63:1173-1190

Hatch TP 1975 Competition between *Chlamydia psittaci* and L cells for host isoleucine pools: a limiting factor in chlamydial multiplication. Infect Immun 12:211-220

Hutchinson GR, Taylor-Robinson D, Dourmashkin RR 1979 Growth and effect of chlamydiae in human and bovine oviduct organ cultures. Br J Vener Dis 55:194-202

Little JM, Centifanto YM, Kaufman HE 1969 Immunoglobulins in human tears. Am J Ophthalmol 68:898-905

Moore JE, Griffiths MS, Pearce JH 1974 Chlamydial infection of conjunctival tissues in culture. Br J Exp Pathol 55:396-405

Moulder JW 1979 Interaction of chlamydiae with host cells. In: Schlessinger D (ed) Microbiology 1979. American Society for Microbiology, Washington, p 105-110

Patterson S, Oxford JS, Dourmashkin RR 1979 Studies on the mechanism of influenza virus entry into cells. J Gen Virol 43:223-229

Ripa KP, Mårdh PA 1977 Cultivation of *Chlamydia trachomatis* in cycloheximide-treated cells. J Clin Microbiol 6:328-331

Valentine RC, Allison AC 1959 Virus particle adsorption 1. Theory of adsorption and experiments on the attachment of particles to non-biological surfaces. Biochim Biophys Acta 34:10-33

General Discussion

Glycolipids in receptor assays

Wallach: There have been several discussions about the use of glycolipids in receptor assays and it is appropriate to ask how such molecules might be presented. Both neutral glycolipids and sialoglycolipids are amphipathic molecules that, in aqueous media, form micelles with sugar groups facing water. The critical micelle concentrations of neutral and acidic glycolipids are $<10^{-5}$ to about 10^{-5} M. Both glycolipid categories can be incorporated into phospholipid liposomes in such a way that, under appropriate conditions, the sugar residues distribute in the plane of the liposomal membrane, with the phospholipids as 'spacers'. For this kind of distribution to be achieved the lipid chains of the glycolipids must mix freely under conditions of assay. Temperature is an important factor in this, because amphipathic lipids undergo sharp temperature-dependent changes from a solid phase to a liquid phase. If the phospholipid chosen as spacer is in a solid phase at the assay temperature, the glycolipid will not be distributed as desired, but segregated into a separate phase. Thus, for an assay temperature of 37 °C one can expect mixing with dimyristoyllecithin (transition temperature 23 °C), but phase segregation with dipalmitoyllecithin (transition temperature 41 °C). One may minimize any tendency for glycolipids to segregate by introducing cholesterol at a molar ratio of 0.3/1.0 (cholesterol/phospholipid); cholesterol 'buffers' chain 'fluidity' and fosters mixing.

Beachey: What is the best lipid mixture to use?

Wallach: One could use dimyristoyllecithin, which is cheap and pure, and cholesterol.

Mirelman: How much glycolipid should be added?

Wallach: This depends on the glycolipid. The trick is to get it into the liposomes in a predictable distribution.

Mirelman: But what ratios would you add?

Wallach: It is hard to generalize because it depends on the size and charge of the sugar moiety. You could certainly include glycolipids at an 0.1/1 molar ratio to lecithin in a system containing cholesterol.

Razin: Would the ganglioside liposome system be suitable for transferring glycolipids to the erythrocyte membrane?

Wallach: Yes. GM_1 has been transferred to pigeon erythrocytes, for example (King et al 1976).

Razin: For phospholipid you really need a transport protein to transfer it from one bilayer to another.

Levine: Is there anything special about the pigeon erythrocytes that the GM_1 ganglioside is put on?

Wallach: The idea was that one could measure the effect of the receptor for cholera toxin because this is low in these cells, although they possess adequate adenylate cyclase.

Feizi: But if you intend to incorporate glycolipids into membranes you have to avoid using carrier lipids. If you want to minimize entry into red cell membranes while you are using a ganglioside to inhibit the binding of an antibody to red cell, you need to incorporate the ganglioside into micelles. Otherwise the lipid may bind to red cells and give enhanced agglutination at ambient or higher temperatures.

Helenius: Many glycolipids have detergent-like properties. A detergent in a mixture with lipids has an equilibrium-free concentration which is lower than in a pure system. When present in a bilayer a detergent will therefore partition into another bilayer more slowly than if it is free in solution.

Candy: One of the original reasons for choosing pigeon erythrocytes to study cholera toxin was that these erythrocytes had very low levels of NAG glycohydrolase activity (Gill 1977).

Levine: That experiment was done because it was shown that, with the low level of adenylate cyclase activity in the pigeon erythrocyte, as one gives increasing doses of cholera toxin one can measure increasing adenylate cyclase activity, which then reaches a plateau (King et al 1976). However when one treats the erythrocytes with GM_1 ganglioside the threshold goes much higher. That shows that when the extra ganglioside is not present subunit B of cholera toxin saturates all the subunit B receptors, so one is limiting the amount of subunit A that could get inside to turn on the adenylate cyclase. You then use exogenous B subunit receptors (GM_1 ganglioside), artificially putting it onto the erythrocyte membrane. By getting more B subunit attached you get more A subunit inside, and the level of adenylate cyclase gets even higher.

Candy: But if you use any cell membrane other than from pigeon erythrocytes you have to keep putting in more NAD, which is an essential cofactor for adenylate cyclase activation by cholera toxin. Other cell membranes rapidly hydrolyse NAD by NAD glycohydrolase activity and become refractory to cholera toxin (Gill 1977).

Levine: In terms of what erythrocytes that GM_1 ganglioside would stick to, I don't know whether the pigeon erythrocyte has a surface that is different from other erythrocytes.

Wallach: It has very low levels of GM_1.

REFERENCES

Gill DM 1977 Mechanism of action of cholera toxin. In: Greengard P, Robison GA (eds) Advances in cyclic nucleotide research. Raven Press, New York, vol 8:93-94

King CA, van Heyningen WE, Gascoyne N 1976 Aspects of the interaction of *Vibrio cholerae* toxin with the pigeon red cell membrane. J Infect Dis 133 (suppl): S75-S81

Functions of surface glycoproteins of myxoviruses and paramyxoviruses and their inhibition

PURNELL W. CHOPPIN, CHRISTOPHER D. RICHARDSON, DAVID C. MERZ and ANDREAS SCHEID

The Rockefeller University, 1230 York Avenue, New York, NY 10021, USA

Abstract Two glycoproteins, HN and F, are present on the surface of paramyxoviruses. HN has receptor-binding and neuraminidase activities. F is involved in viral penetration, cell fusion and haemolysis and is activated by proteolytic cleavage by a host enzyme into two disulphide-bonded subunits (F_1 and F_2). The ability of the virus to initiate infection and undergo multiple cycle replication depends on the presence of an activating protease in the host; thus cleavage of F is a major determinant of pathogenesis. The new N-terminus generated on F_1 by cleavage is involved in biological activity, and the amino acid sequence of this region of F_1 is hydrophobic and highly conserved among paramyxoviruses. In an attempt to design specific inhibitors, oligopeptides analogous to this region were synthesized and found to be highly active, specific inhibitors of viral penetration, cell fusion and haemolysis. Inhibition is amino-acid-sequence-specific and affected by peptide length, steric configuration and addition of groups to the N-terminal and C-terminal amino acids. Replication of influenza virus was also specifically inhibited by oligopeptides resembling the N-terminus of the HA_2 polypeptide. Like that of F_1 protein the N-terminus of HA_2 is generated by a proteolytic cleavage that activates infectivity. These results have provided information on the action of proteins in viral penetration and membrane fusion and they suggest a possible new approach to chemical inhibition of viral replication. Studies with specific antibodies to each of the paramyxovirus glycoproteins have shown that antibodies to the F protein are essential for effective prevention of the spread of infection. Antibodies to the HN protein, although capable of neutralizing released virus, do not prevent spread to adjacent cells through membrane fusion mediated by the F protein. These findings have implications for the design of effective vaccines against paramyxociruses and also provide additional insight into the mechanisms involved in the atypical and severe infections observed in individuals who received inactivated paramyxovirus vaccines and were later infected.

1981 Adhesion and microorganism pathogenicity. Pitman Medical, Tunbridge Wells (Ciba Foundation symposium 80) p 252-269

Evidence obtained in many laboratories with different viruses has established that the membrane-enclosed animal viruses have one or more glycoproteins on their surfaces (reviewed in Compans & Klenk 1979 and Choppin & Scheid 1980). These glycoproteins form spike-like projections from the lipid bilayer of the viral membrane in which their bases are embedded. Although much remains to be learnt about the structure and functions of viral glycoproteins, with some viruses a great deal of information is available, and it is possible to make some generalizations. Viral glycoproteins are involved in the adsorption of the virus to the host cell and, in some cases, in the penetration of the virus or the viral genome into the cell. For most viruses the chemical nature of the receptors has not been determined. However, it has been established that the myxoviruses and paramyxoviruses react with receptors containing neuraminic acid and in recent years much information has been obtained on the specific sites on these viruses that are involved in the early interactions with the host cell. This aspect will be the major focus of this paper. In addition to virus adsorption and penetration, it has also been shown in several virus systems that the immunological reactions important in the prevention of infection involve the viral surface glycoproteins, and that in some instances these proteins may have a direct toxic effect on cells, causing cell lysis, cell fusion or inhibition of cellular biosynthesis.

Structure and function of paramyxovirus glycoproteins

The paramyxovirus family contains a large number of viruses, including the parainfluenza viruses (of which Sendai virus and simian virus 5 [SV5] are prototypes), mumps, Newcastle disease virus (NDV), measles, canine distemper and several others. These viruses cause a wide variety of diseases in humans and lower animals, ranging from mild respiratory infections to chronic neurological diseases. Our laboratory has been engaged for several years in the isolation and biological and biochemical characterization of the paramyxovirus glycoproteins. These studies have been reviewed previously (Choppin & Compans 1975, Scheid et al 1978, Choppin & Scheid 1980) and will be briefly summarized here, with emphasis on recent results on the inhibition of the functions of these glycoproteins (Richardson et al 1980, Merz et al 1980).

Two glycoproteins, designated HN and F, are associated with the paramyxovirus envelope and have been isolated and purified in biologically active form by the use of non-ionic detergents and various chromatographic and sedimentation methods (Scheid et al 1972, Scheid & Choppin 1973, 1974a, b, Tozawa et al 1973, Hsu et al 1979). These proteins form spike-like projections about 10 nm long on the surface of the virion, and are anchored in the viral membrane by a hydrophobic portion at the C-terminal end of the protein.

The HN protein

The HN protein, which has a relative molecular mass (M_r) of 65 000–70 000, depending on the virus strain, is present on the surface of the virion as a dimer, held together by disulphide bonds in the hydrophilic region and hydrophobic bonds at the base of the protein (Scheid et al 1978). The isolated HN protein has receptor binding activity, which is manifested by haemagglutination when the erythrocyte is used as the target cell, and by neuraminidase activity, which is capable of destroying the receptors for the virus (Scheid et al 1972, Scheid & Choppin 1973). Thus, the HN protein is responsible for the first step in infection, the adsorption of the virus to the cell. Under appropriate conditions the hydrophobic portion of the HN protein of SV5 (M_r about 5000) can be removed by a proteolytic enzyme, resulting in a water-soluble protein with an M_r of about 59 000 which retains receptor-binding and neuraminidase activity, although it cannot function as an agglutinin because it is monovalent (Scheid et al 1978). The intact HN molecules aggregate by their bases when detergent is removed, forming rosette-like clusters which are multivalent and thus can cause haemagglutination (Scheid et al 1972).

The F protein

This paramyxovirus glycoprotein has an M_r of about 65 000 and is cleaved by a host cell enzyme to yield two disulphide-bonded polypeptides, F_1 and F_2, with M_r values of about 50 000 and 15 000, respectively (Homma & Ouchi 1973, Scheid & Choppin 1974a, 1976, 1977). The F_1 polypeptide contains the C-terminus which is associated with the viral membrane, and F_2 contains the original N-terminus. The F protein is responsible for several biological activities which involve membrane fusion, i.e. penetration of the virus into the host cell, cell fusion and haemolysis. The isolated F protein has no biological activity as a pure protein; however Hsu et al (1979) recently demonstrated that when pure F protein is reconstituted into a membrane with lipids it is biologically active, if a means of attachment of the reconstituted membrane to the cell is provided. This can be provided by either the HN protein or a lectin, e.g. wheat germ agglutinin. However, the evidence for the biological activity of the F protein came originally from studies showing that the proteolytic cleavage of the precursor protein (F_0), to yield F_1 and F_2, activated virus-induced cell fusion, haemolysis and the initiation of infection (Homma & Ouchi 1973, Scheid & Choppin 1974a, 1976, 1977). This cleavage is normally accomplished by a host protease and some cells lack the appropriate enzyme to cleave the protein of certain viruses. For example, bovine kidney (MDBK) or mouse fibroblast (L) cells lack an enzyme capable of cleaving the Sendai virus F_0, and they produce non-infectious virions which are also incapable of causing cell fusion and haemolysis. These virions can be activated *in vitro* by trypsin. Mutants have been isolated which require a different protease from wild-type viruses for activation and these mutants exhibit a different host range at the levels of both cultured cells and the chick embryo (Scheid & Choppin 1976).

Because the ability of the virus to initiate infection, and therefore to undergo multiple-cycle replication, depends on the activation of the F protein by an appropriate protease supplied by the host, these results indicated that the host range and tissue tropism of the virus, and its ability to spread in the host and ultimately cause disease, depend on the availability of the appropriate activating protease in the host (Scheid & Choppin 1975, 1976, Choppin & Scheid 1977). On the basis of these findings we postulated that virus virulence in the natural animal host would also depend on the susceptibility of the virus to cleavage by host enzymes. We suggested as examples the virulent and avirulent strains of NDV which had been isolated over the years (Scheid & Choppin 1975), because retrospective analysis of the gels published by Lomniczi et al (1971) suggested that F_0 was present in cells infected with the less virulent strains but not the virulent strains.

Proteolytic cleavage was subsequently clearly shown to have a role in viral pathogenesis in adult animals in extensive studies by Nagai et al (1976) and Nagai & Klenk (1977). They examined many strains of NDV which were virulent or avirulent for the chicken; in each case virulence correlated with the ability of chicken fibroblasts to cleave the F protein of the virus.

The biological importance of the cleavage of the F protein led to a study of the structure of the protein around the cleavage site. As a result of the cleavage a new N-terminus is generated on the F_1 protein (Scheid & Choppin 1977). Several lines of evidence suggested that this new N-terminus is involved in the biological activities of the protein. First, the expression of the activities is dependent on the cleavage. Secondly, the amino acid sequences beginning at this N terminus have been determined for three different paramyxoviruses with different natural hosts (SV5, NDV and Sendai virus) and the sequences are highly conserved (Scheid et al 1978, Gething et al 1978, Richardson et al 1980). The first 20 amino acids of these F_1 polypeptides are shown below.

Sen	Phe-Phe-Gly-Ala-Val-Ile-Gly-Thr-Ile -Ala-Leu-Gly-Val-Ala-Thr-Ala-Ala-Gln-Ile -Thr
SV5	Phe-Ala-Gly-Val-Val-Ile-Gly-Leu-Ala-Ala-Leu-Gly-Val-Ala-Thr-Ala-Ala-Gln-Val-Thr
NDV	Phe-Ile -Gly-Ala-Ile -Ile-Gly-Gly-Val-Ala-Leu-Gly-Val-Ala-Thr-Ala-Ala-Gln-Ile -Thr

There is a very high degree of homology in these sequences, with variation occurring at only six positions (underlined).

The importance of this region of the protein was further demonstrated by the finding that a mutant of Sendai virus, which is activated by a different protease from wild-type virus, had the same N-terminal sequence as the wild type, i.e.

Phe-Phe-Gly. There are no charged residues in the above sequences, and the extremely hydrophobic nature of this region of the molecule raised the possibility that it could be involved in hydrophobic interactions with the target cell membrane. Finally, we noted a similarity between these N-terminal sequences of the F_1 polypeptide and an oligopeptide which had been found earlier to inhibit plaque formation, haemolysis and cell fusion by measles virus (Nicolaides et al 1968, Miller et al 1968, Norrby 1971, Graves et al 1978) and SV5 (P. W. Choppin, unpublished experiments).

Because of these indications that the N-terminus of the F_1 polypeptide of paramyxoviruses was involved in the biological activities of the F protein we reasoned that it might be possible to specifically inhibit these activities with oligopeptides which resembled this N-terminus. We therefore synthesized a number of oligopeptides, using the N-terminal region of the F_1 polypeptide of Sendai virus as the primary model, and tested them against Sendai, SV5 and measles viruses. It was shown with some oligopeptides that they could inhibit haemolysis and cell fusion. However, in most experiments we examined inhibition of infectivity at the level of penetration, using plaque formation as the assay method because of its greater sensitivity. In each experiment several concentrations of each polypeptide were tested against the virus and dose–response curves were constructed. From these curves 50% effective concentrations were ascertained, which provide useful values for comparison.

Sendai, SV5 and measles viruses were all inhibited by the appropriate oligopeptides; however, measles virus was much more sensitive than the other two to their action; e.g. the 50% effective concentrations of Z-Phe-Phe-Gly against SV5, Sendai and measles viruses were 400, 320 and 0.2 μM, respectively (Z is used to designate the benzyloxycarbonyl group). Because of its greater sensitivity, detailed structure–activity studies were done with measles virus. These results are described elsewhere (Richardson et al 1980) and are only briefly summarized here. Table 1 lists the 50% effective concentrations of some of the many oligopeptides that we tested to illustrate important points. As can be seen, oligopeptides with the appropriate amino acid sequence were highly active inhibitors. The most active inhibitor (0.02 μM) was a heptapeptide with the sequence of the N-terminal of Sendai virus. A tripeptide (Z-Phe-Phe-Gly) and a dipeptide (Z-Phe-Phe) with the same initial sequence, although quite active, showed progressively less inhibitory activity with decreased length of the peptide chain. Thus the length of the peptide is a significant factor. The most important factor, however, was the correct sequence; Phe-Phe-Gly, for example, was several orders of magnitude more active than Gly-Phe-Phe. This sequence specificity was also strikingly illustrated by a mutant of measles virus that was selected for resistance to Z-Phe-Phe-L(NO_2)Arg. Whereas the wild-type virus was sensitive to both this tripeptide and Z-Phe-Phe-Gly at a concentration of 0.2 μM, this mutant was unaffected by the former peptide at a concentration of >1000 μM but remained sensitive to the latter at 21 μM, thus demonstrating the importance of the third amino acid in activity. Table 1 also shows that the addition of the benzyl-

TABLE 1 Inhibition of measles virus plaque formation by synthetic oligopeptides

Peptide	*50% effective concentration* (μM)
Z-D-Phe-L-Phe-Gly-D-Ala-D-Val-D-Ile-Gly	0.02
Z-D-Phe-L-Phe-Gly	0.20
Z-D-Phe-L-Phe-Gly-(methyl ester)	20
Z-L-Phe-L-Phe-Gly	23
Z-D-Phe-L-Phe	28
Z-L-Phe-L-Phe	42
Z-L-Phe-L-Ser	141
Z-Gly-L-Phe-L-Phe	530
D-Phe-L-Phe-Gly-D-Ala-D-Val-D-Ile-Gly	130
D-Phe-L-Phe-Gly	180

Z: benzyloxycarbonyl group.

oxycarbonyl group to the N-terminus greatly increases activity (cf. Z-D-Phe-L-Phe-Gly and D-Phe-L-Phe-Gly), and the esterification of the C-terminal amino acid decreases activity (cf. Z-D-Phe-L-Phe-Gly and Z-L-Phe-L-Phe-Gly-(methyl ester)). The explanation for the effects of the addition of these groups on inhibitory activity is not clear; however, they could play a role in the positioning of the inhibitor at its site of action, because the benzyloxycarbonyl group contributes additional hydrophobicity at the N-terminus, and the esterification of the C-terminal amino acid would decrease the polarity of the peptide, and these changes could affect the orientation of the peptide. The steric configuration of the N-terminal amino acid is also important, with greater activity obtained with D-phenylalanine as the N-terminal amino acid, rather than with the naturally occurring L-phenylalanine. The possibility was considered that this could be related in part to protection by the D-amino acid from chymotryptic degradation in the assay system; however, it seems more likely from the available data that a steric effect is directly involved in inhibitory activity. The D-phenylalanine-containing peptide was also more active in inhibiting haemolysis, an assay in which chymotryptic action is less likely to be a factor than in plaque formation. It is clear, too, that the D-amino acid-containing peptides do not act by simply inhibiting the metabolic activity of the cell. Indeed, none of the oligopeptides tested appears to be significantly toxic in the system studied. The cells survive for many days in the presence of the peptides, show no detectable cytopathic changes, and can multiply normally.

These results have indicated that oligopeptides with amino acid sequences which resemble the N-terminal region of the F_1 polypeptide of paramyxoviruses are highly active specific inhibitors of viral infectivity and of virus-induced cell fusion and haemolysis, activities which involve fusion of viral and cell membranes. Table 2 summarizes some of the characteristics of this inhibitory activity. The results suggest that the inhibitors competitively interfere with the N-terminal region of the

TABLE 2 Characteristics of the inhibition of paramyxoviruses by oligopeptides resembling the N-terminal region of the viral F_1 polypeptide

Inhibition is amino-acid-sequence-specific and longer oligopeptides are more active
The steric configuration of some amino acids affects activity
The presence of a benzyloxycarbonyl group on the N-terminal amino acid increases activity
Esterification of the C-terminal amino acid decreases activity

F_1 polypeptide. It is not yet clear how, or at precisely what site, these inhibitors act – that is, whether they act on the F_1 polypeptide of the virus or on a cell membrane receptor. However, their availability provides an additional means of not only determining how they act but also of determining the exact site and mechanism by which the F protein initiates infection, and of investigating the biochemical and physical events in virus-induced cell fusion and haemolysis, and membrane fusion in general.

Myxovirus glycoproteins

The myxoviruses (influenza viruses) also possess two glycoproteins on their surface, the haemagglutinin (HA) and neuraminidase (NA) proteins. Thus in myxoviruses, unlike the paramyxoviruses, the receptor-binding (haemagglutinating) and neuraminidase functions reside in different proteins. However, as with paramyxoviruses, there is proteolytic cleavage of one of these glycoproteins (HA) by a host protease (Lazarowitz et al 1971). This cleavage, which can be accomplished by a cellular enzyme or plasmin in serum, yields two disulphide-bonded polypeptides (HA_1 and HA_2), with a new terminus being generated on the HA_2 polypeptide, the C-terminus of which is embedded in the viral membrane (Laver 1971, Lazarowitz et al 1971, 1973, Skehel & Waterfield 1975). Although this cleavage does not affect haemagglutinating activity it does activate the infectivity of the virus, presumably at the level of penetration (Klenk et al 1975, Lazarowitz & Choppin 1975). Furthermore, there is a correlation between the virulence of avian influenza viruses for the chicken and the cleavability of the HA protein by avian cell proteases (Bosch et al 1979). Thus, there is a structural and functional analogy between the F protein of paramyxoviruses and the HA protein of myxoviruses: both are involved in the initiation of infection and are activated by cleavage by a host protease to yield two disulphide-bonded polypeptides. This analogy is strengthened by the fact that the sequence of the first nine amino acids of the new N-terminus that is generated by cleavage on the HA_2 polypeptide of myxoviruses resembles the sequence of the N-terminus of the F_1 polypeptide of paramyxoviruses, except that in myxoviruses an N-terminal glycine precedes phenylalanine, which is the N-terminus of the F_1 polypeptide of paramyxoviruses. The sequence found by Skehel & Waterfield (1975) for the Lee strain of the influenza B is: Gly-Phe-Phe-Gly-Ala-Ile-Ala-Gly-Phe-Leu. With several strains of influenza A virus these authors also found the same sequence, except that leucine was substituted for the first phenylalanine and isoleucine for the terminal leucine. Thus this sequence is highly conserved in the various strains of influenza virus and is strikingly similar to that of the F_1 polypeptide of paramyxoviruses. (Compare the influenza virus HA_2 sequence with those of the paramyxoviruses given above.)

Because of these structural and functional similarities between the HA and F proteins of these two viruses, and the success we had obtained with oligopeptide inhibitors of paramyxoviruses, we synthesized oligopeptides which resembled the N-termini of the HA_2 polypeptides of influenza viruses and tested them for their ability to inhibit the replication of these viruses. With influenza virus also, there was amino-acid-sequence-specific inhibition. Table 3 shows the inhibitory activity of three peptides. Z-Gly-L-Leu-L-Phe-Gly and Z-Gly-L-Phe-L-Phe-Gly, which resemble the N-terminal sequences of influenza A and B viruses, respectively, were highly active against the WSN strain of influenza A virus. However, Z-D-Phe-L-Phe-Gly, which resembles the paramyxovirus sequence, was much less active against influenza virus. As shown above (cf. Table 1), the reverse was true with paramyxoviruses, indicating that with influenza virus also, there is amino-acid-sequence-specific inhibition, with the correct N-terminal amino acid being very important for optimum activity. These findings emphasize the biological importance of this region of the HA protein and the functional relatedness of the HA and F proteins of the two different groups of viruses. They also support the concept that the activation of the infectivity of influenza virus by proteolytic cleavage involves the penetration step and that fusion between viral and cell membranes may be involved in penetration, as it is with the paramyxoviruses. Further studies with these specific oligopeptide inhibitors should provide additional knowledge of the early interactions of both groups of viruses with cell membranes, and reveal whether the use of such inhibitors may provide a new approach to the control of virus infections.

TABLE 3 Inhibition of the WSN strain of influenza A virus by oligopeptides

Peptide	*50% effective concentration (μM)*
Z-Gly-L-Leu-L-Phe-Gly	20
Z-Gly-L-Phe-L-Phe-Gly	53
Z-D-Phe-L-Phe-Gly	290

Z:benzyloxycarbonyl group.

Immunological prevention of the spread of paramyxovirus infections

As indicated in the introductory comments above, it has long been known that the surface glycoproteins of enveloped viruses are involved in the generation of immunity to infection; however, an evaluation of the importance of antibodies to the individual paramyxovirus proteins had to await the preparation of specific antibodies to each protein and evaluation of their role in preventing the initiation and spread of infection. Using the parainfluenza virus SV5 as a model we prepared antibodies against each of the envelope proteins and investigated their inhibitory activities (Merz et al 1980, and unpublished experiments). As expected from previous studies of the biological activities of the isolated proteins, antibodies against the HN protein inhibited haemagglutinating and neuraminidase activities and neutralized the infectivity of the virus by interfering with adsorption in a conventional neutralization test in which virus and antibodies were mixed before inoculation of cells. Similar results had been found with NDV (Seto et al 1974). Antibodies to the SV5 F protein inhibited cell-fusing and haemolysing activities and also neutralized infectivity in the conventional neutralization test by interfering with viral penetration. In addition, however, interesting and significant results were obtained when the ability of the antibodies to inhibit the spread of infection from cell to cell was investigated (Merz et al 1980). In these experiments the antibodies were added to a monolayer of cells after a few cells had been infected. These conditions were designed to mimic, more closely than a conventional neutralization test, the conditions in an individual whose respiratory epithelium has been seeded with a virus inoculum. In such a population of cells virus can spread either by released virus adsorbing to and infecting other cells, or by fusion of the membrane of an infected cell with that of an adjacent cell as a result of the cell-fusing activity of the F protein. It was found that whereas antibodies to the HN protein could inhibit the dissemination of infection by released virus, they could not prevent the spread of infection by membrane fusion. Thus, in cells which were susceptible to the fusing activity of the F protein, there was spread from cell to cell in the presence of anti-HN antibodies. By contrast, antibodies to the F protein completely prevented the spread of infection because they were capable not only of neutralizing released virus but also of preventing cell-to-cell spread by membrane fusion by inhibiting the activity of the F protein.

These results have indicated that effective immunological prevention of the spread of paramyxovirus infections must involve inactivation of the F protein, and that an effective vaccine must induce antibodies to this protein. These findings also provide a possible explanation for the previous failures of formalin-killed paramyxovirus vaccines, and for the atypical and severe infections that have been observed in individuals who received killed measles virus or respiratory syncytial virus vaccines and were subsequently infected with the virus (Fulginiti et al 1967, Chanock et al 1968). Norrby and coworkers (Norrby & Gollmar 1975, Norrby et al 1975, Norrby

& Penttinen 1978) have shown that formalin or Tween–ether-inactivated measles and mumps virus vaccines induced haemagglutinating-inhibiting but not haemolysis-inhibiting antibodies (the latter are now known to be anti-F antibodies). They suggested that the atypical measles in individuals receiving such vaccines might be related to the lack of these antibodies. Others have pointed out that the atypical measles and severe respiratory syncytial infections in individuals who received vaccines had several features of an immunopathological process (Buser 1967, Scott & Bonnano 1967, Chanock et al 1968).

Correlation of our findings (Merz et al 1980) that anti-F antibodies are required to completely prevent spread of paramyxovirus infections with the previous observations on lack of antibodies to F in individuals receiving the vaccines, and with the clinical findings suggesting an immunopathological reaction, has led us to the following hypothetical explanation for the atypical infections. The killed vaccines induce antibodies to the HN (or H, in measles) protein. When these individuals are exposed to the virus, some cells in the respiratory tract are infected and the infection can spread to adjacent cells by fusion, because of the lack of anti-F antibodies. As infection spreads in this manner, virus antigens are produced and released; they then serve as secondary antigenic stimuli, resulting in a hyperimmune response to H and the other viral proteins. Recent studies have shown that the convalescent sera of patients with atypical measles contain high levels of viral antibodies not only to H but also to other viral proteins (Hall et al 1979, W. W. Hall, M. H. Kaplan & P. W. Choppin, unpublished experiments). Therefore there is a situation in which viral replication and spread by fusion is continuing at the same time as a secondary immune response is being mounted to the antigens being produced. This sequence of events, discussed in detail in Merz et al (1980), could explain the clinical and pathological findings observed in atypical measles, and possibly in severe respiratory syncytial virus infections, although it is not yet known which respiratory syncytial virus protein is involved in cell fusion or whether formalin inactivates the antigenicity of that protein, as it does with the other paramyxovirus F proteins.

The previous failures of the formalin-killed paramyxovirus vaccines resulted in a loss of interest in the possibility of developing inactivated vaccines for these viruses. However, now that it is known that anti-F antibodies are essential, this situation should be re-evaluated, particularly since methods are now available for isolating the proteins in biologically active form, which has the dual advantage of avoiding the use of formalin and obtaining a pure protein (Scheid et al 1972, Scheid & Choppin 1974b, Hsu et al 1979, Merz et al 1980). Vaccines consisting of isolated purified glycoproteins alone or in combination with lipids, could be important additions to the available immunization procedures against diseases caused by paramyxoviruses.

Acknowledgements

Research described here was supported by research grants AI-05600 from the National Institute of Allergy and Infectious Diseases, CA-18213 from the National Cancer Institute, and PCM-09091 from the National Science Foundation. C.D.R. is a Canadian Medical Research Council Postdoctoral Fellow and D.C.M. was a Predoctoral Trainee under grant CA-09256 from the National Cancer Institute.

REFERENCES

Bosch FX, Orlich M, Klenk H-D, Rott R 1979 The structure of the haemagglutinin, a determinant for the pathogenicity of influenza viruses. Virology 95:197-207

Buser F 1967 Side reaction to measles vaccination suggesting the Arthus phenomenon. New Engl J Med 277:250-251

Chanock RM, Parrott RH, Kapikian AZ, Kim HW, Brandt CD 1968 Possible role of immunological factors in pathogenesis of RS lower respiratory tract disease. In: Pollard M (ed) Virus-induced immunopathology. Academic Press, New York (Perspectives in virology 6) p 125-139

Choppin PW, Compans RW 1975 Reproduction of paramyxoviruses. In: Fraenkel-Conrat H, Wagner RR (eds) Comprehensive virology. Plenum Press, New York, vol 4:95-178

Choppin PW, Scheid A 1977 The biological role of host-dependent proteolytic cleavage of a paramyxovirus glycoprotein. In: ter Meulen V, Katz M (eds) Proceedings of a conference on slow virus infections. Springer-Verlag, New York, p 129-136

Choppin PW, Scheid A 1980 The role of viral glycoproteins in adsorption, penetration, and pathogenicity of viruses. Rev Infect Dis 2:40-61

Compans RW, Klenk H-D 1979 Viral membranes. In: Fraenkel-Conrat H, Wagner RR (eds) Comprehensive virology. Plenum Press, New York, vol 13:293-408

Fulginiti VA, Eller JJ, Downie AW, Kempe CH 1967 Altered reactivity to measles virus. Atypical measles in children previously immunized with inactivated measles virus vaccine. J Am Med Assoc 202:101-106

Gething MJ, White JM, Waterfield MD 1978 Purification of the fusion protein of Sendai virus: analysis of the NH_2-terminal sequence generated during precursor activation. Proc Natl Acad Sci USA 75:2737-2740

Graves MC, Silver SM, Choppin PW 1978 Measles virus polypeptide synthesis in infected cells. Virology 86:254-263

Hall WW, Lamb RA, Choppin PW 1979 Measles and SSPE virus proteins: lack of antibodies to the M protein in patients with subacute sclerosing panencephalitis. Proc Natl Acad Sci USA 76:2047-2051

Homma M, Ouchi M 1973 Trypsin action on the growth of Sendai virus in tissue culture cells. III. Structural difference of Sendai viruses grown in eggs and in tissue culture cells. J Virol 12:1457-1465

Hsu M-C, Scheid A, Choppin PW 1979 Reconstitution of membranes with individual paramyxovirus glycoproteins and phospholipid in cholate solution. Virology 95:476-491

Klenk H-D, Rott R, Orlich M, Blodorn J 1975 Activation of influenza A viruses by trypsin treatment. Virology 68:426-439

Laver WG 1971 Separation of two polypeptide chains from the haemagglutinin subunit of influenza virus. Virology 45:275-288

Lazarowitz SG, Choppin PW 1975 Enhancement of the infectivity of influenza A and B viruses by proteolytic cleavage of the haemagglutinin polypeptide. Virology 68:440-454

Lazarowitz SG, Compans RW, Choppin PW 1971 Influenza virus structural and non-structural proteins in infected cells and their plasma membranes. Virology 46:830-843

Lazarowitz SG, Goldberg AR, Choppin PW 1973 Proteolytic cleavage by plasmin of the HA polypeptide of influenza virus. Virology 56:172-180

Lomniczi B, Meager A, Burke DC 1971 Virus RNA and protein synthesis in cells infected with different strains of Newcastle disease virus. J Gen Virol 13:111-120

Merz DC, Scheid A, Choppin PW 1980 The importance of antibodies to the fusion glycoprotein (F) of paramyxoviruses in the prevention of the spread of infection. J Exp Med 151:275-288

Miller FA, Dixon GJ, Aynett G, Dice JR, Rightsel WA, Schabel FM, McLean JW 1968 Antiviral activity of carbobenzoxy di- and tripeptides on measles virus. Appl Microbiol 16:1489-1496

Nagai Y, Klenk H-D 1977 Activation of precursors to both glycoproteins of Newcastle disease virus by proteolytic cleavage. Virology 77:125-134

Nagai Y, Klenk H-D, Rott R 1976 Proteolytic cleavage of viral glycoproteins and its significance for the virulence of Newcastle disease virus. Virology 72:494-508

Nicolaides F, DeWald H, Westland R, Lipnick M, Posley J 1968 Potential antiviral agents. Carbobenzoxy di- and tri-peptides active against measles and herpes viruses. J Med Chem 11:74-79

Norrby E 1971 The effect of a carbobenzoxy tripeptide on the biological activities of measles virus. Virology 44:599-608

Norrby E, Gollmar Y 1975 Identification of measles virus-specific haemolysis-inhibiting antibodies separate from haemagglutination-inhibiting antibodies. Infect Immun 11:231-239

Norrby E, Penttinen K 1978 Differences in antibodies to the surface components of mumps virus after immunization with formalin-inactivated and live virus vaccines. J Infect Dis 138:672-676

Norrby E, Enders-Ruckle G, ter Meulen V 1975 Differences in the appearance of antibodies to structural components of measles virus after immunization with inactivated and live virus. J Infect Dis 132:262-269

Richardson CD, Scheid A, Choppin PW 1980 Specific inhibition of paramyxovirus and myxovirus replication by oligopeptides with amino acid sequences similar to those at the N-termini of the F_1 or HA_2 viral polypeptides. Virology 105:205-222

Scheid A, Choppin PW 1973 Isolation and purification of the envelope proteins of Newcastle disease virus. J Virol 11:263-271

Scheid A, Choppin PW 1974a Identification of biological activities of paramyxovirus glycoproteins. Activation of cell fusion hemolysis and infectivity by proteolytic cleavage of an inactive precursor protein of Sendai virus. Virology 57:475-490

Scheid A, Choppin PW 1974b The hemagglutinating and neuraminidase protein of a paramyxovirus: Interaction with neuraminic acid in affinity chromatography. Virology 62:125-133

Scheid A, Choppin PW 1975 Activation of cell fusion and infectivity by proteolytic cleavage of a Sendai virus glycoprotein. In: Reich E et al (eds) Proteases and biological control. Cold Spring Harbor Laboratory, Cold Spring Harbor, p 645-659

Scheid A, Choppin PW 1976 Protease activation mutants of Sendai virus: Activation of biological properties by specific proteases. Virology 69:265-277

Scheid A, Choppin PW 1977 Two disulfide linked polypeptide chains constitute the active F protein of paramyxoviruses. Virology 80:54-66

Scheid A, Caliguiri LA, Compans RW, Choppin PW 1972 Isolation of paramyxovirus glycoproteins, Association of both haemagglutinating and neuraminidase activities with the larger SV5 glycoprotein. Virology 50:640-652

Scheid A, Graves MS, Silver SM, Choppin PW 1978 Studies on the structure and function of paramyxovirus glycoproteins. In: Mahy BWJ, Barry RD (eds) Negative strand viruses and the host cell. Academic Press, London, p 181-193

Scott TF McN, Bonnano DE 1967 Reactions to live-measles virus vaccine in children previously inoculated with killed-virus vaccine. New Engl J Med 277:248-250

Seto JT, Becht H, Rott R 1974 Effect of specific antibodies on biological functions of the envelope components of Newcastle disease virus. Virology 61:354-360

Skehel JJ, Waterfield MD 1975 Studies on the primary structure of the influenza virus hemagglutinin. Proc Natl Acad Sci USA 72:93-97

Tozawa H, Watanabe M, Ishida N 1973 Structural components of Sendai virus. Serological and physicochemical characterization of hemagglutinin subunit associated with neuraminidase activity. Virology 55:242-253

DISCUSSION

Taylor-Robinson: The fact that you can select a mutant which is not the organism causing the disease is something that we have to heed in studies with all micro-organisms.

Elbein: Are the mutants that you isolated with tripeptide different at the N-terminus?

Choppin: I don't know yet. They don't grow very well.

Elbein: The mutants might be susceptible to trypsin or to variations in the amino acid sequence.

Choppin: Presumably the mutation is just to the left of the cleavage site; the F_1 polypeptide has the same N-terminus.

Elbein: It is just at the point of cleavage where they have different amino acids.

Choppin: Yes, presumably one still has to have the same N-terminus for activity. These oligopeptides are inhibitors of the fusion reaction itself, not of cleavage.

Razin: You can label the F protein and isolate it. Perhaps you can isolate the active peptide that is formed by protease action on the labelled F protein. Can this active peptide cause fusion by itself? Can you inhibit binding of the labelled peptide by the tripeptide inhibitor?

Choppin: We can isolate the disulphide bonded protein $F_{1,2}$, and we can isolate the F_1 polypeptide alone. Neither of these has biological activity by itself. To get biological activity, the protein must bc rcconstituted into a membrane. Furthermore the $F_{1,2}$ protein alone in a membrane is still not active because the adsorption mechanism is on the other protein. It can be delivered to the cell by adding either the HN protein or wheat germ agglutinin, and then it is active. With membrane fusion one would not expect activity unless the protein is inserted into a membrane.

Levine: I was fascinated by the explanation for the atypical measles syndrome and the increased illness in the recipients of inactivated vaccines for respiratory syncytial (RS) infection. Chanock et al (1968) also suggested that the striking age-specific attack rate for RS bronchiolitis is mediated by maternally transferred antibody. Do you have serological evidence that certain RS antibodies are passed but that the antibodies against the F protein are not, which might therefore explain RS bronchiolitis? In influenza, killed vaccine with all its drawbacks is nevertheless pretty protective when it resembles the circulating viruses. Why is that?

Choppin: The respiratory syncytial virus analogy is an extrapolation. It is not even known which protein in respiratory syncytial virus is responsible for the fusion. With measles we know the protein glycoproteins and the antibodies to them are differentially produced after injection of formalin-killed vaccine.

Influenza does not induce the same kind of cell fusion in the respiratory epithelium. Even if it did, we probably would not have the same problem as with measles virus, because in influenza virus the receptor-binding protein, the haemagglutinin protein, is also the one that must be cleaved to activate the initiation of infection. Thus the receptor binding protein is also the one that is involved in the next step. In measles and with the other paramyxoviruses, these functions are on different proteins.

Hughes: Could you comment on the curious requirement for the unnatural, D-amino acids in the inhibiting peptides? Secondly, if one made a reagent by putting the synthetic tripeptides onto wheat germ agglutinin or some other lectin, would one then have a very versatile fusing agent, and indeed a very selective cytotoxic agent? The conjugate would bind to the cell surface by the lectin, followed by fusion triggered by the peptide haptens.

Choppin: The question of D versus L puzzled us also. There are at least two different explanations to be considered. A partial explanation would be that we are dealing with an *in vivo* system and the D-amino acid might confer some protection against chymotryptic activity. Therefore the oligopeptide would survive longer in the system, with better activity, than in oligopeptide that was being degraded by proteases in the system. However, the same increased activity with an N-terminal amino acid was found in the haemolysis reaction, and protease digestion is less likely to be a problem there. Another fact that argues against this explanation is that we get the most activity with D-phenylalanine, L-phenylalanine-glycine. If the first phenylalanine is changed to L, the L,L configuration is less active than L,D. In other words, there seems to be some preference for alternating phenylalanines, and that may have something to do with the orientation of the aromatic group around the peptide. A bit more support for this argument is that greatest activity is found with a benzyloxycarbonyl group on the N-terminus, and that is another aromatic group. If the benzyloxycarbonyl group is removed, the activity goes down, and although the inhibitory activity is still amino-acid-sequence-specific, it makes less difference whether the N-terminus is a D or an L. In short, we think the greater activity is related to the steric orientation of the aromatic groups, but that is just a hypothesis.

If you made a synthetic fusogen, you would have to put it into a liposome to get it to work. The entire F protein is not active unless it is reconstituted into a membrane. Another point is that we are interfering with a short peptide. I would predict that one would not get fusion activity unless one synthesized a much longer polypeptide. In the paramyxoviruses, there are 26 or 27 hydrophobic amino acids in a row, and with these viruses there is overt haemolysis and cell fusion, in addition to virus infectivity. In influenza virus, the hydrophobic stretch is only about nine amino acids, and there is no overt fusion or haemolysis. We think that the stretch of hydrophobic amino acids in influenza virus may not be long enough to cause overt fusion and haemolysis, but that is pure hypothesis.

Hughes: It may be significant that you need a stretch of 27 hydrophobic amino acids; that is about the same stretch as the hydrophobic sequence of glycophorin (Furthmayr 1977). Do you think it is simply a question of having a hydrophobic sequence which can span the membrane?

Choppin: The coincidence is there. We don't know whether that is what is necessary, but to make a fusogenic reagent I think we need a lot more equipment than just the tripeptide, which in some way seems to be interfering with the N-terminal of the viral glycoprotein.

Taylor-Robinson: Are the tripeptides active at very low concentrations? Is there a chance that they would work in an animal model?

Choppin: The most active tripeptide has a 50% inhibitory concentration of 0.2 μM, which is an approachable concentration. The heptapeptide I mentioned is even more active, 0.02 μM. For comparison, that is a lot more active than actinomycin. These hydrophobic oligopeptides are very insoluble, and the concentration in solution is unquestionably lower. These are concentrations that probably can be achieved locally, but whether the peptides will work in an animal we don't know yet.

Silverblatt: Perhaps there is a relation between these peptides and naturally occurring peptides such as substance P.

Sharon: Since you say they do not inhibit interaction of the isolated fusogenic agents with membranes, what is the mechanism of action of these peptides?

Choppin: I didn't say that. We think they are inhibiting the interaction of the viral polypeptide with some important site. We don't know yet what that site is. There are two possibilities at least: one, that it is interacting with some site in the plasma membrane of the target cell, and the other that it is interacting with some site on the viral protein itself. We know that as a result of cleavage the protein folds into a different conformation. It is conceivable that this inhibitor reacts with the viral protein and prevents the viral protein from assuming its proper conformation. We expect to have the answer to that soon, by using labelled oligopeptides.

Sharon: So you can demonstrate inhibition in a system which consists of the isolated or cleaved F protein and isolated membrane of the host cell?

Choppin: You can never demonstrate activity of the F with the isolated protein. It has to be reconstituted into a membrane.

Sharon: Does this work with isolated host membranes or do you need intact cells?

Choppin: We have used it for either red blood cells or tissue-cultured cells. We have not done experiments with isolated membrane fragments.

Sharon: Does the isolated HN act as a haemagglutinin?

Choppin: The isolated HN haemagglutinates and has neuraminidase activity. That is a pure protein that is active, but those activities do not depend on membrane fusion.

Sharon: I think this is one of the rare examples of a lectin with enzymic activity.

Choppin: There is no question about that. The isolated monomer HN has receptor binding activity and has neuraminidase activity. It does not haemagglutinate. If you allow it to become multivalent by aggregating, it will then haemagglutinate.

Mirelman: Do you obtain antibodies against a cleaved peptide or against the intact F?

Choppin: We have been unable to demonstrate an antigenic difference in the system between cleaved F and uncleaved F. That doesn't mean it doesn't exist.

Mirelman: Does the antibody against the cleaved F and the antibody against the intact F exert protection?

Choppin: We have not done precisely that experiment, but based on the available information we would expect that it would. We have no evidence that in a natural system this region of the peptide is antigenic.

Mirelman: Obviously the hydrophobic tripeptide sequence of F is very important. Could you have these haptens coupled to a macromolecular carrier for use in immunological studies?

Choppin: We haven't done that.

Feizi: How readily do the F mutants arise during natural infections?

Choppin: We have no idea.

Feizi: In the presence of antibodies to F can the virus escape by producing a different F protein by a phenomenon similar to antigenic variation of parasites? In other words, in the course of an epidemic do mutants appear with antigenically different F proteins?

Choppin: In nature the paramyxoviruses are markedly stable antigenically. Measles and NDV are completely unlike influenza in this regard.

Feizi: Can you detect any antigenic differences among different F proteins?

Choppin: They have not been found in wild-type strains. I have heard from Dr Walter Gerhard that one can make a mutant which is antigenically detectable by monoclonal antibody, but this is apparently not an important natural occurrence.

Feizi: Is that with respect to F?

Choppin: I believe that it is true with respect to either glycoprotein. If this were happening frequently under natural conditions we would see atypical measles, which we don't see.

Taylor-Robinson: Dourmashkin & Tyrrell (1970) by electron microscopy showed fusion of Sendai virus to the cilia of chicken tracheal epithelial cells in organ culture. Later, however, in studies concerned with the attachment of influenza virus to pieces of the chorioallantoic membrane, they could not demonstrate that there was any fusion whatsoever. In this case, they believed that virus entry was by viropexis (Dourmashkin & Tyrrell 1974). Do you think there is fusion with myxoviruses?

Choppin: This is not yet clear. The penetration of paramyxoviruses by fusion was first described by Morgan & Howe in 1968 but electron microscopy is a terrible way to decide how virus particles are getting into the cell. You never know whether the virus particle you see fusing with the cell is the one that is causing infection. The same is particularly true in something like influenza virus, where the best particle-to-infectivity ratio that anyone has claimed is 10:1. With influenza virus many virus particles are phagocytized, and that is the case with paramyxovirus particles also. I don't think we know for sure yet with myxoviruses whether the infecting particle penetrates by fusing at the plasma membrane, or whether it is endocytosed, and then there is a reaction between the viral membrane and the endocytic vesicle membrane. In either case there would presumably be membrane fusion at some point. I would agree that there is not very good electron microscopic evidence for fusion of influenza virus with the plasma membrane.

Helenius: We have been studying Semliki Forest virus, an enveloped animal virus somewhat simpler in structure than the paramyxoviruses and the myxoviruses. In the centre the virus has a spherical nucleocapsid containing 240 copies of protein and one molecule of RNA. The membrane contains spike proteins, each composed of three polypeptides, one of which has haemagglutinating activity. The spike protein unit is anchored to the bilayer by hydrophobic peptide moieties.

When we started to study the mechanism of entry of this virus into cells we were biased in favour of the type of mechanism that Dr Choppin was talking about: membrane fusion at the plasma membrane. It turned out that a membrane fusion reaction does, in fact, occur but it takes place intracellularly after the virus has been internalized by endocytosis. The evidence for this is based partly on morphology and partly on studies using radioactive virus tracers and specific inhibitors. The sequence of events is as follows (Helenius et al 1980).

(1) The virus binds to the cell surface, the preferred binding sites being the microvilli. In mouse and human cells we have shown that the main receptors for the virus are the major histocompatibility antigens, H2 and HLA antigens respectively. Recent studies have, however, shown that these surface glycoproteins cannot be the sole receptors because mutant cells which lack them can in some cases be infected. The number of possible receptors is thus an open question.

(2) After attachment the virus moves along the membrane from the microvilli and gets trapped in the so-called coated pits. These are micro-domains of the cell surface which are specialized for endocytosis. The half-life of attachment of the virus to the cell surface is only 7 min, after which the viruses are found in different endocytotic vacuoles within the cell. These vacuoles then gradually develop into lysosomes and the pH inside drops. The low pH milieu in the lysosomes apparently induces a fusion reaction between the virus and the lysosomal membrane whereby the capsid emerges into the cytoplasmic compartment. This type of fusion can be triggered *in vitro* between the virus and almost any type of membrane. We can, for instance, trigger it at the plasma membrane of cells by lowering the pH of the medium; the capsid then enters through the membrane and the cell is infected.

In view of these results it seems that the entry of Semliki Forest viruses is similar to that of the paramyxoviruses discussed by Dr Choppin, except that under physiological conditions the fusion reaction is delayed until the virus has been endocytosed into the cell. The mechanism of endocytosis is a continuous cellular process, which functions independently of added virus. We find that a BHK-21 cell can take up more than 3000 viruses per minute.

REFERENCES

Chanock RM, Parrott RH, Kapikian AZ, Kim HW, Brandt CD 1968 Possible role of immunological factors in pathogenesis of RS lower respiratory tract disease. In: Pollard M (ed) Virus-induced immunopathology. Academic Press, New York (Perspectives in virology 6) p 125-139

Dourmashkin RR, Tyrrell DAJ 1970 Attachment of two myxoviruses to ciliated epithelial cells. J Gen Virol 9:77-88

Dourmashkin RR, Tyrrell DAJ 1974 Electron microscopic observations on the entry of influenza virus into susceptible cells. J Gen Virol 24:129-141

Furthmayr H 1977 Structural analysis of a membrane glycoprotein: glycophorin A. J Supramol Struct 7:121-134

Helenius A, Kartenbeck J, Simons K, Fries E 1980 On the entry of Semliki Forest virus into BHK-21 cells. J Cell Biol 84:404-420

Morgan C, Howe C 1968 Structure and development of viruses as observed in the electron microscope. IX. Entry of parainfluenza virus type I. J Virol 2:1122-1133

Effect of inhibitors on glycoprotein biosynthesis and bacterial adhesion

ALAN D. ELBEIN, BARBARA A. SANFORD, MARY A. RAMSAY and Y. T. PAN

Departments of Biochemistry and Microbiology, The University of Texas Health Science Center at San Antonio, 7703 Floyd Curl Drive, San Antonio, Texas 78284, USA

Abstract Group B streptococci adhere to influenza-virus-infected canine kidney epithelial cells but not to uninfected cells. For studies of the molecular nature of this interaction the bacteria were radiolabelled and a quantitative binding assay was developed with which the following properties of the system were observed. (1) Adhesion was specific for group B streptococci (GBS); streptococci from other serological groups did not bind and did not inhibit adhesion of radioactive GBS. (2) Binding of GBS to infected kidney cells was inhibited by the addition of cell walls from GBS to the kidney cell monolayers. (3) Preincubation of GBS with free influenza virus prevented their attachment to infected kidney cell monolayers. With a centrifugation type of assay, labelled influenza virus bound to GBS. This binding could be inhibited by several glycoproteins after removal of the terminal sialic acid. Asialo-glycopeptides of the complex type, isolated from these inhibitory glycoproteins, also bound to GBS.

The influenza viral glycoproteins have been partially characterized and shown to contain a glycosylamine type of complex oligosaccharide. This type of oligosaccharide is biosynthesized by means of lipid-linked saccharide intermediates. Several antibiotics such as tunicamycin and streptovirudin, and other inhibitors such as 2-deoxyglucose and glucosamine, inhibit this lipid-linked pathway. These inhibitors also prevent the formation of mature influenza virus as well as the adherence of group B streptococci. Other inhibitors of protein glycosylation should be valuable as tools for improving further our understanding of the mechanism of cell adhesion.

Glycoproteins have been implicated in various types of recognition reactions, such as those occurring in differentiation, tissue development and cell adhesion. Many of the receptor sites for hormones and other small molecules involve glycoproteins. There is also some evidence to suggest that these complex carbohydrates also play a part in bacterial adhesion. For example, Ofek et al (1977) have reported that the adherence of *E. coli* to human mucosal cells is blocked by the addition of mannose

---

1981 Adhesion and microorganism pathogenicity. Pitman Medical, Tunbridge Wells (Ciba Foundation symposium 80) p 270-287

and α-methylmannoside, suggesting that mannose receptors are involved in adhesion. It is not clear from these studies whether the mannose residues are present in the bacterial cells or in the mucosal cells, but mannose has been shown to be present in many mammalian glycoproteins. In studies with Gram-negative organisms, pili have been implicated in mediating the attachment of bacteria to epithelial cells (Punsalang & Sawyer 1973, Swanson 1973). But in the adherence of group A streptococci to oral mucosal cells, the lipoteichoic acid component of the streptococci appears to be one of the interacting macromolecules (Beachey & Ofek 1976). These various studies make it appear likely that these cell:cell interactions include both carbohydrate and protein components.

The bacterial adherence system to be described here is based on the observations that group B streptococci (GBS) adhere to canine kidney (MDCK) epithelial cells once these mammalian cells have been infected with influenza virus, but that these bacteria do not adhere to uninfected kidney cells (Sanford et al 1978, Sanford et al 1979). Additional evidence suggests that GBS are probably recognizing and interacting with glycoproteins on the surface of the influenza virus. Thus, when GBS are mixed with a solution of influenza virus, they bind the virus and remove it from solution (Sanford et al 1980). That is, the absorbed virus suspension is no longer able to agglutinate erythrocytes. Since influenza is a budding virus and contains two known coat glycoproteins, a haemagglutinin and a neuraminidase, the evidence suggests that these viral glycoproteins are synthesized in the MDCK cells and inserted into the cell membrane before mature virus emerges. The budding virus then emerges by taking a portion of the cell membrane as it leaves the cell.

In order to study this system in more detail we developed a quantitative assay for measuring bacterial adherence by using bacteria that had been labelled with a radioactive tag. The streptococci could be labelled in either of two ways, with essentially the same results. The bacteria could be grown in $[^{14}C]$fructose to label the cells uniformly, or they could be incubated with a surface label, 4,4′-$[^{3}H]$diisothiocyano-1,2-diphenylethane-2,2′-disulphonic acid, which forms covalent bonds with free amino groups at the surface of the cells. The assay system used to test the ability of these bacteria to adhere was as follows. MDCK cells were grown in plastic tissue culture dishes in Eagle's medium with 10% calf serum, 0.03% glutamine and 20 μg neomycin/ml. When the cells had reached confluency they were inoculated with a dilution of influenza A/NWS/33 virus that contained eight haemagglutination (HA) units/0.25 ml. The virus–cell mixtures were incubated at 25 °C for 1 h, with gentle rotation. After aspiration of the virus suspension and addition of 2 ml growth medium the cell cultures were incubated at 34 °C for 48 h. The medium was removed by aspiration and the monolayers were washed well with phosphate-buffered saline. The radioactively labelled group B streptococci were then added to the monolayers in the saline solution and the mixture was allowed to incubate for about 60 min. The 'unbound' streptococci were then removed by aspiration and the monolayers were washed several times with saline to

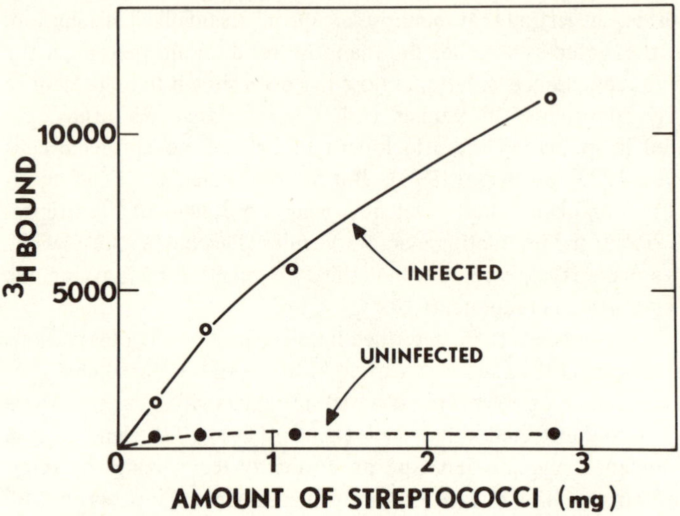

FIG. 1. Effect of concentration of group B streptococci on the adherence of bacteria to canine kidney cell monolayers infected with influenza virus. The binding of ^{3}H-labelled streptococci to virus-infected monolayers (○-○) and to uninfected monolayers (●- - -●) was examined. Ordinate, c.p.m.

remove any free or loosely bound bacteria. The monolayers were detached from the plates with trypsin–EDTA and the cell mixture was placed in scintillation vials so that the number of radioactive bacteria adhering to these monolayers (Pan et al 1979) could be determined.

Fig. 1 shows the results obtained in this assay. Radioactive GBS bound very well to the MDCK cells infected with influenza virus and this binding was proportional to the number of bacteria added to the monolayers. The binding also showed saturation when increased amounts of bacteria were added to the monolayers. However, the GBS did not adhere to uninfected MDCK cells. In order to determine the specificity of this binding for the bacteria we tested various streptococci from serological groups other than group B for their ability to inhibit the binding of radioactive GBS to the infected MDCK cells. Unlabelled streptococci (*S. pyogenes, S. mitis,* etc.) were mixed, in various amounts, with the labelled streptococci and this mixture was tested for its ability to bind to infected MDCK cells. As expected, unlabelled GBS markedly inhibited the adherence of radioactivity, with about 40% inhibition occurring when equal amounts of unlabelled and labelled cells were mixed and about 80% inhibition when a fivefold excess of unlabelled cells was used. However, the unlabelled streptococci from other groups showed much less effect on binding, only inhibiting at fairly high concentrations, probably as a result of non-specific binding by the large number of bacteria present. Thus, the binding appears to be fairly specific for GBS.

TABLE 1 Effect of various sugars and glycoproteins on bacterial adherence

Compound added to streptococci	*Concentration or amount used (mM or mg)*	*Radioactive streptococci bound (c.p.m.)*
None (control)	–	4317
Lactose	1 mM	5164
	10	5071
	100	5115
Methyl-D-mannoside	1 mM	5677
	10	4939
	100	5259
Heparin	0.05 mg	4505
	0.25	5773
	1.0	5005
	2.5	5183
None	–	4706
LDL	0.3 mg	8902
	0.6	5881
	1.2	3106
LDL treated with neuraminidase	0.3 mg	2756
	0.6	2325
	1.2	1893
	2.4	1544
LDL treated with neuraminidase and then β-galactosidase	0.6 mg	6224
	1.2	6256
	2.0	6266

LDL: low density lipoprotein.

It is also possible to block binding of GBS by first incubating the MDCK cell monolayers with isolated cell walls from GBS, indicating that the binding or recognition component is at the cell surface of the bacteria.

Since preliminary evidence suggests that the GBS are recognizing and adhering to viral coat glycoproteins, we tested the effect of various sugars and glycoproteins on the binding. Table 1 shows some of the results of these studies. Various amounts of glycoprotein or sugar were mixed with the radioactively labelled group B streptococci and after incubation for 30 min the mixture was tested for its ability to adhere to virus-infected kidney cell monolayers. The rationale was that if the glycoprotein or sugar had a similar recognition site to that in the infected kidney cells it should interact with the streptococci and block their adherence. Human low density lipoprotein inhibited binding to some extent but this inhibition was greatly enhanced when the low density lipoprotein was first treated with neuraminidase

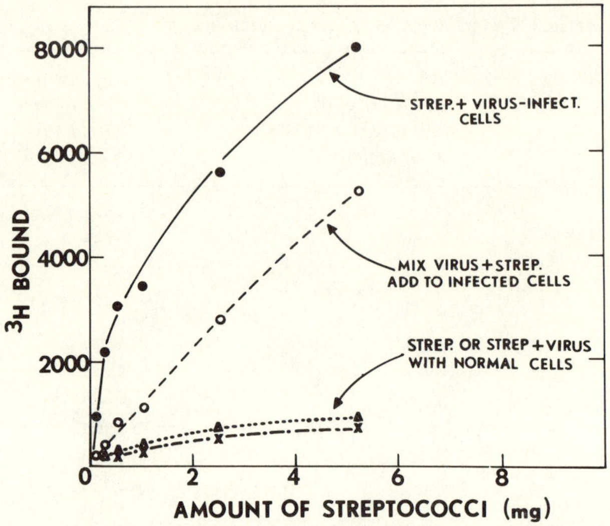

FIG. 2. Inhibition of streptococcal adherence by free influenza virus. In control experiments the binding of ^{3}H-labelled streptococci to infected (●—●) or uninfected (▲- - -▲) kidney cell monolayers was examined. In another experiment the ^{3}H-labelled streptococci were first mixed with free influenza virus (haemagglutinin titre of 256) and this mixture was tested for its ability to adhere (o - - - o) to the cells. x— - — - — - x, ^{3}H-labelled streptococci and free virus on uninfected cells.

to remove terminal sialic acid residues. Low density lipoprotein contains an asparagine-linked oligosaccharide of the complex type composed of the sugars mannose, *N*-acetylglucosamine (GlcNAc), galactose and sialic acid (Swaminathan & Aladjem 1976). Removal of sialic acids by neuraminidase has been shown to expose terminal galactose residues on this glycoprotein. Thus, this experiment could indicate that galactose has a role in the recognition process. However, none of the simple sugars, including the disaccharide lactose, had any effect on adherence. One problem with the lipoprotein studies could be that removal of sialic acid allows the molecules to aggregate, since they no longer contain negative charges, and these aggregated lipoproteins might entrap the bacteria non-specifically. Studies using other glycoproteins and glycopeptides are in progress.

Since the streptococci appear to bind to viral glycoproteins at the MDCK cell surface, we tested the effect of free virus on bacterial adhesion. Purified influenza virus was mixed with various amounts of ^{3}H-labelled GBS and these mixtures were tested for their ability to bind to the kidney cells. Fig. 2 shows the results of such an experiment. As shown in Fig. 1, the bacteria adhered very well to MDCK cells infected with influenza virus but did not bind to uninfected cells. However, when the streptococci were first mixed with free virus and then tested, there was considerable inhibition of binding. The experiment suggests that the streptococci

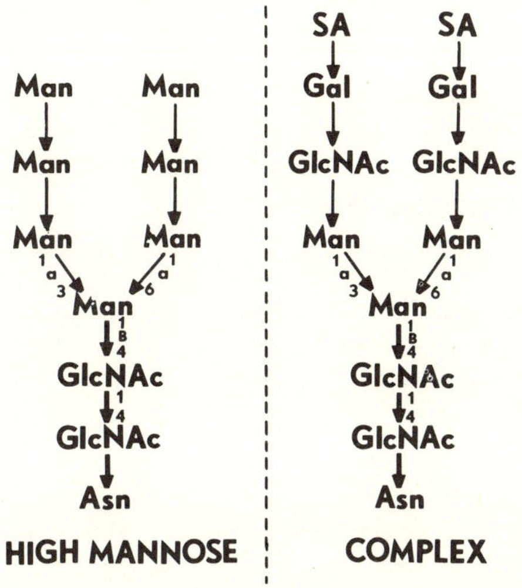

FIG. 3. General structure of GlcNAc→asparagine-linked glycoproteins. Both the 'high mannose' and 'complex' types of oligosaccharides are shown.

recognize and adhere to free virus and are therefore not available to bind to kidney cells. These results agree with a previous study in which group B streptococci inhibited haemagglutination by influenza virus (Sanford et al 1980). In the experiment shown in Fig. 2 the amount of influenza virus used was not sufficient to tie up all the streptococcal receptors but larger amounts of virus would result in greater inhibition of binding.

The influenza virus glycoproteins have been partially characterized and shown to have the general structure presented in Fig. 3 (Collins & Knight 1978). Two types of oligosaccharide are found in these viruses and both kinds are attached to protein in a GlcNAc–asparagine linkage. One type, called the high mannose type, contains only mannose and GlcNAc, while the second or complex type contains mannose, GlcNAc, galactose and sialic acid. Both oligosaccharides have a common core region composed of two GlcNAc residues linked in an *N,N′*-diacetylchitobiose group to which are attached three mannose units in a branched structure. The first mannose unit is linked to the GlcNAc in a β1–4 bond while the next two mannoses are linked in α1–3 and α1–6 bonds to the first mannose (Kornfeld & Kornfeld 1976). Although the influenza viral coat glycoproteins have been fairly well characterized as high mannose and complex oligosaccharides the details of their anomeric configuration, glycosidic linkage and branching are not completely known.

Studies in the past 10 or 15 years (Waechter & Lennarz 1976, Elbein 1979) have shown that lipid-linked saccharide intermediates participate in the biosynthesis of the core regions of these glycosylamine types of oligosaccharides. The postulated

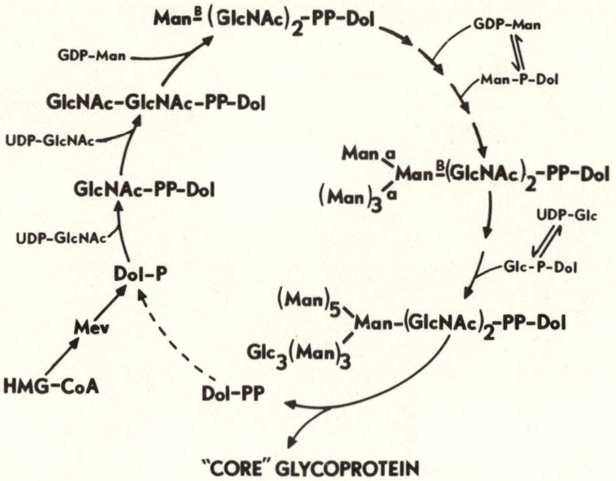

FIG. 4. Series of reactions in the biosynthesis of the core region of GlcNAc-asparagine types of glycoproteins. The lipid-linked saccharide intermediates involve the formation of an oligosaccharide-lipid containing *N*-acetylglucosamine (GlcNAc), mannose (Man) and glucose (Glc) that takes part in the glycosylation of proteins having an oligosaccharide attached to asparagine.

pathway of reactions for the biosynthesis of these oligosaccharides is shown in Fig. 4. The lipid molecule participating in these reactions is a membrane-bound lipid called dolichol. This is a long-chain polyisoprenol containing 100–110 carbons with an α-saturated isoprene unit at one end to which a phosphate group is attached. The pathway of synthesis is initiated by the transfer of GlcNAc-1-P from UDP-GlcNAc to dolichyl-P to form the first lipid intermediate, GlcNAc-pyrophosphoryl-dolichol. A second GlcNAc is then added to this lipid to form the di-GlcNAc lipid, *N,N'*-diacetylchitobiosyl-pyrophosphoryl-dolichol. This $(GlcNAc)_2$-linked lipid then serves as the acceptor for a number of mannose residues to form a large lipid-linked oligosaccharide. Some of the mannose units come directly from GDP-mannose; others apparently come from another lipid intermediate, mannosyl-phosphoryl-dolichol. Once the lipid-linked oligosaccharide is formed the oligosaccharide is transferred to the polypeptide chain to form the 'core' glycoprotein. The evidence indicates that this transfer to protein occurs while the polypeptide chain is still attached to membrane-bound polysomes. In some animal systems, especially those that are producing viral glycoproteins, glucose is also added to the lipid-linked oligosaccharide to form an oligosaccharide with the composition $(Glc)_3(Man)_9$-$GlcNAc_2$. When this oligosaccharide has been transferred to protein the 'preglycoprotein' undergoes several processing reactions before the protein reaches its final state. In these reactions all the glucoses and a number of mannoses are removed by specific hydrolytic enzymes. Other sugars such as GlcNAc, galactose and sialic

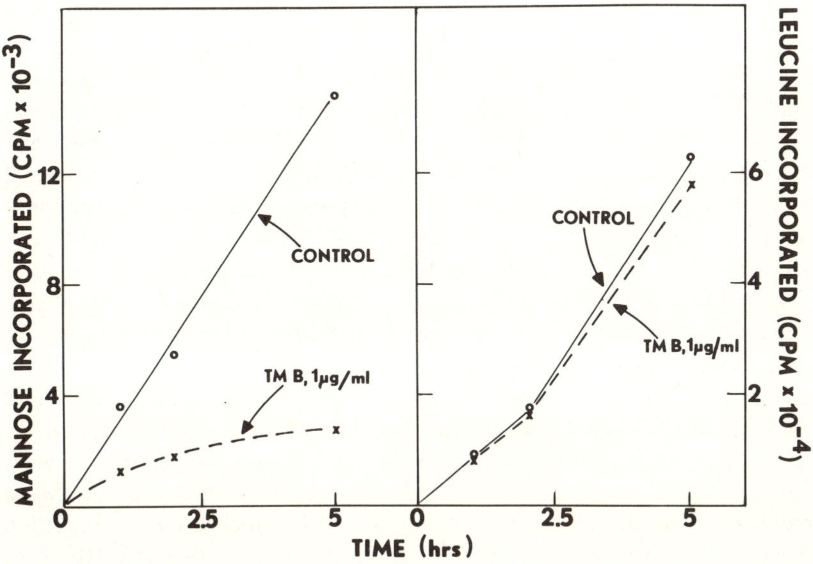

FIG. 5. Effect of tunicamycin peak B (TM B) on the incorporation of [^{3}H]mannose and [^{3}H]leucine into protein by MDCK cells. Monolayers were incubated for 2 h in 2 μg of tunicamycin in modified Eagle's medium containing 2% calf serum; 5 μCi of [^{3}H]mannose or [^{3}H]leucine was then added to each dish and the cells were incubated for the times shown. Cells were removed by trypsin treatment and washed several times in 5% trichloroacetic acid. Insoluble residue was placed in scintillation vials to determine its radioactive content. Incorporation of [^{3}H]mannose is a measure of glycosylation; incorporation of [^{3}H]leucine is a measure of protein synthesis.

acid may then be added to the glycoprotein to form the final complex oligosaccharide.

Several antibiotics inhibit the formation of glycoproteins by blocking the lipid-linked saccharide pathway at various steps. For example, tunicamycin and streptovirudin are antibiotics that inhibit the GlcNAc-1-P transferase which catalyses the formation of GlcNAc-pyrophosphoryl-dolichol from UDP-GlcNAc and dolichyl-P (Tkacz & Lampen 1975, Ericson et al 1977, Elbein et al 1979). Since this reaction is the first step in the lipid-linked saccharide pathway, the formation of GlcNAc-pyrophosphoryl-dolichol is necessary for the formation of lipid-linked oligosaccharide. If the GlcNAc lipid is not formed the lipid-linked saccharide cannot be synthesized. Under these conditions the protein is not glycosylated.

As indicated above, the influenza virus glycoproteins contain a high mannose and a complex type of oligosaccharide and are therefore likely to be synthesized by the pathway shown in Fig. 4. It was therefore of interest to study the effect of antibiotics such as tunicamycin and streptovirudin on the formation of virus particles and on the binding of GBS to infected MDCK cells. In order to first show that tunicamycin (and streptovirudin) inhibited protein glycosylation in these cells we examined the incorporation of [^{3}H]mannose and [^{3}H]leucine into protein in MDCK cells in the presence and absence of tunicamycin (Fig. 5). Since tunicamycin is produced as a complex of several different, but closely related, compounds, and since this complex has also been shown to inhibit protein synthesis, we separated the tunicamycin complex into its individual components, using high performance liquid chromatography (HPLC). Four major peaks were obtained and each was inhibitory to the GlcNAc-1-P transferase. Since the log of the retention time of these fractions on HPLC falls on a straight line the components are probably homologues of each other, differing by one carbon atom in the fatty acid moiety (Keenan & Elbein unpublished). Fig. 5 shows the effect of one of these peaks (labelled peak B) with the MDCK cells. In the control cells (without tunicamycin) the incorporation of mannose into protein was linear, indicating that the proteins were being glycosylated. When tunicamycin peak B (TM B) was added to these cells one hour after inoculation with virus, the incorporation of mannose was almost completely inhibited, demonstrating that this antibiotic blocked the glycosylation step. However, the antibiotic had little effect on the incorporation of [^{3}H]leucine into the protein, showing that the purified tunicamycin fraction did not inhibit protein synthesis. Similar results were obtained with other tunicamycin fractions and also with the streptovirudin fractions separated by HPLC.

We next examined the effect of tunicamycin on the formation of virus and on the binding of streptococci, as shown in Table 2. In this experiment the cells were inoculated with virus and one hour later various concentrations of tunicamycin were added. After incubation for 24–48 h to allow the virus to replicate, the kidney cell monolayers were tested for haemadsorption and for their ability to bind GBS. Haemadsorption of red blood cells is a measure of the presence of viral glycoproteins in the MDCK cell membrane since one of the viral glycoproteins is a haemagglutinin. The medium was tested for free influenza virus by haemagglutination. As Table 2 shows, when the concentration of tunicamycin was 0.3 μg/ml or higher the MDCK cells lost their ability to adsorb red blood cells and free virus no longer appeared in the medium. At this concentration of antibiotic the cells also lost their ability to bind GBS. Thus, this experiment further supports the previous studies indicating that GBS recognize and bind to the viral glycoproteins. However, these studies do not prove that the carbohydrate portion of the viral glycoprotein is necessarily involved in recognition and adhesion. The carbohydrate may be necessary for the viral glycoprotein to be properly inserted into the mammalian cell membrane, so in the absence of glycosylation the viral proteins may remain in the cytoplasm and not be available as bacterial receptors.

When streptovirudin was tested in place of tunicamycin essentially the same results were obtained. That is, streptovirudin inhibited the formation of virus particles and the adherence of GBS to the MDCK cell monolayers. However, in

TABLE 2 Effect of tunicamycin on influenza virus infection and binding of group B streptococci (GBS)

Tunicamycin (μg/ml)	*Haemadsorption plaque assay*[a] *(% of control)*	*Haemagglutinin titre*[b] *(supernatant)*	*Binding of GBS*[c] *(% of control)*
0	100	1:128	100
0.078	100	1:16	71
0.156	50	1:11	6
0.32	7.7	–[d]	2
0.625	0	–	0
1.25	0	–	0
2.5	0	–	0
5.0	0	–	0
10.0	0	–	0

[a]Haemadsorption by kidney cells. The control (100%) is equal to the mean number of haemadsorption plaques on virus-infected monolayers that have not been pretreated with tunicamycin.
[b]Haemagglutinin released from cells as measured by agglutination of erythrocytes.
[c]Binding of streptococci as measured by microscopic observation. The control (100%) is equal to the mean number of bacterial adherence plaques on virus-infected monolayers that have not been pretreated with tunicamycin. This assay correlated well with the ^{3}H-binding assay done in other experiments.
[d]Dash indicates undiluted supernatant which was haemagglutination-negative.

TABLE 3 Effects of cycloheximide on glycoprotein synthesis and bacterial adherence

Cycloheximide (μg/ml)	*[^{3}H]Mannose incorporated (% of control)*			*Bacterial adherence (% of control)*
	M-P-Dol	*Lipid-linked oligo*	*Glycoprotein*	
0	100	100	100	100
0.05	68	45	43	73
0.2	57	28	25	42
1.0	39	18	8	29
5.0	33	23	5	22
10.0	28	14	3	23

these experiments larger amounts of streptovirudin were required (1–2 μg/ml), perhaps because this antibiotic is not as pure as the tunicamycin or perhaps because it is not taken up so well by whole cells.

We have also tested the effects of a number of other antibiotics and inhibitors on the biosynthesis of glycoproteins and on bacterial adherence. For these experiments duplicate sets of MDCK cell monolayers were infected with virus and one

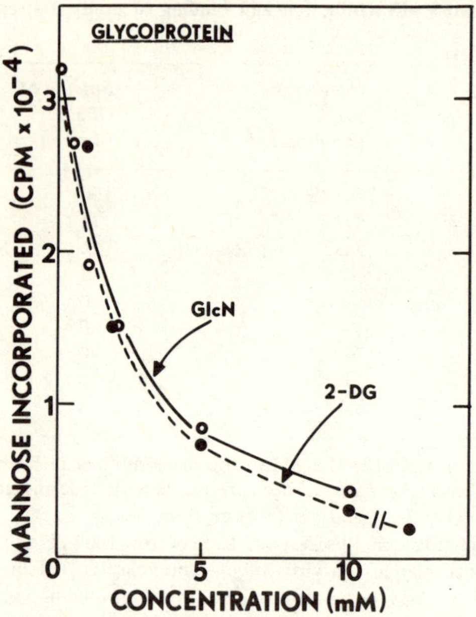

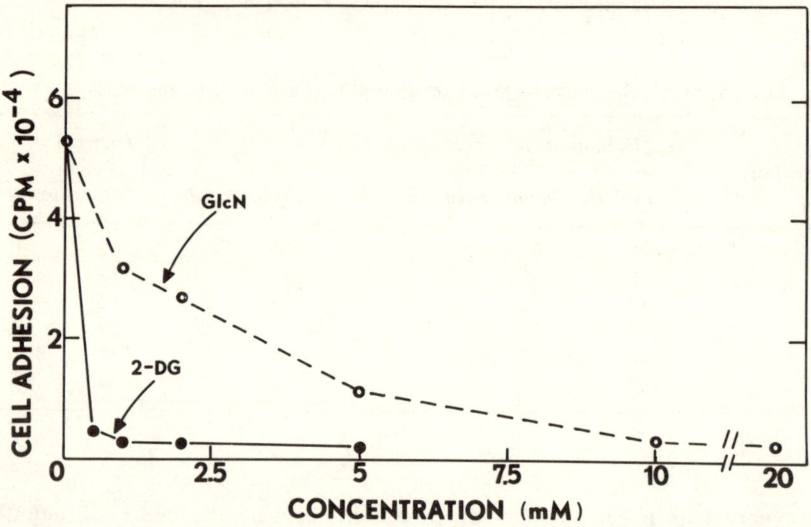

FIG. 6. The effect of various concentrations of 2-deoxyglucose (2-DG) and glucosamine (GlcN) on protein glycosylation and adherence of group B streptococci in MDCK cells. Various concentrations of these inhibitors were incubated with infected MDCK cells and the incorporation of [^{3}H] mannose into protein (above) or the binding of ^{3}H-labelled streptococci (below) was examined.

hour after infection various concentrations of the inhibitors were added to the monolayers. The cells were allowed to incubate for three hours with the inhibitor and $[^3H]$mannose was then added to one set of kidney cells to determine the incorporation of mannose into the various lipid-linked saccharide intermediates (see Fig. 3) and into glycoprotein. The other set of cells was allowed to incubate for 48 h with the inhibitors so that virus replication could occur; the cells were then tested for their ability to bind GBS. Table 3 shows the results of one such experiment where cycloheximide, a known inhibitor of protein synthesis, was tested. As shown, 1 μg cycloheximide/ml was sufficient to inhibit the formation of glycoprotein, and at this concentration bacterial adherence was also prevented. These results were as expected, since cycloheximide should block the synthesis of virus proteins. However, this experiment indicated that this type of assay should be valuable for locating inhibitors of protein synthesis or protein glycosylation.

Several other antibiotics have been shown to inhibit the formation of lipid-linked saccharides in cell-free particulate enzyme preparations from porcine aorta. These include the polypeptide antibiotics amphomycin (Kang et al 1978) and tsushimycin, and the nucleoside antibiotic showdomycin (Kang et al 1979). However, when these compounds were tested in the above assay with MDCK cells they inhibited neither protein glycosylation nor bacterial adherence. Probably these peptide and nucleoside antibiotics cannot penetrate the cell membrane and therefore do not enter the cells.

We have tested two other inhibitors, 2-deoxyglucose and glucosamine, on the MDCK cell system. Both these compounds have been reported to inhibit glycoprotein synthesis and viral replication in several different types of cells (Datema & Schwarz 1979). These two compounds blocked the incorporation of $[^3H]$mannose into lipid-linked saccharides in infected MDCK cells, with optimum inhibition of lipid-linked oligosaccharides requiring about 5 mM of either of these sugars. Glycoprotein synthesis (i.e. mannosylation) was also mostly inhibited at 5 mM-2-deoxyglucose or 5 mM-glucosamine (Fig. 6). Interestingly enough, although both sugars also inhibited bacterial adherence, 2-deoxyglucose appeared to be much more potent in inhibiting binding at concentrations as low as 0.5 mM.

The studies described here demonstrate that group B streptococci recognize and adhere to influenza virus glycoproteins that have been synthesized in infected MDCK cells and inserted into the cell membrane. The experiments with various antibiotics and other inhibitors of protein glycosylation have given valuable information on the role of glycoproteins in this process. However, it is still not clear which macromolecule in the streptococcal cell wall takes part in the recognition of viral glycoprotein. Nor is it clear yet which portion of the viral glycoprotein participates in the adherence process. Studies are being designed to isolate and characterize the individual components in this system.

Acknowledgements

These studies were supported by grant HL 17783 from the National Heart, Blood and Lung Institute and a grant from the Robert A. Welch Foundation.

REFERENCES

Beachey EH, Ofek I 1976 Epithelial cell binding of group A streptococci by lipoteichoic acid on fimbriae denuded of M protein. J Exp Med 143:759-771

Collins JC, Knight CA 1978 Purification of the influenza hemagglutinin glycoprotein and characterization of its carbohydrate components. J Virol 26:457-467

Datema R, Schwarz RT 1979 Interference with glycosylation of glycoproteins. Inhibition of formation of lipid-linked oligosaccharides in vivo. Biochem J 184:113-123

Elbein AD 1979 The role of lipid-linked saccharides in the biosynthesis of complex carbohydrates. Annu Rev Plant Physiol 30:239-272

Elbein AD, Gafford J, Kang MS 1979 Inhibition of lipid-linked saccharide synthesis. Comparison of tunicamycin, streptovirudin and antibiotic 24010. Arch Biochem Biophys 196:311-318

Ericson M, Gafford J, Elbein AD 1977 Tunicamycin inhibits GlcNAc-lipid formation in plants. J Biol Chem 252:7431-7433

Kang MS, Spencer JP, Elbein AD 1978 Amphomycin inhibition of mannose and GlcNAc incorporation into lipid-linked saccharides. J Biol Chem 253:8860-8866

Kang MS, Spencer JP, Elbein AD 1979 The effect of showdomycin on glycolipid formation. Inhibition of glycosyl-phosphoryl-dolichol on aorta and stimulation of glycosyl-ceramide in yeast. J Biol Chem 254:10037-10043

Kornfeld R, Kornfeld S 1976 Comparative aspects of glycoprotein structure. Annu Rev Biochem 45:217-237

Ofek I, Miselman D, Sharon N 1977 Adherence of *Escherichia coli* to human mucosal cells mediated by mannose receptors. Nature (Lond) 265:623-625

Pan YT, Schmitt JW, Sanford BA, Elbein AD 1979 Adherence of bacteria to mammalian cells: inhibition of tunicamycin and streptovirudin. J Bacteriol 139:507-514

Punsalang AP Jr, Sawyer WD 1973 Role of pili in the virulence of *Neisseria gonorrhoeae.* Infect Immun 8:255-263

Sanford BA, Shelokov A, Ramsay MA 1978 Bacterial adherence to virus-infected cells: a cell culture model of bacterial superinfection. J Infect Dis 137:176-181

Sanford BA, Smith N, Shelokov A, Ramsay MA 1979 Adherence of group B streptococci and human erythrocytes to influenza A virus-infected MDCK cells. Proc Soc Exp Biol Med 160:226-232

Sanford BA, Smith N, Shelokov A, Ramsay MA 1980 Adsorption of influenza A viruses by group B streptococci. J Infect Dis, in press

Swanson J 1973 Studies of gonococcus infection. IV. Pili: their role in attachment of gonococcus to tissue culture cells. J Exp Med 137:571-578

Swaminathan N, Aladjem F 1976 The monosaccharide composition and sequence of the carbohydrate moiety of human serum low density lipoproteins. Biochemistry 15:1516-1522

Tkacz J, Lampen J 1975 Tunicamycin inhibition of polyisoprenyl-N-acetylglucosaminyl pyrophosphate formation in calf-liver microsomes. Biochem Biophys Res Commun 65:248-257

Waechter CJ, Lennarz WJ 1976 The role of polyprenol-linked sugars in glycoprotein synthesis. Annu Rev Biochem 45:95-112

DISCUSSION

Mirelman: You mentioned that streptococci bind the virus. We observed that Gram-negative organisms also bind viruses, and for example influenza virus is bound by a mannose-binding *E. coli*. Could 'hitch-hiking' of viruses on bacteria serve as a sort of reservoir of infection, or perhaps assist in penetration? Is anything known about the carriage of viruses on the backs of bacterial cells?

Choppin: Viruses seem to get there on their own without requiring bacteria to deliver them.

One striking thing about the influenza virus carbohydrate portion compared to the normal cellular glycoproteins is that there is no terminal neuraminic acid on the viral carbohydrate chain because of the viral neuraminidase. That would fit with your observations. One way of testing that would be to use another virus grown in the same cell, such as VSV or the Semliki Forest virus that Dr Helenius mentioned, which does not have neuraminidase as a component. Then you would have the same carbohydrate portion, but with neuraminic acid on the end.

Elbein: In our system we know that the influenza virus does not have neuraminic acid as a terminal sugar. We can label that virus with galactose oxidase and NaB^3H_4, indicating that galactose residues are exposed.

Choppin: So the infected cells would have a reduced sialic acid content.

Elbein: Some of the influenza viruses don't have neuraminidase.

Choppin: There are neuraminidase-minus mutants, which is another way that one could look at the specific requirements.

In terms of the agglutination phenomenon there is something else that might be helpful. It was shown some time ago by Becht et al (1972) that cells infected with a variety of viruses, not just influenza, which don't have neuraminic acid are agglutinable by lectins such as concanavalin A and wheat germ agglutinin, but the uninfected cell is not. The virus particles grown in those cells are agglutinated by those lectins. Presumably the carbohydrate portion is now on the spike-like projection on the surface of the cell, where it is more accessible to the lectin than it is in a normal cell. The other way that agglutinability can be produced is by transforming the cell to a malignant state. The accessibility of the carbohydrate receptors by virtue of their being on a viral glycoprotein might be playing a role in your system.

Freter: When first reading about it, I thought that bacteria may adhere to virus-infected cells because the viral neuraminidase removes neuraminic acid and negative charges from the cell surface, thereby facilitating bacterial adhesion. Subsequent results have shown that the presence of viral haemagglutinin is a more likely explanation, but is there still a possibility that additional factors might add to bacterial adhesion?

Elbein: With streptococci we tried to do some experiments where we removed the sialic acid. The streptococci have an antigen that has sialic acid, that is true, but neuraminidase has no effect on binding of streptococci.

Freter: How about the cells?

Elbein: You can't treat these cells with neuraminidase easily. If you fiddle around with the kidney cells too much they lift off the plate and then you are in trouble with the assay.

Freter: Does viral neuraminidase remove sialic acid from the kidney cells?

Elbein: I don't know. Apparently this strain of influenza virus has little or no neuraminidase.

Choppin: It has a very unstable neuraminidase, so if the virus is used as an enzyme, activity may not be very good. However, the enzyme does remove the neuraminic acid from the virus itself, and certainly has an active neuraminidase when it comes out of the cell.

Wallach: How effective is 25-hydroxycholesterol in blocking dolichol synthesis and glycosylation, compared to its effect on cholesterol synthesis?

Elbein: We haven't done anything on that but others have looked at the formation of dolichol itself. Several people feel that the regulation of the whole lipid-linked pathway is due to the continual synthesis of dolichol and that levels of dolichol phosphate control the pathways. 25-Hydroxycholesterol blocks HMG-CoA reductase. In some cases it seems to inhibit dolichol formation and prevent glycosylation of proteins to some extent, but the results are conflicting. It may depend on the age of the cells, when they grow, and so on.

A new inhibitor of cholesterol, compactin, seems to block dolichol synthesis much better (Brown et al 1978). Some people have used compactin in sea urchin embryos for developing systems (Carson & Lennarz 1979). It blocks glycosylation of the proteins necessary for development. That inhibitor could be useful in these studies.

Sharon: To what extent has the structure of the carbohydrate unit of your lipoprotein been worked out?

Elbein: Not very far. It is a complex oligosaccharide. I don't know how many oligosaccharide chains it has. We have prepared the glycopeptide by pronase digestion and, after removal of sialic acid, labelled the galactose with ^{3}H. The asialo–[^{3}H] galactose-labelled glycopeptide binds like the low density lipoprotein. It doesn't seem to be quite as good an inhibitor of streptococci binding to kidney cells. One of the two assays we use is the binding of very active group B streptococci (GBS) to kidney cells. One way we can test that is by using inhibitors that block that binding. In the other assay we use a radioactive component such as LDL glycopeptide, bind it to unlabelled GBS, and then centrifuge the mixture. Asialo-fetuin also binds to GBS.

Sharon: You mentioned the effect of tunicamycin on influenza virus. Is a protein synthesized which then remains intact?

Elbein: Schwarz et al (1976) have evidence that the influenza viral protein is synthesized in the unglycosylated form in the presence of tunicamycin, but the protein turns over much more rapidly than the normal glycoprotein.

Sharon: So one may conclude that glycosylation protects the polypeptide against degradation. The non-glycosylated protein may not reach the outer membrane.

Elbein: Yes.

Feizi: These discussions show that we need to determine whether it is the viral glycoprotein or cellular glycoproteins that are involved in the bacterial adhesion.

Elbein: We only get binding of streptococci to virus-infected monolayers.

Feizi: If your assays were done with cells in suspension it would be possible to investigate the effect of trypsinization or neuraminidase treatment *per se* on the adhesion of bacteria.

Elbein: Except that it makes the quantitative assay that much more difficult, since some cells, but not necessarily all, come loose.

Helenius: Epithelial cells in culture frequently grow like real epithelia with tight junctions between neighbouring cells. Two plasma membrane regions are seen: the apical region with the brush border and the basolateral surface. These are totally different in protein composition and function. Boulan & Sabatini (1978) have shown that some viruses such as influenza virus bud from the apical membrane, others only from the basolateral membrane. The glycoproteins of influenza virus are presumably present only on the brush border surface of the cell. In this case one would expect the influenza virus neuraminidase activity to be localized on the apical surface. The polarity of the cells used may thus in part complicate interpretations which depend on the use of monolayer cells.

Feizi: But it is possible that the viral infection and trypsinization and neuraminidase treatment alter the polarity of the epithelial cells, and thus they may alter the clustering ability and accessibility of lectin binding sites.

Elbein: One could always argue that all these cells are virus-infected and the virus protein on the surface has exposed something else with a similar structure. Whatever is involved GBS apparently recognize at least a portion of a complex oligosaccharide. I wouldn't want to say that what they recognize is galactose. Whatever the sequence is, something like a galactose *N*-acetylglucosamine group and maybe involving some of the mannose residues is necessary also. The initial recognition factor may be a simple sugar but I would be surprised if a simple sugar was the whole adherence factor.

Sharon: It is again a problem of what you want to know.

Beachey: Have you any clues as to the composition of the surface adhesins of the GBS?

Elbein: Not really. We have tried some of the group B antigens. For example we have treated streptococci with trypsin and all those sorts of things. We have not been able to find any way to inhibit the binding. We have had a difficult time making cell wall preparations of these bacteria. We got some cell walls with a Ribi disintegrator but the yield was very poor. We have tried trypsin digestion, neuraminidase, trichloroacetic acid extraction, and so on, and haven't been able to find the binding component.

Hughes: Do wall-less organisms bind? You could treat with lysozyme to remove the cell wall.

Elbein: We don't know.

Candy: I would like to support Dr Feizi's contention that the alternative hypothesis for increased bacterial adhesion to virus-infected cells is a modification of the existing cell wall by the virus. We studied the effect of rotavirus infection in bovine kidney cells on the adhesion of colonization factor I and colonization factor II strains. The adhesion of both strains was enhanced in the virus-infected cells, compared to controls. However there was some degree of adhesion to the uninfected cell cultures. It wasn't an all-or-none phenomenon like you described, Dr Elbein. I learnt with interest from Dr Choppin that globoside is present in the plasma membrane of these cells and that the viral infection actually enhances the synthesis of carbohydrate chains on the glycolipid of the cell surface, and therefore this could be one mechanism by which virus infection enhances bacterial adhesion, by inducing the synthesis of cell membrane receptor molecules.

Taylor-Robinson: There are well known associations of bacterial infections with influenza virus infections: *Staph. aureus* and *Haemophilus influenzae.* Could your results explain those infections? Or have you not looked at the range of bacteria which could tell you that?

Elbein: We haven't carried it that far. In many influenza virus infections there are secondary bacterial infections. The strains of streptococci may have some potential significance but I don't really know. There are many possible mechanisms whereby virus infections of cells could lead to secondary bacterial infections. In Gram-negative bacteria, as Dr Mirelman indicated, *E. coli* may bind to infected cells. We have tried to make clinical correlations and there is some evidence suggesting that bacterial infection could result from surface receptors which are virus-induced.

Taylor-Robinson: There are, of course, other possible explanations for associated bacterial infections. Elliot Larson (Larson et al 1977) has shown that polymorphonuclear leucocyte function is diminished in influenza virus infections and this could be one reason for the proliferation of bacteria.

Elbein: We have not specifically looked. Dr Sanford has looked at staphylococcal adhesion and there is some indication that this could also be altered in influenza.

REFERENCES

Becht H, Rott R, Klenk H-D 1972 Effect of concanavalin A on cells infected with enveloped RNA viruses. J Gen Virol 14:1-8

Boulan R, Sabatini D 1978 Asymmetric budding of virus in epithelial monolayers: a model system for study of epithelial polarity. Proc Natl Acad Sci USA 75:5071-5075

Brown MS, Faust JR, Goldstein JL, Kaneko I, Endo A 1978 Induction of 3-hydroxy-3-methylglutaryl coenzyme A reductase activity in human fibroblasts incubated with compactin a competitive inhibitor of the reductase. J Biol Chem 253:1121-1128

Carson DD, Lennarz WS 1979 Inhibition of polyisoprenoid and glycoprotein biosynthesis causes abnormal embryonic development. Proc Natl Acad Sci USA 76:5709-5713

Larson HE, Parry RP, Gilchrist C, Luquetti A, Tyrrell DAJ 1977 Influenza viruses and staphylococci *in vitro:* some interactions with polymorphonuclear leucocytes and epithelial cells. Br J Exp Pathol 58:281-288

Schwarz RT, Rohrschneider JM, Schmidt MFG 1976 Suppression of glycoprotein formation of Semliki Forest influenza, and avian sarcoma virus by tunicamycin. J Virol 19:782-791

Sublethal concentrations of antibiotics and bacterial adhesion

EDWIN H. BEACHEY, BARRY I. EISENSTEIN and ITZHAK OFEK

Veterans Administration Medical Center, 1030 Jefferson Avenue, and University of Tennessee College of Medicine, Memphis, Tennessee 38104, USA

Abstract Various antibiotics in sublethal concentrations markedly impair adhesion of *Streptococcus pyogenes* and *Escherichia coli* to human cells. In streptococcal cells penicillin G caused an enhanced loss of lipoteichoic acid, the ligand (adhesin) that binds the organism to host cells, with consequent loss of their adhesive properties. In *E. coli* sublethal concentrations of penicillin prevented the surface expression of the mannose-specific adhesins by distorting cell wall biosynthesis. In contrast to streptococci, *E. coli* cells could not be made to lose their adhesins once they had been formed. Streptomycin in subinhibitory concentrations similarly suppressed the acquisition of mannose-binding and adhesive activities in several strains of antibiotic-sensitive *E. coli* but not in isogenic derivatives with ribosomal mutation to high-level streptomycin resistance, *rpsL*, or in bacteria in the stationary phase of growth, suggesting that streptomycin exerted its sublethal suppressive effects by classic mechanisms of action on the bacterial ribosome. Strain VL2, derived from one streptomycin-resistant mutant, retained a high level (1000 μg/ml) of resistance to streptomycin but reacquired sensitivity to the sublethal effect; growth in 30 μg streptomycin/ml suppressed mannose-sensitive haemagglutination (< 1% of control) as well as mannose-sensitive adhesion to epithelial cells (42%) or leucocytes (7%). Although these streptomycin-treated bacteria demonstrated an unaltered degree of fimbriation their fimbriae were significantly longer than those on the untreated bacteria. Furthermore, in contrast to the untreated bacteria, the fimbriae isolated from the drug-treated bacteria were found to lack mannose-binding activity as measured by haemagglutination. It therefore appears that streptomycin can cause even resistant bacteria to produce an aberrant fimbrial protein, presumably by causing misreading in 'competent' ribosomes. These studies indicate that the use of sublethal doses of certain antibiotics whose mode of action is well known may shed light on the genetic and chemical modulation of bacterial factors involved in mucosal colonization.

It has been assumed that antibiotic therapy clears pathogenic bacteria from the host by the known mechanisms of action of the administered drugs. Some antibiotics are bactericidal and are thought to clear microorganisms simply by killing them; others

1981 Adhesion and microorganism pathogenicity. Pitman Medical, Tunbridge Wells (Ciba Foundation symposium 80) p 288-305

are bacteriostatic and are thought to clear microorganisms by rendering them more susceptible to the defence mechanisms of the host. Supplementing this rather simplistic view of microbial clearance mediated by antibiotics are results which suggest that, in subinhibitory concentrations, antibiotics may act in more subtle ways to alter host–parasite interactions without interfering with the growth and replication of the infecting microorganisms. Several studies have demonstrated that sublethal doses of antibiotics selectively interfere with the synthesis or expression of certain excreted (Boethling 1975, Piovant et al 1978, Shibl & Al-Sowaygh 1979) as well as somatic (Hirashima et al 1973, Horne et al 1977, Horne & Tomasz 1977, 1979, Alkan & Beachey 1978, Eisenstein et al 1979, 1980a) components of bacteria without altering total protein synthesis. The implications of these findings are obvious: it may be possible to selectively suppress the virulence properties of potentially pathogenic microorganisms.

Recent work in our laboratories has indicated that subinhibitory doses of antibiotics interfere with the production or retention of the adhesive ligands (or adhesins) that bacteria need for attaching themselves to mucosal surfaces (Alkan & Beachey 1978, Eisenstein et al 1979, 1980a, Ofek et al 1979). Since the adherence of bacteria to mucosal surfaces has been gaining recognition as an important initial step in the pathogenesis of bacterial infections (Ofek & Beachey 1980a,b) it has been of special interest to determine the effects of certain antibiotics on the formation of these adhesive ligands. It is the purpose of this paper to describe studies of the effects of subinhibitory concentrations of antibiotics on the ability of *Streptococcus pyogenes* and *Escherichia coli* to adhere to human epithelial cells.

Methods

The strains of group A streptococci used have been described previously (Alkan & Beachey 1978). A streptomycin-susceptible strain (M-Str^s or VL1) of *E. coli* with a minimum inhibitory concentration (MIC) of 30 μg/ml was obtained from a clinical specimen and an isogenic streptomycin-resistant mutant (M-Str^r or VL10 – MIC > 1000 μg/ml) of the clinical strain was selected by cloning a spontaneous mutant that grew in high concentrations of streptomycin, as described by Eisenstein et al (1979). An additional streptomycin-resistant mutant (Str^{r1} or VL2) strain, derived from strain VL10, which retained a high level of antibiotic resistance but grew faster than VL10 (doubling time, 49.3 min for strain VL2 as compared to 59.8 min for strain VL10), was also used in these studies.

The sensitive and resistant strains were grown in the presence or absence of antibiotics in brain-heart infusion broth under static conditions for 24 to 48 h as previously described (Eisenstein et al 1979). Epithelial cell, leucocyte or erythrocyte-binding activities of the bacteria were quantitated as described in detail (Alkan & Beachey 1978, Ofek & Beachey 1978, Ofek et al 1979). We assayed the mannose-binding activity of the *E. coli* strains by aggregometry and agglutination, using

TABLE 1 Effect of subinhibitory concentrations of penicillin or streptomycin on the adhesive properties of S. pyogenes and E. coli in the resting phase

Bacteria[a]	*Antibiotic (μg/ml)*	*Adhering ability (% of control*[b]*)*
S. pyogenes	Penicillin G (1)	30
E. coli	Penicillin G (27)	100
E. coli	Streptomycin (10)	100

[a]Resting-phase bacteria grown in brain–heart infusion broth and harvested after 48 h of growth were washed and resuspended in 1/20th the original volume of fresh broth containing the specified antibiotics. The suspensions were then incubated for 3 h at 37 °C, washed and exposed to human buccal epithelial cells.
[b]Expressed as: (number of adherent bacteria per epithelial cell in the presence of antibiotic/ number of bacteria per epithelial cell in the absence of antibiotic) × 100.

mannose-containing yeast cells as described previously (Ofek & Beachey 1978). The morphology, number and length of fimbriae were estimated by electron microscopy (Novotny et al 1969, Eisenstein et al 1979, 1980a). Fimbriae were isolated from strain VL2 grown in the presence or absence of streptomycin and partially purified by ammonium sulphate precipitation and sucrose density gradient centrifugation, as described by Silverblatt (1979) and Silverblatt & Cohen (1979). The ability of the isolated fimbriae to bind mannose residues was determined in assays of the agglutination of guinea-pig erythrocytes in the absence and the presence of 2.5% α-methyl mannoside as described by Salit & Gotschlich (1977).

Results and discussion

Initially we investigated the effects of subminimal inhibitory concentrations (SIC) of penicillin G or streptomycin on the abilities of resting-phase cultures of *S. pyogenes* or *E. coli* to adhere to epithelial cells. For this purpose 48-h broth cultures of bacteria were washed once in 0.02 M-phosphate–0.15 M-NaCl, pH 7.4, and resuspended in 5% of the original volume of fresh broth with or without added antibiotics. As shown in Table 1 the addition of SIC penicillin to resting-phase *S. pyogenes* resulted in a marked reduction (70%) in the ability of the bacteria to adhere to epithelial cells, whereas similar treatment of resting-phase *E. coli* with either penicillin or streptomycin, a ribosomally active antibiotic, had no effect.

The loss of adhering ability in resting-phase streptococci was attributed to a penicillin-induced loss of lipoteichoic acid (Alkan & Beachey 1978, Horne & Tomasz 1977, 1979), a surface ligand centrally involved in the adhesion of *S. pyogenes* to epithelial cells (Ofek et al 1975, Beachey & Ofek 1976). The retention of adhering

TABLE 2 Loss of adhering ability of E. coli exposed to sublethal antibiotics during the growth phase[a]

Antibiotic (μg/ml)	*Adhering ability (% of control[b])*
Penicillin G (27)	10
Streptomycin (10)	20
Streptomycin (2)	33
No antibiotic	100

[a]*E. coli* strain VL1 was subcultured 1:100 from a stationary phase broth culture to early visible growth at 37 °C. Samples (0.05 ml) were further subcultured in fresh broth (5 ml) containing the specified antibiotics and incubated for 48 h at 37 °C. Adherence assays were done as described (Eisenstein et al 1979).
[b]Values expressed as described in footnote to Table 1.

ability by resting-phase *E. coli* incubated with penicillin or streptomycin indicated that the antibiotics were unable to release the adhesive ligands of the organisms. In contrast, when growing cultures of *E. coli* were treated in the same way with SIC penicillin or streptomycin their ability to adhere to epithelial cells was markedly reduced (90% and 80% respectively) (Table 2). Thus, reduced adhering ability in *E. coli* appeared to be due to prevention of the formation of surface adhesin by the organisms rather than to the loss of adhesin once it had been formed.

Since the *E. coli* strains used in these studies bind via mannose-specific adhesins to mannose residues on animal cells (Ofek & Beachey 1978, 1980a,b), subsequent adhesion assays included control mixtures containing 2.5% mannose so that we could assess the mannose sensitivity of binding. In addition, mannose-binding activity was assessed by yeast cell agglutination assays (Ofek & Beachey 1978).

The effect of subminimal inhibitory concentrations of penicillin on the mannose binding of *E. coli* was found to correlate with the transformation of rod-shaped bacteria into filaments devoid of fimbriae, as determined by electron microscopy (Fig. 1). The filamentous organisms lacked the ability to agglutinate yeast cells or to adhere to epithelial cells (see Table 2). The mechanism of the association between filament formation and loss of surface adhesins might be inferred from a similar association observed in a temperature-sensitive mutant of *E. coli*, strain PAT 84 (Mirelman et al 1976).

When PAT 84 was grown at temperatures restrictive for carboxypeptidase activity it formed filaments and lacked fimbriae; when it was shifted to permissive temperatures the mutant formed normal rods and surface fimbriae and it regained its ability to adhere to epithelial cells and to agglutinate yeast cells (Table 3). The same organisms grown in SIC penicillin at permissive temperatures assumed properties identical to those of organisms grown at restrictive temperatures without penicillin

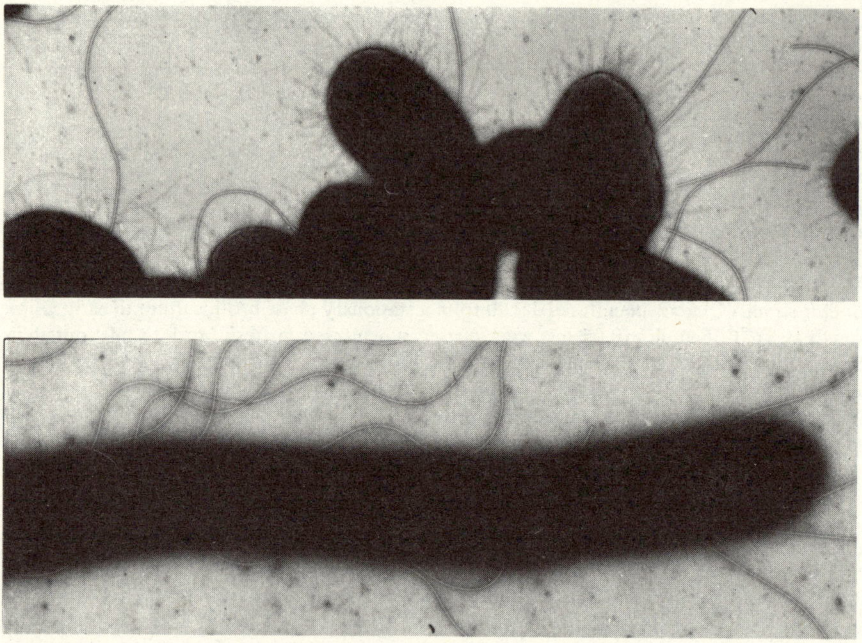

FIG. 1. Filament formation and loss of fimbria in *E. coli* (strain VL1) grown in sublethal concentrations of penicillin G. When grown for 24 h in the absence of penicillin the organisms from normal rods and are fully fimbriated (A); when grown for the same time in about half the minimum inhibitory concentration of penicillin G (6 μg/ml) the organisms fróm long filaments devoid fimbriae (B).

(Table 3). Subminimal inhibitory concentrations of penicillin inhibit the carboxypeptidase activity needed for normal septation (Mirelman et al 1976). Thus, penicillin induces filaments and has an effect similar to that of growing PAT 84 at a temperature restrictive for carboxypeptidase activity. Although these results suggest that a causal relationship exists between abnormal septation which is affected by carboxypeptidase activity and the formation of fimbrial adhesins (Ofek et al 1979), this remains to be proved.

Like penicillin, the aminoglycosides neomycin, gentamicin and streptomycin markedly suppressed the development of mannose-binding activity of strain VL1 in a dose-dependent fashion (Fig. 2). Spectinomycin and tetracycline were also suppressive but to a lesser degree. Chloramphenicol was only minimally suppressive at 50% MIC (Fig. 2). The remaining antibiotics tested, including rifampin, clindamycin, erythromycin, novobiocin and nalidixic acid, had no reproducible effects on the acquisition of mannose-binding activity (results not shown). In general, the

TABLE 3 Relationship between filament formation and loss of fimbriae, yeast cell agglutination and epithelial cell adhesion of a temperature-sensitive mutant, PAT 84, of E. coli[a]

Culture conditions	*Fimbriae*	*Filaments*	*Yeast-cell-agglutinating power*	*Adhering ability (% of control)*
Permissive temperature (30 °C)	+	0	100	100
Restrictive temperature (39 °C)	0	+	<10	10
Permissive temperature (30 °C) + penicillin G (27 μg/ml)	0	+	<10	10

[a]*E. coli* strain PAT 84 was subcultured 1:100 from a stationary-phase broth culture to early visible growth at 30 °C. Samples (0.05 ml) were further subcultured into fresh broth (5 ml) and further incubated for 48 h at 39 °C (restrictive) or 30 °C (permissive) with and without penicillin G. Organisms were collected, washed and examined for fimbriae and filament formation by electron microscopy, and for adhering ability as described in footnote to Table 1. Yeast-cell-agglutinating power was calculated according to a modification (Eisenstein et al 1979) of the method of Duguid & Gillies (1957) as 10^{11}/minimal number of bacteria per millilitre needed to agglutinate yeast cells. The values are expressed as % of control.

suppression of the acquisition of mannose-binding activity, as measured by yeast cell agglutination, paralleled suppression of the ability of *E. coli* grown in subminimal inhibitory concentrations of antibiotic (Table 4) to adhere to epithelial cells.

It should be emphasized that our studies included *E. coli* strains that exhibited mannose-sensitive interactions with eukaryotic cells. Thus the antibiotic affected the formation of mannose-specific adhesin on the bacterial surface. Sandberg et al (1979), using ampicillin-induced filaments of certain uropathogenic *E. coli*, reported findings similar to ours. In contrast Vosbeck et al (1979), using an *E. coli* strain exhibiting mannose-resistant binding to tissue culture cells, were unable to show inhibition of adhesion by either β-lactam or the aminoglycoside antibiotics. It appears, therefore, that the suppressive effects of sublethal concentrations of antibiotics on adhering ability depend on the chemical composition and the mechanism of expression of the surface adhesin of a particular microorganism.

Streptomycin exerts its known effects by binding to streptomycin receptor(s) on bacterial ribosomes; single-step mutants that are highly resistant to streptomycin are thought to lack these receptors (Ozaki et al 1969). In an attempt to determine whether the suppressive effect of SIC streptomycin was mediated through the classic target site on ribosomes we examined the effect of streptomycin on mannose binding and epithelial cell adherence in an isogenic streptomycin-resistant mutant of the VL1 strain of *E. coli* used in the previous studies. The highly resistant mutant (VL10) was selected and cloned from a large population of the sensitive strain grown on agar plates containing streptomycin greatly in excess of the MIC. The purified, cloned derivative strain VL10 was found to have a degree of surface

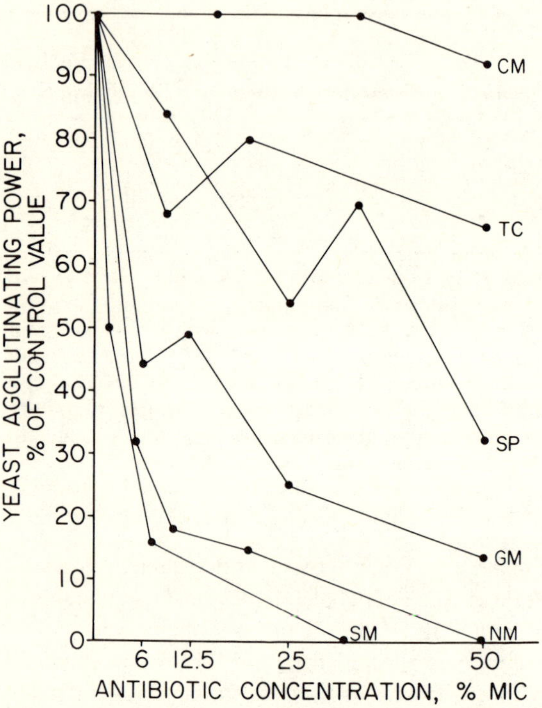

FIG. 2. Effect of increasing concentrations of antibiotics on the yeast-agglutinating power of *E. coli*, strain VL1. The yeast-cell-agglutinating power (see footnote to Table 3) of cells grown for 48 h in subinhibitory concentrations of various antibiotics was compared to that of cells grown without drug. Results are expressed as the means of at least four separate experiments. CM, chloramphenicol; TC, tetracycline; SP, spectinomycin; GM, gentamicin; NM, neomycin; SM, streptomycin. (Reproduced by permission from Eisenstein et al 1980a.)

fimbriation and yeast-cell-agglutinating (mannose-binding) and epithelial-cell-adhering abilities similar to those of the parent strain. Growth of the resistant strain in enormous concentrations (up to 5000 μg/ml) of streptomycin had little effect on mannose binding or epithelial-cell-adhering ability (Table 5). Fimbriae were present but because of electron-dense deposits on the surface of the bacteria these were difficult to quantitate at the higher concentrations of streptomycin (Eisenstein et al 1979). Similar contrasting effects were noted with streptomycin-sensitive (*rpsL*$^+$) and streptomycin-resistant (*rpsL*) derivatives of *E. coli* strain K12 with defined alleles for the ribosomal receptor for streptomycin, protein S12. These results suggested, therefore, that the suppressive effects of SIC streptomycin on the

TABLE 4 Loss of the ability of E. coli VL1 grown in SIC streptomycin to agglutinate yeast cells or to adhere to buccal epithelial cells[a]

Streptomycin in culture (μg/ml)	*Yeast cell agglutination (% of control)*	*Epithelial cell adhesion (% of control)*
0	100	100
0.5	50	100
2	5	30
10	<1	10

[a]See footnotes to Tables 1 and 3 for methods and derivation of values. Each value represents the mean of at least three experiments. (Data from Ofek et al 1979.)

TABLE 5 Effects of streptomycin on adherence and yeast cell agglutination of a streptomycin-resistant strain (VL10) of E. coli grown for 48 h with or without antibiotic

Streptomycin in culture (μg/ml)	*Yeast cell agglutination (% of control)*	*Epithelial cell adherence (% of control)*
0	100	100
5000	100	100
10 000	32	68
20 000	No growth	

surface properties of *E. coli* are exerted through the classical streptomycin target site on bacterial ribosomes.

Because other investigators had shown that the binding of type 1 fimbriae to eukaryotic cells is mannose-sensitive (Salit & Gotschlich 1977) we studied the relationship between mannose-binding activity and the degree of fimbriation of *E. coli* VL1 in various phases of growth as well as in the presence and absence of SIC streptomycin. We observed a time-dependent increase in the proportion of fimbriated organisms; at 6 h the mean percentage of fimbriated organisms was 48, at 24 h it was 91, and at 48 h it was 97%. Moreover, the acquisition of fimbriae paralleled the time-dependent increase in yeast cell agglutination: yeast-cell-agglutinating power (see footnote to Table 3) at 6 h was 100; at 24 h it was 6050 ± 30 and at 48 h it was 9700 ± 830 (Eisenstein et al 1979). *E. coli* strain VL1 grown in 2 μg streptomycin/ml for 48 h showed a reduction in fimbriation that was positively correlated with reduced mannose-binding activity ($r = 0.35, n = 52, P < 0.05$).

Serial passage of VL10 in broth resulted in the selection of a faster growing strain, VL2 (generation time of 49.3 min compared to 59.8 min for the parent

TABLE 6 Susceptibility of a streptomycin-resistant strain of E. coli to the suppressive effects of SIC streptomycin on mannose-sensitive interactions with animal cells

			Effect of growth of organisms in 30 μg streptomycin/ml on:		
Isogenic strains	*Streptomycin MIC (μg/ml)*	*Generation time (min)*	*Epithelial cell adhesion*[a]	*Leucocyte adhesion*[a,b]	*Haemagglutinating power*[a,b]
VL10	>1000	59.8	100	95	100
VL2	>1000	49.3	42	7	<10

[a]All values are expressed as % of control values obtained with the organisms grown without antibiotic. (Data from Eisenstein et al 1980b.)

[b]Attachment to leucocytes was determined by the monolayer technique of Bar-Shavit et al (1977). Haemagglutination assays used a 2% suspension of guinea-pig erythrocytes and serial dilutions of test bacteria in microtitre wells, as previously described (Eisenstein et al 1980a). Haemagglutinating power was calculated by the method of Duguid & Gillies (1957). (See also footnote to Table 3.)

VL10), which retained its high level of resistance to streptomycin. In contrast to the total resistance to streptomycin of the haemagglutinating, epithelial-cell-adhering and leucocyte-attaching abilities of the parent VL10 strain, each of these activities was suppressed in VL2 grown in an antibiotic concentration of less than 0.03 of its MIC (Table 6).

It was of particular interest to examine fimbriation in the streptomycin-resistant strain, VL2. Electron microscopy revealed that drug-treated organisms of this strain were as fimbriated as the untreated organisms but that the fimbriae on the drug-treated organisms were significantly longer (Fig. 3). These results suggested that the fimbriae of the drug-treated organisms must have been altered in some way that made them lose their adhesive properties.

In order to examine the possibility that the fimbriae of drug-treated VL2 *E. coli* lacked adhesiveness we took fimbriae isolated and partially purified from VL2 grown for 48 h with 30 μg streptomycin/ml and compared their ability to agglutinate guinea-pig erythrocytes with that of fimbriae from VL2 grown without streptomycin. As shown in Table 7 the fimbriae from the streptomycin-treated organisms totally lacked agglutinating activity at a concentration 15-fold greater than the minimal agglutinating concentration of the fimbriae from untreated VL2. From these results it appears that streptomycin can cause even resistant bacteria to produce aberrant fimbrial protein, presumably by causing misreading in 'competent' ribosomes (Gorini 1974). Our results suggest that strain VL10 acquired a second mutation, possibly analogous to *ram* (Gorini 1974), which enabled the double mutant strain VL2 to interact in this manner with streptomycin.

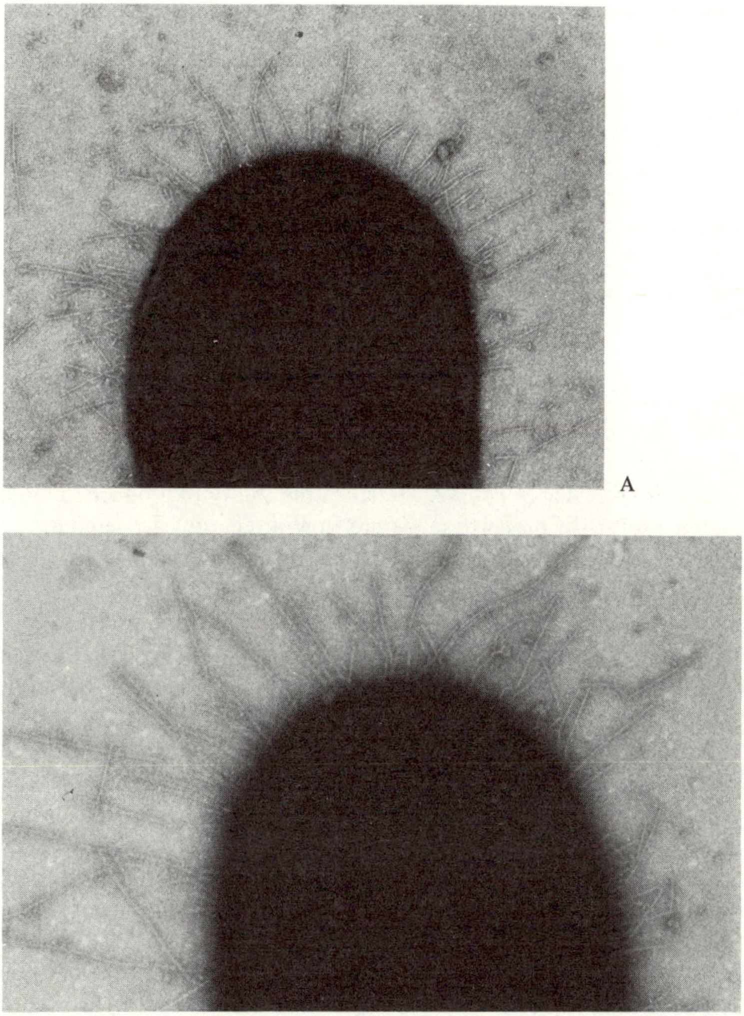

A

B

FIG. 3. Electron micrographs of streptomycin-resistant *E. coli*, strain VL2, grown for 48 h without (A) and with (B) 30 μg/ml of streptomycin. Note longer, thicker fimbriae of the cell grown in streptomycin. (Reproduced by permission from Eisenstein et al 1980a.)

Streptomycin inhibits or distorts protein synthesis by binding to the S12 protein of sensitive ribosomes (Ozaki et al 1969), thereby interfering with the translation of mRNA (Gorini 1974). Subinhibitory concentrations of streptomycin have been known for some time to distort rather than inhibit protein synthesis, with the resultant production of aberrant proteins by the treated bacteria (Edelman & Gallant

TABLE 7 Streptomycin-induced loss of haemagglutinating activity of fimbriae of a streptomycin-resistant (MIC>1000 μg/ml) strain (VL2) of E. coli

Fimbriae prepared from strain VL2 grown[a] with:	*Minimal amount (μg/ml) of protein needed to cause haemagglutination[b]*
No drug	35
Streptomycin, 30 μg/ml	>500

[a]Bacteria were cultured with or without antibiotic as described in footnote to Table 2.
[b]See footnote, Table 6.

1977). The notion that the second, more subtle effect of streptomycin was responsible for the diminished mannose-binding activity of *E. coli* in our studies is based on the finding that drug-treated *E. coli* VL2 retained the ability to produce fimbrial protein with altered adhesive properties. We are now attempting to demonstrate mRNA misreading in drug-treated bacteria by showing alterations in the primary structure of the fimbrial proteins.

The penicillin-induced loss of the lipoteichoic acid adhesin from the surface of streptococci is a rather novel effect of this antibiotic. Perturbation in peptidoglycan synthesis had been the only known mode of action of penicillin. Recently, however, several investigators (Horne & Tomasz 1977, 1979, Alkan & Beachey 1978) have reported an additional effect: penicillin seems to enhance the release of lipid-containing substances, including lipoteichoic acid, from whole bacteria by an ill-defined mechanism. Thus penicillin may act by two mechanisms to reduce the ability of bacteria to adhere. The first mechanism, demonstrated here in certain strains of *E. coli*, involves the classic target site of penicillin action resulting in the formation of filaments denuded of adhesins; the second mechanism, operating in streptococci, involves the enhanced release of lipid-containing materials, some of which constitute the surface adhesins of the microorganisms. Whether one or more of the known penicillin-binding proteins are involved in the second mechanism as they are in the first remains to be elucidated.

Taken together, our studies indicate that subinhibitory levels of several antibiotics, especially those acting on the cell wall or on ribosomes, are clearly capable of suppressing the adhesive properties of bacteria without manifestly altering bacterial growth. Antibiotics can act selectively on the virulence properties of bacteria, presumably by interfering with the synthesis of protein subunits (streptomycin) or with the arrangement of these subunits into functional organelles (penicillin). Thus the use of sublethal doses of certain antibiotics whose mode of action is well known may shed light on the genetic and chemical modulation of bacterial factors involved in mucosal colonization.

Acknowledgements

These studies were supported by programme-directed research funds from the US Veterans Administration and by Research Grants AI-13550 and AI-10085 from the US Public Health Service. E.H.B. is the recipient of a Medical Investigatorship Award from the US Veterans Administration. B.I.E. is the recipient of a Clinical Investigatorship No. AM-00686 from the US Public Health Service.

We thank L. Hatmaker, V. Long, D. Dodd and J. Smith for excellent technical assistance.

REFERENCES

Alkan ML, Beachey EH 1978 Excretion of lipoteichoic acid by group A streptococci: influence of penicillin on excretion and loss of ability to adhere to human oral mucosal cells. J Clin Invest 61:671-677

Bar-Shavit Z, Ofek I, Goldman R, Mirelman D, Sharon N 1977 Mannose residues on phagocytes as receptors for the attachment of *Escherichia coli* and *Salmonella typhi.* Biochem Biophys Res Commun 78:455-460

Beachey EH, Ofek I 1976 Epithelial cell binding of group A streptococci by lipoteichoic acid on fimbriae denuded of M protein. J Exp Med 143:759-771

Boethling RS 1975 Regulation of extracellular protease secretion in *Pseudomonas maltophilia.* J Bacteriol 123:954-961

Duguid JP, Gillies RR 1957 Fimbriae and adhesive properties in dysentery bacilli. J Pathol 74:397-411

Edelman P, Gallant J 1977 Mistranslation in *E. coli.* Cell 10:131-137

Eisenstein BI, Ofek I, Beachey EH 1979 Interference with the mannose binding and epithelial cell adherence of *Escherichia coli* by sublethal concentrations of streptomycin. J Clin Invest 63:1219-1228

Eisenstein BI, Beachey EH, Ofek I 1980a Influence of sublethal concentrations of antibiotics on the expression of the mannose specific ligand of *Escherichia coli.* Infect Immun 28:154-159

Eisenstein BI, Ofek I, Beachey EH 1980b Dissociation between mannose-sensitive hemagglutination and mannose binding activity in a strain of *Escherichia coli.* In: Nelson JD, Grass C (eds) Current chemotherapy and infectious disease. American Society for Microbiology, Washington DC, vol 2:780-781

Gorini L 1974 Streptomycin and misreading of the genetic code. In: Nomura M et al (eds) Ribosomes. Cold Spring Harbor Laboratory, Cold Spring Harbor, NY, p 791-803

Hirashima A, Childs G, Inouye M 1973 Differential inhibitory effects of antibiotics on the biosynthesis of envelope proteins of *Escherichia coli.* J Mol Biol 79:373-389

Horne D, Tomasz A 1977 Tolerant response of *Streptococcus sanguis* to beta-lactams and other cell wall inhibitors. Antimicrob Agents Chemother 11:888-896

Horne D, Tomasz A 1979 Release of lipoteichoic acid from *Streptococcus sanguis:* stimulation of release during penicillin treatment. J Bacteriol 137:1180-1184

Horne D, Hackenback R, Tomasz A 1977 Secretion of lipids induced by inhibition of peptidoglycan synthesis in streptococci. J Bacteriol 132:704-717

Mirelman D, Yashouv-Gan Y, Schwarz U 1976 Peptidoglycan biosynthesis in a thermosensitive division mutant of *Escherichia coli.* Biochemistry 15:1781-1790

Novotny C, Carnahan J, Brinton CC Jr 1969 Mechanical removal of F pili, type I pili, and flagella from Hfr and TRF donor cells and the kinetics of their reappearance. J Bacteriol 98:1294-1306

Ofek I, Beachey EH 1978 Mannose binding and epithelial cell adherence of *Escherichia coli.* Infect Immun 22:247-254
Ofek I, Beachey EH 1980a Bacterial adherence. In: Stollerman GH (ed) Advances in internal medicine. Year Book Medical Publishers, Chicago, vol 25:503-532
Ofek I, Beachey EH 1980b General concepts and principles of the adherence of bacteria to animal cells. In: Beachey EH (ed) Bacterial adherence. Chapman & Hall, London (Receptors and recognition, series B) vol 6:1-27
Ofek I, Beachey EH, Jefferson W, Campbell GL 1975 Cell membrane-binding properties of group A streptococcal lipoteichoic acid. J Exp Med 141:990-1003
Ofek I, Beachey EH, Eisenstein BI, Alkan M, Sharon N 1979 Suppression of bacterial adherence by subminimal inhibitory concentrations of β lactam and aminoglycoside antibiotics. Rev Infect Dis 1:832-837
Ozaki M, Mizuchima S, Nomura M 1969 Identification and functional characterization of the protein controlled by the streptomycin-resistant locus in *E. coli.* Nature (Lond) 222:333-339
Piovant M, Varenne S, Pages JM, Lazdunski C 1978 Preferential sensitivity of synthesis of exported proteins to translation inhibitors of low polarity in *Escherichia coli.* Mol Gen Genet 164:265-274
Salit IE, Gotschlich EC 1977 Hemagglutination by purified type I *Escherichia coli* pili. J Exp Med 146:1169-1181
Sandberg T, Stenqvist K, Svanborg Edén C 1979 Effects of subminimal inhibitory concentrations of ampicillin, chloramphenicol, and nitrofurantoin on the attachment of *Escherichia coli* to human uroepithelial cells *in vitro.* Rev Infect Dis 1:838-844
Shibl AM, Al-Sowaygh IA 1979 Differential inhibition and bacterial growth and hemolysin production by lincosamide antibiotics. J Bacteriol 137:1022-1023
Silverblatt F 1979 Ultraviolet irradiation disrupts somatic pili structure and function. Infect Immun 25:1060-1065
Silverblatt F, Cohen LS 1979 Antipili antibody affords protection against experimental ascending pyelonephritis. J Clin Invest 64:333-336
Vosbeck K, Handschin H, Menge E, Zak O 1979 Effects of subminimal inhibitory concentrations of antibiotics on adhesiveness of *Escherichia coli in vitro.* Rev Infect Dis 1:845-851

DISCUSSION

Silverblatt: Just because a pilus is intact doesn't necessarily mean that it is able to bind. We often find that organisms whose pili look perfectly normal in the electron microscope have lost the ability to bind. Perhaps another determinant of adhesion might be a subtle change in the tertiary or quaternary structure of the pili subunits that obscures the binding site.

Beachey: We suggested the possibility of misreading because streptomycin is known to distort the translation of mRNA, resulting in amino acid substitutions to aberrant proteins which then lose their function (Ozaki et al 1969). We haven't excluded the possibility that longer fimbriae have something on their surfaces that may mask adhesive structures on these organelles.

Silverblatt: Pili expression is a multi-genetic factor. There is not only a structural gene but also at least one assembly gene.

Vosbeck: Our pet strain (*E. coli* SS142) is a mannose-resistant adherer. We found a dose-dependent inhibition of adhesion when this strain was grown in the presence of subinhibitory concentrations (1/16 and 1/4 of the MIC) of several antibiotics, including tetracycline, clindamycin and trimethoprim–sulphonamide combinations. Streptomycin, chloramphenicol and certain β-lactam antibiotics did not inhibit under our conditions (Vosbeck et al 1979).

We also looked at the piliation of *E. coli* SS 142. In its natural state, grown without antibiotics, it is heavily piliated with very thin pili. If one grows it in tetracycline, adhesion is inhibited and almost no pili are demonstrable by electron microscopy. Clindamycin inhibits a little less than tetracycline and a few more pili occur than with tetracycline. We did these experiments several times and we did them blindly, so we are convinced that there is loss of piliation associated with growth in subinhibitory concentrations of some antibiotics.

We then tested several antibiotics on 10 different *E. coli* strains in the monolayer adhesion assay I described earlier (p 138). With streptomycin, chloramphenicol, tetracycline, clindamycin and trimethoprim–sulphametrole adhesion was reduced with most strains. The adhesion of *E. coli* SS142 was not inhibited by streptomycin, as an exception.

What do these results mean? It appears that inhibition of adhesion, or of the synthesis of some adhesion factor, can be observed with many different antibiotics. This effect does not seem to correlate with the mechanism of action of these antibiotics. The antibiotics inhibitory in our assay system are mostly protein synthesis inhibitors, except for trimethoprim–sulphametrole which inhibits tetrahydrofolate reductase and as such inhibits C-1 fragment metabolism. Others, including Dr Svanborg Edén, have found similar effects with subinhibitory concentrations of ampicillin. Dr Wadström's group (Tylewska et al 1980) found that rifampicin, which inhibits RNA polymerase, inhibits adhesiveness in their system (hydrophobic interaction chromatography). So it really seems that no specific mechanism is involved in the inhibition of bacterial adhesion. It is an open question what this means in terms of clinical effects and whether decreasing adhesion has something to do with the clinical efficacy of antibiotics.

Svanborg Edén: Interference of a small amount of drug with binding to the mucous surface might be of interest in prophylaxis against infection but not in treatment of established infection, where the bactericidal effects may be more useful. *In vitro* the combination of ampicillin treatment of bacteria and addition of antibodies, for example against pili or another bacterial surface structure, decreases adhesion synergistically. When prophylaxis is needed *in vivo* a low amount of antibody may be present on the mucosal surface, and may act together with the antibacterial agent to protect the patient.

Beachey: Did the bacteria form filaments under those conditions?

Svanborg Edén: Yes. At 1/4 of the MIC.

Beachey: Did you look for pili?

Svanborg Edén: The antiserum did not contain antibody to anything else but pili. The increased efficiency of the antibodies thus suggests a decreased amount of pili present.

Beachey: This would be a good strain to look at under the electron microscope to see whether the fimbrial protein had assembled into fully formed organelles or just knobs on the surface. Using ferritin-labelled antibodies, one may be able to detect vestiges of fimbriae as blebs on the surface.

Vosbeck: We did a similar study to the one I mentioned earlier (p 138) with subinhibitory concentrations of β-lactam antibiotics. We tested the same 10 *E. coli* strains with ampicillin, penicillin G, mecillinam and nalidixic acid, four antibiotics which induced morphological changes. In all cases we found an apparent great increase in adhesion. With nalidixic acid, this increase in adhesion could be verified by independent methods (Vosbeck et al 1979). With the β-lactam antibiotics the situation seems to be more complex. [^{14}C]Acetate is perhaps not an ideal labelling agent for this kind of study, since it is incorporated into the cell wall lipids; the lipids may be shed under the influence of antibiotics acting on the cell wall and what we find may be binding of lipid fragments, not of bacteria. We shall have to look at that in more detail before we reach a final conclusion about what is happening with β-lactam antibiotics.

Also, our culture conditions are really quite different from those that other people are using. Dr Svanborg Edén used a four-hour incubation in her experiments with ampicillin, whereas we incubate for 18 h. Her bacteria were in the logarithmic growth phase whereas ours were in the stationary phase.

Razin: Do β-lactam antibiotics inhibit piliation?

Vosbeck: We haven't looked at that.

Beachey: Did you look at these tissue culture cells under the microscope to see whether adherence of the *E. coli* correlated with the uptake of radioactivity by these cells?

Vosbeck: We are doing that now to make sure of what we are seeing by measuring radioactivity.

Hughes: Dr Beachey, could you clarify how you see these β-lactam antibiotics and cell wall inhibitors working? It seems to me that there are two possibilities. If streptococci are treated with penicillin you get reduced adhesion, so you could say that the cell wall polymer itself is the adhesin. The other possibility is that the cell wall polymer is needed to stick on the real adhesin.

Beachey: The second suggestion is feasible. If lipoteichoic acid is the adhesin there is always the question of how we visualize the orientation of the lipoteichoic acid molecule in the cell. If the model proposed by Wicken & Knox (1977) is correct, the lipid moiety of lipoteichoic acid should be buried in the bacterial cell membrane and the other end should be sticking through the peptidoglycan layer to the outside, where it is in a hydrophilic environment. We now have some evidence (Beachey et al 1980) that liopteichoic acid under the right conditions forms fairly stable complexes with certain cell surface proteins of the streptococcus. Such complexing may form a bridge or a network composed of lipoteichoic acid and protein on the bacterial surface, to form what we originally called fimbriae – we now call them fibrillae because they don't have a distinct structure as do the fimbriae of Gram-negative bacteria. Somehow these complexes are held together so that the lipid ends of the lipoteichoic acid molecule can flip out and react with eukaryotic cell membranes. We also know that the lipid portion of the molecule is important for interaction with the cell membrane (Ofek et al 1975).

Hughes: I would be interested in what Dr Helenius thinks about lipids popping out of one hydrophobic environment and inserting themselves into another.

Helenius: There are cases where hydrophobic or amphiphilic molecules can move directly from one membrane to the other but they are very few. Cholesterol is one example; for instance it moves easily from liposomes into red blood cells. Some phospholipids can do this too, but much more slowly.

Hughes: I think you need special protein carriers for phospholipids. That is a special case. In any exchange of components between the eukaryotic cell surface membrane and the bacterial cytoplasmic membrane as envisaged by Dr Beachey there is the cell wall between them to be considered. That is really quite a large space for a hydrophobic molecule to negotiate.

Beachey: It is well established that group A streptococci excrete lipoteichoic acid into the culture medium (Alkan & Beachey 1978). As these molecules pass through the cell wall it is possible that they are protonated so that they can form hydrogen bonds with the surface proteins of the streptococcus. In this way the lipoteichoic acid would be at least transiently anchored in the proper orientation on the surface.

Hughes: Is anything specific known about the attachment sites of the fimbriae, pili or flagella to bacterial surfaces? Flagella go right through the peptidoglycans and interact with the cytoplasmic membrane, but what about the pili?

Svanborg Edén: In the electron microscope the long filaments produced after ampicillin treatment had pili. We observed that most of the circumference of the bacteria was bare and then a tuft of pili was seen coming out in one place.

Mirelman: One of the points of using a β-lactam such as penicillin is that it is possible to alter the shape of the Gram-negative bacteria. It is very difficult, however, to do reproducible experiments with morphologically different bacteria: you have to hit the correct concentrations of penicillin to get a reproducible change. Cephalexin on the other hand is a very good antibiotic, in the sense that over a large range of concentrations it gives very nice filamentous bacteria. These spaghetti-like bacteria don't have pili and they appear similar to those shown by Ed Beachey. We used radioactively labelled, soluble mannan and tried to see whether the filamentous bacteria could still bind to mannose. Although they don't show pili, they bound radioactive mannan quite well, and they attached quite well to epithelial cells and macrophages (Bar Shavit et al 1980). The appearance of pili on the outside therefore does not mean that they are the component that binds mannose residues. A small stub that cannot be seen in the electron microscope might be enough for binding.

Beachey: Dr Barry Eisenstein and I have been looking at binding of radiolabelled mannose by bacteria that either adhere or don't adhere. All the strains that are chemoreceptor-positive, whether or not they adhere to epithelial cells, and whether or not they have fimbriae, bind the same amount of mannose. The amount of mannose that is bound by the fimbriae, therefore, must be so small that with the high background of the chemoreceptor binding, you wouldn't be able to detect it. So your mannan may be binding to chemoreceptors. I wonder whether chemoreceptor-negative mutants bind mannan.

Mirelman: The problem is that the binding of mannan is reversible with α-methylmannoside. If the binding of mannose by the chemoreceptor goes inside the cell can it be reversed with α-methylmannoside?

Beachey: All our binding experiments were done at 4 °C with 0.02% sodium azide, and in these conditions the binding of the radiolabelled mannose was completely reversed by adding an excess of unlabelled mannose but not by α-methyl-D-mannoside.

Mirelman: If you were measuring binding through a chemoreceptor you would not be able to remove it by adding exogenous α-methylmannoside to remove the radioactive mannan.

Howard: Could the very low level of antibiotics used cause a structural defect in the synthesis of a small proportion of all proteins in these bacteria? The structural requirements for intermolecular association and assembly of pilus proteins into a functional pilus might be of such stringency that a low frequency of defects in pilus protein structure might preclude pilus formation. In contrast, a low frequency of defects in other proteins might be tolerated much more easily. As you can identify the pilus protein immunochemically it might be of interest to biosynthetically label bacteria that have been treated with sublethal antibiotic levels and lack pili and then investigate whether there is an internal pool of pilus protein in these cells.

Taylor-Robinson: Some of the points that have been made have important implications for those who incorporate antibiotics in various cell systems designed to test the association of bacterial adhesion with pathogenicity. I am thinking in my own case of the organ culture system where it is convenient to be able to use a selective antibiotic to suppress unwanted organisms. Clearly, however, the antibiotic although not suppressing the bacterium under test may have an inhibitory effect on its attachment to cells. We have been conscious of this and although we have often treated organ cultures with antibiotics initially, we have usually removed them before inoculation of bacteria. I wonder whether others recognize this problem? Do you put antibiotics into your organ cultures with impunity, Dr Candy?

Candy: We don't do long-term organ cultures, and antibiotics are not needed for such short-term preparations.

Beachey: In favour of selective suppression of the assembly of surface proteins by subinhibitory antibiotics was the observation in our study that flagella were formed, even though fimbriae were not, in the presence of penicillin (see Fig. 1, p 292).

REFERENCES

Alkan ML, Beachey EH 1978 Excretion of lipoteichoic acid by group A streptococci: influence of penicillin on excretion and loss of ability to adhere to human oral mucosal cells. J Clin Invest 61:671-677

Bar Shavit Z, Goldman R, Ofek I, Sharon N, Mirelman D 1980 Mannose binding activity of Escherichia coli: a determinant of attachment and ingestion by macrophages. Infect Immun 29:417-424

Beachey EH, Simpson WA, Ofek I 1980 Interaction of surface polymers of *Streptococcus pyogenes* with animal cells. In: Rutter P (ed) Microbial adhesion to surfaces. Ellis Horwood, Chichester, in press

Ofek I, Beachey EH, Jefferson W, Campbell GL 1975 Cell membrane-binding properties of group A streptococcal lipoteichoic acid. J Exp Med 141:990-1003

Ozaki M, Mizuchima S, Nomura M 1969 Identification and functional characterization of the protein controlled by streptomycin-resistant locus in *E. coli.* Nature (Lond) 222:333-339

Tylewska SA, Wadström T, Hjerten S 1980 The effect of subinhibitory concentrations of penicillin and rifampicin on bacterial cell surface hydrophobicity and on binding to pharyngeal epithelial cells. Antimicrob Agents Chemother 17:292-294

Vosbeck K, Handschin H, Menge EB, Zak O 1979 Effects of subminimal inhibitory concentrations of antibiotics on adhesiveness of *Escherichia coli* in vitro. Rev Infect Dis 1:845-851

Wicken AJ, Knox KW 1977 Biological properties of lipoteichoic acids. In: Schlessinger D (ed) Microbiology 1977. American Society for Microbiology, Washington, p 360-365

Final general discussion

Streptococcal adherence

Beachey: Group A streptococci (*S. pyogenes*) are residents of the mucosal surfaces of the nasopharynx and of the skin of humans. Other than the human host, no natural reservoir for these organisms is known. To survive, the streptococci produce surface adhesins that bind the organisms to epithelial surfaces. They produce other surface substances which protect virulent organisms from being ingested when they invade through the epithelial barriers.

By scanning electron microscopy *S. pyogenes* has been shown to adhere in large numbers to the surface of isolated oral epithelial cells. The adherence of the streptococci to certain areas and not to others on the epithelial cell surfaces suggested the presence of special areas of recognition that are unequally distributed throughout the cell membranes. Transmission electron microscopy of ultrathin sections of epithelial cells containing adherent streptococci revealed fibrillar structures radiating from the surface of the bacteria to the membranes of the epithelial cells (Beachey & Ofek 1976). Previous studies had indicated that both M protein (Swanson et al 1969) and lipoteichoic acid (LTA) (Beachey & Ofek 1976) are associated with these surface fibrils.

Since it was known that LTA binds spontaneously to a variety of mammalian cells (Wicken & Knox 1975), Dr Ofek and I investigated the possibility that LTA may mediate the adherence of streptococci to oral (Beachey 1975, Beachey & Ofek 1976) and skin (Alkan et al 1977) epithelial cells. For this purpose, various surface substances of streptococci, including M protein, C carbohydrate, peptidoglycan and lipoteichoic acid, were purified and tested for their inhibitory effects on adherence. Of the purified substances tested, only LTA was inhibitory in a dose-related fashion.

Further evidence that LTA was centrally involved in adherence was obtained in studies of the inhibitory effects of antibodies to LTA. Treatment of streptococci with 1:100 dilutions of antisera to LTA blocked adherence of the organisms to epithelial cells. The inhibitory effect could be abolished by absorbing the antisera with erythrocytes coated with LTA. Such absorption of antisera containing a mixture of antibodies to LTA, M protein and C carbohydrate removed only LTA antibodies and at the same time abolished the inhibitory effect on adherence (Beachey & Ofek 1976), indicating that the antibodies to other surface substances had no effect on adherence. Finally, as I showed in my presentation, the loss of cell-associated LTA induced by incubating the streptococci in the presence of penicillin was paralleled by a loss of epithelial-cell-adhering ability, indicating that the retention of a critical amount of cell-associated LTA was important for the retention of adhering ability. These studies all pointed to a central role for LTA in the adherence process.

A number of binding studies in our laboratory with radiolabelled LTA have indicated that blood platelets (Beachey et al 1977), erythrocytes (Beachey et al 1979, Chiang et al 1979), lymphocytes (Beachey et al 1979) and epithelial cells (Simpson et al 1980a) all possess specific binding sites or receptors for the binding of LTA. Binding to these cell membranes has been shown to be dependent on the lipid moiety of the LTA molecule. Dr Andrew Simpson in our laboratories has recently shown that LTA binds to the fatty acid binding sites of serum albumin (Simpson et al 1980b) and that albumin may serve as a receptor analogue for the binding of LTA to cell membranes (Simpson et al 1980c).

These findings left us with a dilemma. If the lipid moiety of LTA mediates the binding of streptococci to receptors on eukaryotic cells, how does the LTA molecule remain anchored to the streptococcal surface? According to the model proposed by Wicken & Knox (1977), the LTA molecule is anchored via its lipid moiety to the plasma membrane of the streptococcus. Because of this problem, we investigated the possibility that LTA may form complexes via the hydrophilic polyglycerol-phosphate end of the molecule with surface components of streptococci.

We found that both LTA and its deacylated derivative could be quantitatively precipitated in the presence of purified M protein at low pH ranges, with optimal precipitation at pH 3.7. Once formed, the LTA–M protein complexes were resistant to solubilization at neutral pH either in water or a high concentration of NaCl (Beachey et al 1980). To test the possibility that lipid moieties of LTA remain free to interact with albumin or cell membrane receptors, insoluble complexes of LTA and M protein were exposed at neutral pH to solutions of albumin. These experiments indicated that intact LTA complexed with M protein could still bind to albumin whereas complexes formed by deacylated LTA could not.

Since our studies have strongly suggested that the lipid moieties of LTA are somehow available on the surface of streptococci to interact with receptors on cell membrane, one might speculate that the LTA molecule may be protonated during its transit through the cell wall, and would therefore be able to form hydrogen bonds with surface protein molecules. The hypothesis is supported by studies showing that M protein can be extracted from whole streptococci with non-ionic detergents (Fischetti et al 1976) or guanidine (Russell & Facklam 1975), reagents known to dissociate hydrogen bonds. One could speculate further that LTA complexed to proteins may form micromicelles with the lipid moieties of other LTA molecules in order to protect these hydrophobic groups. As the streptococcal cell approaches a host cell membrane, the lipid ends of the lipoteichoic acid molecule may flip open and attach to specific receptor sites, their hydrophilic ends remaining firmly anchored to the surface proteins.

So far, this remains a speculative view of the spatial orientation of the lipoteichoic acid molecule on the surface of *S. pyogenes*. I would like to know from the lipid chemists whether it is a workable hypothesis.

Elbein: Can you block the binding with things like heparin? Have you tried polyanions?

Beachey: We tried lipopolysaccharides from *Serratia marcescens* and *E. coli*. None of them worked.

Elbein: I was thinking of other negatively charged macromolecules that have lots of negative charges due to sulphate or phosphate. The teichoic acids have a lot of phosphate.

Mirelman: Streptococci also sometimes have glycerolteichoic acid but not lipoteichoic acid. In other words they lack the lipid. Have you tested whether ribitol or glycerol polymers would inhibit binding?

Beachey: Group A streptococci produce only glycerolteichoic acid. It is true that they secrete a mixture of lipoteichoic acid and the deacylated form. A deacylase has recently been isolated from the membranes of group A streptococci (Kessler et al 1979). In some strains all the teichoic acid that is secreted is deacylated (Kessler & Shockman 1979), probably because of high deacylase activity. In any case, the deacylated polymer is ineffective as an inhibitor and it does not bind to cell membranes.

Choppin: If a bacterium is bound to the surface of the cell can you then deacylate and elute it?

Beachey: We haven't tried that but we can block binding by treating the bacteria with hydroxylamine, a chemical deacylating reagent.

Terminology

Taylor-Robinson: Our second topic concerns terminology. Should we use a single term instead of pili, fimbriae, fibrils and so on?

Beachey: I agree with Professor Freter that the surface projections of Gram-negative bacteria that are not flagella and are not involved with the transfer of DNA should be called fimbriae. The term pili should be reserved for those structures involved in the transfer of genetic information, the sex pili. For the structures on the surfaces of Gram-positive bacteria, which have been called variously hair, fuzz and other names, I prefer the terms fibrils or fibrillae. The latter are really indefinite structures that are difficult to measure.

Svanborg Edén: The Ørskovs, at the WHO *E. coli* reference centre, use fimbriae 1,2,3,4 etc., also including CFA antigens K88, K99 etc. F1 = type 1 fimbriae, F2 = K88, F3 = K99, F4 = 987, F5 = CFA/I, F6 = CFA/II, F7 = one type of fimbriae on UTI pathogenic *E. coli*, F8 = another such type (Ørskov, personal communication).

Razin: I was surprised to find how little we know about the pili or fimbriae. Our knowledge of Gram-negative outer membranes is very extensive now, and there are many mutants both with respect to the lipopolysaccharide component and with regard to the outer membrane proteins. It is important to encourage research on the origin of these pili. Are they really associated with the plasma membrane? The experimental system for investigating this problem is available. I agree with Dr Beachey on the terminology proposed for these structures.

Freter: Early immunology provides an example of the horrible things that happen when we are too specific in defining things before we know what they are. I would prefer to have a term that is all-inclusive and to go by taxonomic precedents. 'Fimbriae', as far as I know, was the term first applied by Duguid. We should wait 10 years to see whether there are subdivisions before we use more specific terms.

Mirelman: The main reason pili have not been researched well is that all the people who are working on outer membranes use for their studies bacteria which are in the logarithmic growth phase and they never see pili in these cells. It may come as a surprise to them that stationary phase cells have pili or fimbriae.

Taylor-Robinson: Other terms that need discussion are adhesion, adherence, sticking, binding and attachment. Dr Pearce, you used the word association.

Pearce: Association is a neutral term for organisms which may be extra- or intracellular. Adherence seems appropriate for organisms where it is ecologically important that they stick to certain surfaces. It is perhaps less appropriate for obligate intracellular parasites such as viruses or chlamydiae.

Choppin: The generally accepted term among virologists is adsorption!

Candy: The word cohesion brings in the idea that the substrate to which the bacteria adhere is important too. What we are investigating is not really bacterial adhesion so much as cohesion between the bacteria and surfaces or cells.

Sharon: I don't really see the difference between association, adhesion, adherence and attachment.

Rutter: For some test systems, such as K88-positive *E. coli* and pig brush border cells, I feel that the term adherence is accurate. But I agree with Professor Freter that association might be more appropriate for organisms in the mucus that are not attached to the intestinal cells.

Freter: If one wants to be non-specific one can call the whole thing 'association'. Adherence then is a special case of association; trapping in the mucus gel would be another example of association.

Newell: It is rather different from the use of fimbriae, where one is inventing a specific term. These are all words in the English language which can be used simply with their defined meanings. It doesn't matter whether we use one word or the other. Each system will have its own preference.

Sussman: I strongly agree. There is a marked difference between adherence and attachment as normally used in non-technical speech. When we are talking about structures we ought to wait until the mechanism is clearly defined. Then we may find that one or other of a variety of terms may be applicable to the phenomona we are trying to describe. We ought not to be too fussy yet but go on allowing some elegant variation.

Taylor-Robinson: What about the terms receptor site on the eukaryotic cell and binding site on the organism? Are they accepted?

Razin: Receptor is accepted by most of us.

Sharon: But 'binding site' is not a good term because it has a particular meaning for immunologists and enzymologists. And you don't isolate a site, you isolate substances.

Ligand is a poor term in a sense because it has a clear meaning in enzymology. It is usually a small molecule, e.g. a metal ion that binds to protein. In the field of adherence some people are using ligand to refer to a large molecule that binds to a receptor. Since the two fields are so close to each other, using the same terms for two different substances is wrong. The term lectin is appropriate as long as it is confined to a sugar-binding protein (Goldstein et al 1980).

Lectins are cell-agglutinating in that operational definition. We should find a general term to replace binding site and ligand, and adhesin may be appropriate.

Tramont: We really want a new word because any old word we choose is going to mean something else to others. Adhesion doesn't bother me.

Sharon: Adhesin is a substance responsible for adhesion.

Howard: You can also be too specific here. In our field we refer to erythrocyte receptors and merozoite receptors and we are not claiming any dominant or sub-dominant relationship of one to the other as the terms receptor and binding site do. There may be considerable information exchange when two plasma membranes of complex cells higher than bacteria come together. Components of one membrane could be activated to recognize components on another membrane in a series of steps like a cascade. This situation, which may apply to the interaction of merozoites and erythrocytes, is clearly different to lectin-mediated bacterial adhesion which appears to be a less dynamic interaction. Our terminology must therefore be extremely broad in its implications.

Razin: Operationally I like the terms receptor and binding sites because they let one distinguish easily between the receptor which is on the eukaryotic cell and the binding site which is on the prokaryotic cell.

Howard: You could keep the term binding site for the particular molecular determinant(s) involved in the interaction, without using it for the molecule of which the binding site is but a part.

Friend: I don't think the term 'site' should be used if the molecule is moving around in the membrane.

Sharon: There has been a suggestion that the term 'affinitin' might be used for anything that can bind to cells. I personally like the terms 'cognor' and 'cognon', recently proposed by Ballou and his coworkers (Burke et al 1980). A cognor is a substance such as an antibody or lectin that recognizes, and a cognon is a substance or group, usually smaller, that is recognized.

Newell: Surely the word receptor has dominance built into it: you don't receive the Queen, the Queen receives *you!*

Receptors

Razin: There is a report about fibronectin binding to staphylococci (Kuusela 1978) but we haven't discussed this topic yet. Can any generalization be made about the type of bonds responsible for adhesion? For example do fimbriae really act by overcoming the electrostatic repulsion between the negatively charged bacterium and the negatively charged eukaryotic cell?

Sharon: In lectin–carbohydrate interactions there are hydrogen bonds; there are hydrophobic interactions but rarely electrostatic interactions. For example, in the interaction of concanavalin A and mannose residues there are no electrostatic forces, since the sugar is uncharged. Electrostatic bonds may however participate in the interaction of wheat germ agglutinin and sialic acid residues.

With enzymes, electrostatic effects may be of greater importance, for example with trypsin, specific for the positively charged lysine and arginine residues. Chymotrypsin on the other hand is specific for hydrophobic side chains, such as those of tyrosine. In interactions with bacterial surfaces one or other type of bond may predominate.

Hughes: The data simply aren't there for bacteria. There is some evidence that pseudomonads recognize fibronectin, and that may be involved in the adherence of *Pseudomonas,* but I don't know of any other evidence. The theory should be easy to test. There are very good monoclonal and other antibodies directed against fibronectin which could well be used as reagents to test its role in bacterial adhesion.

Feizi: My comments refer to the differences in reactivities of microbial lectins with different indicator cells and to the chemical basis of these differences. First, I shall discuss how a plant lectin (*R. communis* agglutinin 120) and an animal lectin (calf heart lectin) which may both be described as 'lactose-sensitive' react well with trypsinized rabbit erythrocytes but differ in their reactivities with untrypsinized rabbit erythrocytes and with human erythrocytes. Secondly, I shall discuss why such differences may occur by using as models monoclonal anti-carbohydrate antibodies (human monoclonal anti-I and i antibodies) which have many features resembling lectins.

Comparison of reactivities of R. communis agglutinin 120 and β-galactosyl-binding lectin of calf heart

Bob Childs and I have compared the inhibitability of *R. communis* agglutinin 120 and calf heart lectin with oligosaccharides and have shown that the binding of both lectins to trypsinized rabbit erythrocytes is inhibited by lactose (Galβ1→4Glc) and by Galβ1→4GlcNac and Galβ1→3GlcNac. However, only *R. communis* lectin is inhibited by Galβ1→6GlcNAc (Childs & Feizi 1979). Therefore one might say that the plant lectin is less discriminating than the animal lectin.

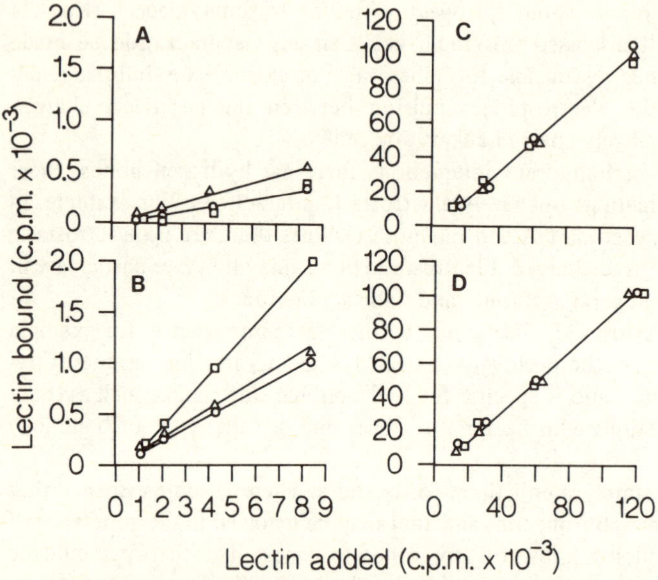

FIG. 1 (Feizi). Binding of ^{125}I-labelled calf heart lectin (A, B) and *R. communis* agglutinin (C, D) to human erythrocytes (A, C) and rabbit erythrocytes (B, D). Symbols: ○, native erythrocytes; □, trypsin-treated erythrocytes; △, neuraminidase-treated erythrocytes.

We have also studied the binding of the radioiodinated lectins to native, trypsin-treated and neuraminidase-treated human and rabbit erythrocytes (Fig. 1). The plant lectin reacted equally well with all the cells while the calf lectin reacted well with trypsinized rabbit erythrocytes and relatively poorly with the other cells. It can be concluded that lectin binding sites for *R. communis* agglutinin are optimally expressed on both human and rabbit erythrocytes and their accessibility is not improved by the enzyme treatment. However, with the bovine lectin, human erythrocytes reacted poorly even after enzyme treatment, and rabbit erythrocytes reacted well only after trypsinization. As β-galactosyl structures are present on both rabbit and human erythrocytes, it seems that the bovine lectin requires more than the presence of these sugars on cells for detectable binding to occur. Enhanced binding of the bovine lectin to trypsinized rabbit erythrocyte is presumably a consequence of removing glycoproteins or peptides which hinder the binding of this lectin to the untreated cells.

Thus it is possible that lectins which behave similarly with simple sugars may differ substantially in their reactivities with receptors on the cell surface.

The differing fine specificities of monoclonal antibodies directed against carbohydrate antigens I and i

In the autoimmune haemolytic disorder known as cold agglutinin disease there occur, in the patients' sera, high titre monoclonal antibodies the majority of which are IgM proteins directed against the I or i antigens of erythrocytes. The I or i specificities are assigned to these antibodies according to their relative reactions with adult (rich in I) or fetal (rich in i) erythrocytes. We have recently defined the carbohydrate sequences on which these antigens reside and have made some observations on the fine specificities of the individual monoclonal antibodies. The knowledge gained is relevant to those interested in studying the fine specificities of closely related lectins.

Anti-i antibodies recognize antigenic determinants on a linear oligosaccharide sequence consisting of two *N*-acetyllactosamine (Galβ1→4GlcNAc) units joined by β1→3 linkage (Niemann et al 1978). This structure is shown below as the oligosaccharide moiety of the erythrocyte glycosphingolipid lacto-N-*nor*-hexaosyl ceramide:

Galβ1→4GlcNAcβ1→3Galβ1→4GlcNAcβ1→3Galβ1→4Glcβ→Cer

A part of this antigenic determinant is found on the shorter chain glycosphingolipid paragloboside:

Galβ1→4GlcNAcβ1→3Galβ1→4Glc→Cer

When these two purified glycosphingolipids were incorporated into cholesterol/lecithin micelles and tested as inhibitors of six anti-i antibodies by radioimmunoassays, the long chain glycolipid was a potent inhibitor of five of these antibodies (0.02–0.08 nmol required for 50% inhibition) while the short chain glycosphingolipid was not active at the highest level tested (0.8 nmol) (Niemann et al 1978). In separate experiments with oligosaccharides free in solution, it was found that 30–400 nmol of long chain oligosaccharides from mucins (Wood et al 1981) and 800–1600 nmol of the synthetic trisaccharide Galβ1→4GlcNAcβ1→3Gal (Wood & Feizi 1979) were required to give 50% inhibition. These data demonstrate (a) the profound effect of multivalence (on micelles) and (b) the relative lack of inhibitory activity of the short oligosaccharide chain with these antibodies.

Antibodies termed anti-I recognize antigenic determinants on a branched oligosaccharide chain formed by the addition of Galβ1→4fGlcNAcβ1→6 branch to the i active chain (Watanabe et al 1979, Feizi et al 1979). This structure is shown below as the carbohydrate moiety of the erythrocyte glycosphingolipid termed lacto-N-*iso*-hexaosyl ceramide:

Galβ1→4GlcNAcβ1↘
6
Galβ1→4GlcNAcβ1→3Galβ1→4Glcβ→Cer
↗3
Galβ1→4GlcNAcβ1

TABLE 1 (Feizi) Comparison of the inhibitability of two monoclonal anti-I antibodies obtained from donors Ma and Woj with a synthetic oligosaccharide and two glycosphingolipids (adapted from Wood & Feizi 1979 and Feizi et al 1979)

	Anti-I from:	
	Ma	*Woj*
Inhibitor	*(nmol giving 50% inhibition)*	
Galβ1→4GlcNAcβ1→6Gal	12	16
Galβ1→4GlcNAcβ1↘6 / Galβ1→4GlcNAcβ1↗3 Galβ1→4GlcNAcβ1→3Galβ1→4Glcβ→Cer	0.2	0.2
Galα1→3Galβ1→4GlcNAcβ1↘6 / Galβ1→4GlcNAcβ1↗3 Galβ1→4GlcNAcβ1→3Galβ1→4Glcβ→Cer	0.1	Inactive at 0.2

This glycosphingolipid when incorporated into cholesterol/lecthin micelles was a potent inhibitor of the monoclonal anti-I antibodies of 10 out of 11 donors. However, these monoclonal antibodies recognize different oligosaccharide domains on this branched structure. This was revealed by doing radioimmunoassays with analogues of the branched structure. Three types of anti-I specificity were revealed; the first involves the 1→4, 1→6 branch, the second the 1→4, 1→3 branch and the third requires both branches to be accessible. However, even those antibodies recognizing the 1→4, 1→6 branch are not necessarily identical. We have compared the inhibitability of two such anti-I antibodies (from patients Ma and Woj) with a synthetic oligosaccharide and two glycosphingolipid analogues (Table 1). The amounts of the trisaccharide required to inhibit the two antibodies were comparable (Wood & Feizi 1979). With the glycosphingolipid containing the unsubstituted 1→4, 1→6 chain the amounts required to give inhibition were much smaller (due to their incorporation into micelles) but again were comparable for the two antibodies. However, when the 1→4, 1→6 chain was substituted with α1→3 linked galactose only anti-I Ma was inhibited by the resulting branched structure.

The differences in the fine specificities of these monoclonal anti-carbohydrate antibodies now account for their often differing reactivities with complex glycoproteins (mucins) isolated from different sources (Feizi & Kabat 1972). When we first observed these differences we thought that there were several different 'types' of I antigen. It now seems that the differing reactivities we were observing reflected (a) the narrow specificities of the monoclonal antibodies and (b) differences in the extent and nature of additional glycosylations of the branched I oligosaccharides occurring on various glycoproteins. On cell membranes, profound effects could also be exerted by the presence of neighbouring molecules which could impair the accessibility of a given carbohydrate ligand, as illustrated above with calf heart lectin and rabbit erythrocyte membranes.

Hughes: This really brings up the question of what are the receptors for the MS lectins. I shall not be surprised to see that the mannose-sensitive activities of different bacterial strains turn out to be lectins binding to different sorts of polysaccharide sequences. It is probable that all of these lectins will be binding ultimately to *N*-glycans of the high mannose type that Dr Elbein was talking about and widely distributed in mammalian cell glycoproteins. However, the fact that high concentrations of mannose inhibit these lectins does not rule out the possibility that different lectins bind to different regions of the high mannose polysaccharides. The mannose-binding plant lectin, concanavalin A, is well known to bind most avidly to mannosyl-oligomers of defined structures and the same may turn out to be true for the analogous bacterial lectins (Baenziger & Fiete 1979).

Following on this point, Dr Beachey said that *E. coli* strains with mannose-sensitive activities bind to specific areas of squamous epithelial cells, raising the question of whether mannose-rich receptors are distributed in particular domains of the cell surface. Indeed there is now evidence that the *N*-glycan structures of glycoproteins are different in different domains of epithelial cells. It looks as though in mucus-secreting epithelial cells, for example, the concentration of high mannose-type glycans is highest in the crypts. As the glycoprotein migrates to the apical surface the chains are matured by the addition of sialic acid (Weiser 1973). Therefore, when we talk about bacterial adhesion to epithelial cell surfaces we should be looking for exactly where they bind on that surface, whether it is the crypt or somewhere else. That may tell us a lot about the type of carbohydrate chain that the bacterial lectin is recognizing.

Taylor-Robinson: Dr Howard talked about the specificity of adherence to erythrocytes in malaria and I would like to add something about the specificity of gonococcal adherence. Alan Johnson, Zell McGee and I for a number of years have studied the effects of gonococci in Fallopian tube organ cultures. We have used the Fallopian tube because *in vivo* it is susceptible to the gonococcus, it is normally sterile, and above all the epithelium is ciliated so that we can measure ciliary activity using it as an index of cell viability (McGee et al 1976). Organ cultures inoculated with the virulent type 1 piliated gonococci lose ciliary activity faster than cultures infected with type 4 gonococci, despite the marginally better growth of type 4 organisms in the cultures. On the other hand, loss of ciliary activity in pig, rabbit and bovine oviduct organ cultures, in which the gonococci multiplied, was no greater than the loss of activity in uninoculated control cultures.

Histologically, we found that the gonococci were closely allied to the surface of the human epithelial cells but not to the cells of the various animal species (Johnson et al 1977). Clearly there must be receptors on cells in the human genital tract which are not present or not exposed on cells in oviducts from the animal species we have tested. Apparently contrary to our findings, Tebbutt et al (1976) reported that there was attachment of gonococci to guinea-pig urogenital tissues. The results of our recent studies with guinea-pig urogenital tissues show that this is the case. However, we find that the organisms don't adhere to the surface of the cells but become trapped in crevices between cells and in the overlying mucus. So we still believe that there is a specific adherence of gonococci to human tissue.

Sterile filtrates of the media from the gonococci-infected Fallopian tube organ cultures inoculated into donor uninfected cultures cause loss of ciliary activity. We are now fairly confident that this loss is due to gonococcal endotoxin which also acts in a specific way. In other words, it affects only the human genital tissues and not those of cattle, pigs and so on. This is a good reason, though perhaps not the only one, for the gonococcus only infecting humans naturally and chimpanzees experimentally. We have not had an opportunity of doing the studies I mentioned on chimpanzee tissues.

Bredt: Organisms such as chlamydiae and gonococci preferentially infect the genital epithelia and the conjunctiva. Is there any factor common to both types of epithelium which could explain the preference for this site?

Taylor-Robinson: There certainly seems to be preferential involvement although

TABLE 2 (Friend) Interaction between potatoes and Phytophthora infestans

Race of Phytophthora infestans	*Potato genotype*					
	r	R_1	R_2	R_3	R_4	R_1R_3
Race 1	S	S	R	R	R	S
Race 2	S	R	S	R	R	R
Race 3	S	R	R	R	R	S
Race 4	S	R	R	R	S	R
Race 1, 3, 5	S	S	R	S	R	S

R = resistant reaction; S = susceptible reaction

one must remember that the eyes are exposed and have a good opportunity of becoming infected as a baby passes through the infected birth canal. But the idea that every organism which infects the genital tract is equally capable of infecting the eye does not seem to be true because we have found that ureaplasmas which are present in the genital tract and sometimes the oropharynx of humans and various animal species do not involve the eye, or affect it rarely.

Bredt: There must be different receptors.

Rutter: Did you say that the cilia stopped beating or that they disappeared?

Taylor-Robinson: They stop beating as the cells die and then there is a loss of ciliated epithelial cells. It is rare for cilia to be detached from the epithelial cells.

Rutter: Do the gonococci attach to the cilia?

Taylor-Robinson: No, they attach to non-ciliated cells, which is quite different from the situation seen with *Bordetella pertussis* in tracheal organ cultures where the organisms attach specifically to the cilia and surfaces of ciliated cells (Muse et al 1977).

Friend: The interaction between potatoes and *Phytophthora infestans,* the fungus which causes late blight disease, is an example of an interaction of the type known by plant pathologists as 'gene-for-gene' relationship. Potato cultivars can contain one or more dominant major genes for resistance (*R* genes) designated R_1, R_2, R_3 etc. Such cultivars are resistant to all races of the fungus except those which contain the corresponding virulence gene; the races are numbered according to the virulence gene or genes they contain. Potato cultivars which do not contain any major resistance genes are susceptible to all races of the pathogen. The interaction is explained in Table 2.

The hypersensitive resistance reaction is characterized by rapid death of cells adjacent to those actually penetrated, followed by death of the fungus (Kitazawa et al 1973, Shimony & Friend 1975). Cell death also acts as a trigger for the biosynthesis of antifungal terpenoid and phenolic compounds (Sato et al 1971, Shimony & Friend 1975).

This genetically determined pattern of interactions works with live fungus and potato leaves or potato tuber slices or discs. When the fungus is homogenized it is possible to isolate fractions apparently derived from the fungal cell wall, known as elicitors; these elicitors will give resistance reactions with all potato cultivars irrespective of whether or not they are genotypically resistant or susceptible to the fungal race from which the elicitor has been isolated. Furthermore all fungal races appear to contain similar non-specific elicitors (Varns et al 1971). The elicitor from *P. infestans* seems to be a carbohydrate; its activity is destroyed by periodate oxidation (Marcan 1978).

We have recently examined the effects of carbohydrates on the resistance reaction, caused by crude fungal elicitors, by measuring resistance as browning of potato discs; the intensity of browning is correlated with hypersensitive cell death (Marcan et al 1979). We found that the β-glucosides laminaribiose and methyl β-glucoside were the only sugars which would inhibit browning and cell death caused by elicitors.

We have accordingly proposed that the β-glucosides could be acting as haptens and so inhibiting the reaction between a non-specific fungal carbohydrate elicitor and a lectin-like receptor in the potato. This hypothetical receptor in the potato would appear to be different from the known glycoprotein lectin whose haemagglutinating activity is inhibited by tri-*N*-acetylglucosamine (Desai & Allen 1979).

The possible importance of β-1,3-glucans in controlling the reaction between *P. infestans* and potatoes is further emphasized by the finding that the fungus itself contains molecules which can suppress activity of the fungal non-specific elicitor. These fungal suppressor components were partially characterized as glucans containing β-1,3 and β-1,6 linkages and 17 to 23 glucose units (Doke et al 1979). They resemble the so-called mycolaminarins which have been previously isolated from other *Phytophthora* species.

Doke & Tomiyama (1980) have published evidence that water-soluble glucans which act as 'suppressor' molecules and which have been isolated from *P. infestans* are race-specific in their action against the non-specific elicitor. For example glucans from race 1 will suppress the hypersensitive reaction of an R_1 potato and the glucans from race 4 suppress the reaction of an R_4 potato.

At first sight these results could explain the specific interactions of different races of *P. infestans* with potato genotypes containing different *R* genes. Each race would contain both a non-specific elicitor and a race-specific suppressor. However, the interaction needs to be examined in more detail at the molecular level.

One way of proceeding would appear to be to isolate the receptor molecules from different potato genotypes to determine whether they can both recognize non-specific elicitors and differentiate between mycolaminarins which may have only small chemical differences, such as the position of 1,6-glucosidic linkages in a β-1,3-glucan 'backbone'. If such a situation did occur then one could postulate that the primary gene product of a fungal virulence gene would probably be a highly specific glucosyl transferase.

Other factors affecting adhesion

Candy: When children have acute diarrhoea due to rotavirus and then excrete enteropathogenic *E. coli*, the condition may become protracted (Carr et al 1976). Veterinary workers see similar mixed infections between rotavirus and *E. coli* in piglets, for example (Mebus 1976). These findings stimulated our work with rotavirus-infected cell cultures, in which we were able to increase the adhesion of the *E. coli.*

Taylor-Robinson: Are there examples of mixed bacterial infections, where one species stimulates or inhibits the adherence of another?

Freter: Entamoeba histolytica does not infect germ-free guinea-pigs; one must infect them with bacteria first.

Taylor-Robinson: There are many examples of stimulation but what about blocking and the effect on receptors?

Mirelman: Amoebae become less virulent if you grow them without bacteria for prolonged periods. If you grow them with bacteria they regain their virulence. What they get from the bacteria that makes them virulent is not known.

Newell: The amoebae eat bacteria, so if they have no bacteria in germ-free gut they can't multiply because they simply have no food.

Tramont: For example, the attachment of gonococci can be blocked with meningococci in the buccal cell system.

Taylor-Robinson: Is that a physical blocking, due to the buccal cells being loaded with meningococci?

Tramont: Prior incubation with meningococci blocks the attachment of gonococci. On the other hand, preincubation with *E. coli* had no effect on the attachment of gonococci. A staphylococcal species isolated from a patient also interfered with the optimal attachment of gonococci.

Candy: Pseudomonas aeruginosa and *Klebsiella pneumoniae* compete for buccal cell receptors too (Johanson et al 1979).

Choppin: In virus systems it is possible to saturate all the receptors for one type of virus with either the homologous virus or related virus, but not affect the binding of other viruses. For example, Coxsackie B viruses compete with each other but not with Coxsackie A in polioviruses. There are a few surprising examples in which viruses of different groups, e.g. a rhinovirus and an adenovirus, can compete for receptors.

Taylor-Robinson: I know too that non-pathogenic *Mycoplasma gallinarum* will inhibit the damaging effect of *Mycoplasma gallisepticum* on chicken tracheal epithelium and it would be interesting to know whether there is competition for receptors by these two distinct avian mycoplasmas.

Models

Bredt: Buccal cells are mostly dead. On the other hand even the erythrocyte reacts quite violently if something is done to it. So the difference between dead and live models shouldn't be forgotten. The membrane of the living cell is fluid. It has a membrane potential which may drive the protein out or let it sink in. And there is membrane turnover. These things don't occur in dead cells. Although dead cells may be good for finding a phenomenon, we need to move on to something that corresponds more to the *in vivo* situation.

Candy: Doctors are putting all sorts of inert surfaces such as heart valves, hips and intravenous cannulae into patients. Colonization of these implants by bacteria may involve specific adhesive mechanisms and can be disastrous for the patients. Thus I would support those workers studying adhesion to inert surfaces.

Sussman: Adhesion of *E. coli* to polystyrene is mannose-sensitive; the MR pili don't attach to polystyrene. These preliminary studies were done in buffer, which is a highly artificial system (A. W. Asscher & M. Harbor, personal communication).

Bredt: In the body you would expect a bit more serum or protein.

Choppin: I don't want to defend the use of dead buccal cells, but in the myxovirus–paramyxovirus systems, in which we know what the receptor is, the virus will adsorb not only to a dead cell, but to one that has been fixed with glutaraldehyde or formalin. One can couple a receptor-containing protein to Sepharose, and the virus will still absorb to it. Thus in some receptor interactions it doesn't matter whether the cell is alive or dead, and one can also demonstrate the same specificity with the isolated receptor molecule.

Sharon: But, as Professor Sussman pointed out, bacteria bind in a mannose-sensitive manner to a surface which certainly doesn't have mannose. So it is dangerous to conclude that if a bacterium binds to a cell in a mannose-sensitive manner there are mannose receptors on the cell. The inhibition by mannose may act in an allosteric manner, and not by direct competition at the binding site.

Razin: There are data about the effect of glutaraldehyde fixation of erythrocytes on their attachment to mycoplasmas. With *M. pneumoniae,* adherence goes down by about 40% once we fix the erythrocytes (Banai et al 1980). Heat-killed *M. pneumoniae* failed to adhere to tracheal epithelium (Powell et al 1976). With *M. gallisepticum* we found that only 10% of the population adhered to erythrocytes (Banai et al 1978). Many of the non-adhering bacteria may represent dead cells.

Beachey: Dead cells in the mouth seem to be several layers deep. We have studied the viability of cells obtained from several successive scrapings from different people. No matter how long we scrape, 95% or more of the epithelial cells are dead. Thus, the study of the adhering properties of the dead cells may be of practical significance. The oral cavity is lined with dead cells capable of adsorbing bacteria.

Mirelman: In talking about models we should also pay attention to the metabolic state of the bacteria. We are usually working with cells that are in the stationary phase but bacteria should also be studied in synchronously dividing cell cultures. Receptors are probably formed at a defined stage in the cell cycle and this could be important if we want to know what happens at the molecular level. With non-synchronous bacteria some cells are older than others and we may have situations where various types of bacteria bind differently, and perhaps even to different receptors, during their life cycle.

Bredt: If we harvest mycoplasmas up to a certain age at a pH below 7.0 we get a decrease in adherence.

Freter: The earlier discussion of the direction of further studies generated two camps: those who advocated studies of the molecular biology of adhesion and those who wanted to test the adhesive properties of bacteria isolated from patients. Both types of studies ought to be done, of course, but in pathogenicity we are dealing with a phenomenon that consists of complex sequential reactions. The effect on the overall outcome of one reaction depends very much on what other reactions occur before and after it. Even if we could know all there is to know about each of these reactions, we still could not predict the whole of pathogenicity until we put everything together and determined how one reaction influences the next. So we must do all these things that we have talked about, but we must also remember to put them together somehow.

The only way I can think of accomplishing such a synthesis is in animal models. These models have their drawbacks and there are none that tell us exactly what happens in humans, but it is still necessary to use them, even if we are studying an experimental disease which has no exact counterpart in human disease. Studying how one phenomenon affects another will give us a list of important reactions and their interdependence, and will give us some idea of what we should eventually look for in human disease.

Beachey: I would agree with that if you can come up with a good model. I am not sure, though, how useful the information gained from an unrepresentative model would be.

Freter: Let me give you an example. Since its discovery in the last century we have learnt a lot about bacterial chemotaxis, including much of its molecular biology. Yet, to my knowledge, there was no demonstration of an *in vivo* role of bacterial chemotaxis before ours. This is not to suggest that this finding was a remarkable intellectual accomplishment – on the contrary it was, in fact, a very obvious conclusion from the data on hand. However, it became obvious only because our study was of the *sequences* of reactions which eventually lead to bacterial adhesion. Such sequences of reactions can be studied best in animals. It is conceivable, still, that chemotaxis has no function in human cholera. But, having learnt that chemotaxis can promote the penetration of motile bacteria into mucus gel (and to what extent and under what conditions it can do that), we can now look into other systems. Perhaps it will turn out eventually that chemotaxis is important in the association of *Proteus* with the human bladder. Animal systems do not resemble precisely any specific human disease but they can predict much about the factors one should look for in human disease. The nature of such factors is quite often unexpected and one would not necessarily decide to study them in humans without prior guidance from animal experiments.

Silverblatt: The ability to adhere to host tissue may not always be in the best interests of the bacteria. *E. coli*, as we know, interact not only with epithelial cells but also with phagocytic cells. We have shown that heavily piliated *E. coli* are much more susceptible to phagocytosis than non-piliated *E. coli*, in the absence of serum opsonins. Pili-mediated phagocytosis results in bacterial killing. However, this is a less efficient bactericidal mechanism than the usual opsonin-mediated phagocytosis (Silverblatt et al 1979). Pili-mediated phagocytosis is a very potent stimulus for leucocyte degranulation but is a less effective inducer of the leucocyte respiratory burst than opsonin-mediated phagocytosis (Pryor et al 1980).

Pili-mediated phagocytosis may be important *in vivo* as well, especially in circumstances where mucosal attachment is not important and where normal opsonic activity is impaired. For example, in a model of haematogenous pyelonephritis we found that non-piliated strains of *Proteus* were much more virulent than heavily piliated variants (Silverblatt & Ofek 1978).

Stickiness therefore may be a two-edged sword. In certain situations it may promote colonization or invasiveness but in others it may be a negative virulence factor.

Sussman: I agree entirely. Piliation in the context of interaction with polymorphs should not be regarded as the parallel of stickiness. Mannose-resistant pili do not stick to polymorphs (S. H. Parry, unpublished work). This raises the interesting question of whether systemic or non-mucosal infections such as peritonitis, wound infection and so on may be associated with MR or non-piliated strains rather than with MS strains.

Silverblatt: These were mannose-sensitive strains. We have looked at *Proteus mirabilis,* which of course bears mannose-insensitive pili, and found that their pili also promoted phagocytosis.

Mirelman: In patients, how common is it to find bacteria that are not opsonized? Binding mechanisms such as MR or MS would be for the non-opsonized cases and I imagine that whenever there is infection one would find opsonins.

Silverblatt: In certain body sites, serum opsonins are either deficient or inhibited, as for example in the renal medulla, the central nervous system, or certain serosal surfaces. This is especially true in the initial stages of infection when little inflammatory response has occurred.

Clinical implications

Mirelman: Adherence to so-called inert materials has already been mentioned at this symposium but its clinical importance hasn't been discussed. Dr Shmuel

Katz, the surgeon who has been working in our laboratory, has brought us a whole series of questions, one of which concerns the materials surgeons use. Sutures can sometimes cause serious infections. So we wanted to see whether there are any differences in the adherence of microorganisms to different manufactured materials. We tested sutures by dipping them into a solution containing 5×10^7 bacteria, for example *Staphylococcus aureus* from a superficial wound, which is one of the most common surgical infections. We then dipped the sutures into other tubes containing saline solutions to get the non-adhering bacteria off. We tested braided Dexon sutures, cat gut, nylon and silk sutures with *Staph. aureus* from a wound and with a mannose-sensitive *E. coli*. I was surprised to learn that synthetic polymers were good receptors for bacteria. Some sutures are monofilaments and others are braided, the second kind being the most commonly used. There were differences in the numbers of cells that bound to a 5 cm length of suture (see Table 3). (We can't measure by mass or by surface area.)

Bacteria also adhere to the synthetic grafts used in artificial arteries or in neurosurgical shunts. We used mice for testing these, making an incision on the animal's back. Putting 5×10^7 bacteria directly onto this area does not cause an abscess but a suture dipped in bacteria produces an abscess within six days.

The presence of bacteria alone doesn't cause infection but if they adhere to the suture they produce a local infection which consequently needs treatment with antibiotics. We asked ourselves whether, to prevent this infection, the sutures could be treated before they go into the body. We hope that we will manage to develop that idea.

TABLE 3 (Mirelman) Numbers of bacteria adhering to four types of suture (data of S. Katz, M. Izhar, D. Mirelman)

	Adhered bacteria (cells/5 cm)	
Suture type	*Staph. aureus*	*E. coli*
Chronic (cat gut)	7.2×10^5	3.6×10^6
Dexon (polyglycolic acid)	9.8×10^6	1.2×10^7
Silk (natural fibre)	1.2×10^6	5.0×10^6
Nylon (polyamide 6,6)	1.2×10^5	1.2×10^6

Sussman: This is a beautiful experimental demonstration of the stitch abscesses that every surgeon sees.

Silverblatt: In clinical practice the sutures are sterile when they go into the patient.

Sussman: The bacteria are on the host and the host is happy with them until the stitch goes in.

Silverblatt: It is well recognized that the presence of a foreign body promotes infection, but is adherence necessary for the infection to occur? Did you put in a suture and then seed the wound, which would be similar to what happens in clinical practice?

Mirelman: Yes, that causes an infection too but it is not very reproducible. You have to put the bacteria in very close to the suture and in very small volumes so that the suspension remains on the suture. In those conditions the bacteria probably adhere to the suture too, so what comes first?

Sussman: There is inhibition of phagocytosis.

Bredt: I think it is more a matter of the foreign body sheltering the staphylococcus. In such a situation the infective dose can go down from 100 000 to 100 staphylococci. Whether they have to adhere to the material is not known, but the phagocytes probably cannot get there.

Mirelman: We removed the sutures after six days and counted the bacteria that remained. The number had fallen noticeably. In other words we think that the bacteria have been phagocytized to some extent.

Levine: Do they stick to stainless steel sutures?

Mirelman: We haven't tested that yet.

Beachey: When *Staph. epidermidis* is cultured from the blood of hospital patients it is usually regarded as a contaminant but one of my colleagues, Dr G. Christensen, found an unusual pocket of *Staph. epidermidis* isolates from blood cultures of patients with intravenous catheters. Some of those patients had signs of infection. Dr Christensen radiolabelled several of the isolates and incubated them with 1-cm pieces of intravenous catheter. The isolates associated with infection and fever adhered in much larger numbers to the intravenous catheters than did isolates from patients without signs of infection. Did you compare pathogenic and non-pathogenic strains, Dr Mirelman?

Mirelman: These were pathogenic.

Bredt: Did you leave the sutures in the buffer or in protein-containing liquid? Inside the body protein is certainly involved.

Mirelman: If we pretreat the suture by soaking it in serum it binds even more bacteria.

Freter: Is there any pattern in the type of material or anything that explains the different degrees of adhesion?

Mirelman: These are initial experiments. We don't see any rationale behind it at the moment.

Taylor-Robinson: It is an interesting phenomenon and it would be useful to bring such revealing pictures to the attention of all involved in surgical procedures as a means of stimulating even greater attention to asepsis.

Tramont: An antibody has to bind before it can function. With organisms such as parasites that produce toxins one of the first steps is their ability to attach to the host. Blocking or inhibition of attachment or adhesion or association by the host is one of the key mechanisms of local immunity. There are other mechanisms as well. For example cancer researchers are interested in antibody-mediated cytotoxicity. I think that the blocking of attachment by local antibody is going to be very important for any organism with a tendency to colonize mucosal surfaces.

Silverblatt: Let me sound another warning that the human response may not be all good either. We have found that anti-pili antibody blocked the ability of the leucocyte to phagocytize the bacteria, so it was acting as an anti-opsonin in this case (R. Weinstein & F. J. Silverblatt, unpublished work).

Sharon: That is also a risk one has to keep in mind in talking about receptor analogues. If you saturate the receptor sites you may also inhibit phagocytosis.

REFERENCES

Alkan ML, Ofek I, Beachey EH 1977 Adherence of pharyngeal and skin strains of group A streptococci to human skin and oral epithelial cells. Infect Immun 18:555-557

Baenziger JU, Fiete D 1979 Structural determinants of concanavalin A specificity for oligosaccharides. J Biol Chem 254:2400-2407

Banai M, Kahane I, Razin S, Bredt W 1978 Adherence of *Mycoplasma gallisepticum* to erythrocytes. Infect Immun 21:365-372

Banai M, Razin S, Bredt W, Kahane I 1980 Isolation of binding sites to glycophorin from *Mycoplasma pneumoniae* membranes. Infect Immun, in press

Beachey EH 1975 Binding of group A streptococci to human oral mucosal cells by lipoteichoic acid (LTA). Trans Assoc Am Physicians 88:285-292

Beachey EH, Ofek I 1976 Epithelial binding of group A streptococci by lipoteichoic acid on fimbriae denuded of M protein. J Exp Med 143:759-771

Beachey EH, Chiang TM, Ofek I, Kang AH 1977 Interaction of lipoteichoic acid of group A streptococci with human platelets. Infect Immun 16:649-654

Beachey EH, Dale JB, Simpson WA, Evans JD, Knox KW, Ofek I, Wicken AJ 1979 Erythrocyte binding properties of streptococcal lipoteichoic acids. Infect Immun 23:618-625

Beachey EH, Simpson WA, Ofek I 1980 Interaction of surface polymers of *Streptococcus pyogenes* with animal cells. In: Rutter P (ed) Microbial adhesion to surfaces. Ellis Horwood, Chichester, in press

Burke D, Mendonça-Previato L, Ballou CE 1980 Cell–cell recognition in yeast: purification of *Hansenula wingei* 21-cell sexual agglutination factor and comparison of the factors from three genera. Proc Natl Acad Sci 77:318-322

Carr ME, McKendrick DW, Spyridakis T 1976 The clinical features of infantile gastroenteritis due to rotavirus. Scand J Infect Dis 8:241-243

Chiang TM, Alkan ML, Beachey EH 1979 Binding of lipoteichoic acid of group A streptococci to isolated human erythrocyte membranes. Infect Immun 26:316-321

Childs RA, Feizi T 1979 Calf heart lectin reacts with blood group Ii antigens and other precursor chains of the major blood group antigens. FEBS (Fed Eur Biochem Soc) Lett 99:175-179

Desai NN, Allen AK 1979 The purification of potato lectin by affinity chromatography on an N, N′, N″-triacetylchitotriose-Sepharose matrix. Anal Biochem 93:88-90

Doke N, Tomiyama K 1980 Suppression of the hypersensitive response of potato tuber protoplasts to hyphal wall components by water-soluble glucans isolated from *Phytophthora infestans.* Physiol Plant Pathol 16:177-186

Doke NA, Garas N, Kuc J 1979 Partial characterization and aspects of the mode of action of a hypersensitivity-inhibiting factor (HIF) isolated from *Phytophthora infestans.* Physiol Plant Pathol 15:127-140

Feizi T, Kabat EA 1972 Immunochemical studies on blood groups. LIV. Classification of anti-I and anti-i sera into groups based on reactivity patterns with various antigens related to the blood group A,B,H, Le^a, Le^b and precursor substances. J Exp Med 135:1247-1258

Feizi T, Childs RA, Watanabe K, Hakomori S 1979 Three types of blood group I specificity among monoclonal anti-I autoantibodies revealed by analogues of a branched erythrocyte glycolipid. J Exp Med 149:975-980

Fischetti VA, Gotschlich EC, Siviglia G, Zabriskie JB 1976 Streptococcal M protein extracted by nonionic detergent. I. Properties of the antiphagocytic and type-specific molecules. J Exp Med 144:32-53

Goldstein IJ, Hughes RC, Monsigny M, Osawa T, Sharon N 1980 What should be called a lectin? Nature (Lond) 285:66 (letter)

Johanson WG Jr, Woods DE, Chaudhuri T 1979 Association of respiratory tract colonization with adherence of gram-negative bacilli to epithelial cells. J Infect Dis 139:667-673

Johnson AP, Taylor-Robinson D, McGee ZA 1977 Species specificity of attachment and damage to oviduct mucosa by *Neisseria gonorrhoeae.* Infect Immun 18:833-839

Kessler RE, Shockman GD 1979 Precursor-product relationship of intracellular and extracellular lipoteichoic acids of *Streptococcus faecium.* J Bacteriol 137:869

Kessler RE, van de Rijn I, McCarty M 1979 Characterization and localization of the enzymatic deacylation of lipoteichoic acid in group A streptococci. J Exp Med 150:1498-1509

Kitazawa K, Inagaki K, Tomiyama K 1973 Cinephotomicrographic observations on the dynamic responses of protoplasm of a potato plant cell to infection by *Phytophthora infestans.* Phytopathol Z 76:80–86

Kuusela P 1978 Fibronectin binds to *Staphylococcus aureus.* Nature (Lond) 276:718-720

McGee ZA, Johnson AP, Taylor-Robinson D 1976 Human Fallopian tubes in organ culture: preparation, maintenance, and quantitation of damage by pathogenic microorganisms. Infect Immun 13:608-618

Marcan HM 1978 Aspects of the mechanism of hypersensitivity of the potato to *Phytophthora infestans* (Mont.) de Bary. PhD Thesis, University of Hull

Marcan H, Jarvis MC, Friend J 1979 Effect of methylglucosides and oligosaccharides on cell death and browning of potato tuber discs induced by mycelial components of *Phytophthora infestants.* Physiol Plant Pathol 14:19

Mebus CA 1976 Calf diarrhea induced by coronavirus & reovirus-like agent. In: Proceedings of the minisymposium on neonatal diarrhea in calves and pigs. University of Saskatchewan, Saskatoon, Saskatchewan, Canada, p 13-25

Muse KE, Collier AM, Baseman JB 1977 Scanning electron microscopic study of hamster tracheal organ cultures infected with *Bordetella pertussis.* J Infect Dis 136:768-777

Niemann H, Watanabe K, Hakomori S, Childs RA, Feizi T 1978 Blood group i and I activities of 'Lacto-N-*nor*hexaosyl-ceramide' and its analogues: the structural requirements for i-specificities. Biochem Biophys Res Commun 81:1286-1293

Powell DA, Hu PC, Wilson M, Collier AM, Baseman JB 1976 Attachment of *Mycoplasma pneumoniae* to respiratory epithelium. Infect Immun 13:959-966

Pryor EP, Willen J, Silverblatt FJ 1980 Piliated *Escherichia coli* activate polymorphonuclear leukocyte killing mechanisms. In: Current chemotherapy and infectious disease. American Society for Microbiology, Washington, p 817-818, in press

Russell H, Facklam RR 1975 Guanidine extraction of streptococcal M protein. Infect Immun 12:679-686

Sato N, Kitazawa K, Tomiyama K 1971 The role of rishitin in localizing the invading hyphae of *Phytophthora infestans* in infection sites at the cut surfaces of potato tubers. Physiol Plant Pathol 1:289-295

Shimony C, Friend J 1975 Ultrastructure of the interaction between *Phytophthora infestans* and leaves of two cultivars of potato (*Solanum tuberosum* L.) Orion and Majestic. New Phytol 74:59-65

Silverblatt FJ, Ofek I 1978 Influence of pili on virulence of *Proteus mirabilis* in experimental hematogenous pyelonephritis. J Infect Dis 138:664-667

Silverblatt FJ, Dreyer JS, Schauer S 1979 Effect of pili on susceptibility of *Escherichia coli* to phagocytosis. Infect Immun 24:218-223

Simpson WA, Ofek I, Sarasohn C, Morrison JC, Beachey EH 1980a Characteristics of the binding of streptococcal lipoteichoic acid to human oral epithelial cells. J Infect Dis 141:457-462

Simpson WA, Ofek I, Beachey EH 1980b Binding of streptococcal lipoteichoic acid to the fatty acid binding sites on serum albumin. J Biol Chem, in press

Simpson WA, Ofek I, Beachey EH 1980c Fatty acid binding sites of serum albumin as membrane receptor analogues for streptococcal lipoteichoic acid. Infect Immun, in press

Swanson J, Hsu KC, Gotschlich EC 1969 Electron microscopic studies of streptococci. I. M antigen. J Exp Med 130:1063-1091

Tebbutt GM, Veale DR, Hutchison JGP, Smith H 1976 The adherence of pilate and non-pilate strains of *Neisseria gonorrhoeae* to human and guinea-pig epithelial tissues. J Med Microbiol 9:263-273

Varns JL, Currier WW, Kuc J 1971 Specificity of rishitin and phytuberin accumulation by potato. Phytopathology 61:968-971

Watanabe K, Hakomori S, Childs RA, Feizi T 1979 Characterization of a blood group I-active ganglioside. Structural requirements for I and i specificities. J Biol Chem 254:3221-3228

Weiser M 1973 Intestinal epithelial cell surface membrane glycoprotein synthesis. I. An indicator of cellular differentiation. J Biol Chem 248:2536-2541

Wicken AJ, Knox KW 1975 Lipoteichoic acids: a new class of bacterial antigen. Science (Wash DC) 187:1161-1167

Wicken AJ, Knox KW 1977 Biological properties of lipoteichoic acids. In Schlessinger D (ed) Microbiology 1977. American Society for Microbiology, Washington, p 360-365

Wood E, Feizi T 1979 Blood group I and i activities of straight chain and branched synthetic oligosaccharides related to the precursors of the major blood group antigens. FEBS (Fed Eur Biochem Soc) Lett 104:135-140

Wood E, Hounsell EF, Feizi T 1981 Preparative affinity chromatography of blood group Ii active sheep gastric mucins and release of antigenically active oligosaccharides by alkaline borohydride degradation. Carbohydr Res, in press

Closing remarks

D. TAYLOR-ROBINSON

Division of Communicable Diseases, MRC Clinical Research Centre and Northwick Park Hospital, Watford Road, Harrow, Middlesex, HA1 3UJ, UK

Chairmen of some previous Ciba Foundation symposia provided good summaries of what was said, but I understand that many of them summarized things after the meeting rather than at the time. I am also comforted by the fact that some chairmen didn't summarize at all! They shied away from distilling almost three days of intensive discussion down to a crystallized form which is easy to digest. I hope your expectations of me in this regard are not too high. The best I can do is to comment briefly and in general terms on some of the points that impressed me. In doing this I shall return to the themes I mentioned in my opening remarks.

The first of these themes concerned the mechanisms of adherence. It was clear from the beginning of this meeting that there were likely to be a number of different mechanisms because of the range of microorganisms we were attempting to cover. The contents of Table 1, constructed by mutual effort, emphasize this and summarize some of the other points that have been made.

However, apart from differences in mechanisms, there are quite a lot of similarities even for diverse microorganisms. These had not been brought home to me before this meeting, perhaps because I hadn't really thought about them. For example, a carbohydrate forms the chemical basis for many of the eukaryotic cell receptors. I knew about mannose receptors for *E. coli* and for some other microorganisms, and the carbohydrate receptors for certain mycoplasmas, but I was unaware of the importance of carbohydrate as a receptor for *E. histolytica.*

The second question concerned the events which are stimulated by the adherence phenomenon. We did not discuss this aspect in a formal way but, on the other hand, we did not totally disregard it. Quite obviously we were concerned with the primary event – that is, adhesion, but we also discussed at some length the validity of models designed to relate the primary event to the final outcome, that is disease. It seems to me that in a series of events, A, B, C and D, which are linked, it would be foolhardy to necessarily relate A (adhesion) directly to D (disease) and forget all about B and C. The point was emphasized by Dr Freter and it is one that I and others should not forget. I have sympathy with Mark Richmond's ideas of examining the problems at a molecular level, but to apply this approach to all microorganisms would not seem possible at this stage and I don't believe that he was saying that. The problems encountered by one group of workers with a particular microorganism may be quite different and at a different level from those faced by other workers with a different microorganism. The models which are chosen may be different but whichever is used the question of whether it realistically relates adhesion to pathogenicity should be continually kept in mind.

The final point posed originally concerned the ways in which microbial adherence may be prevented. We covered a number of aspects of this and it is encouraging to hear that there are indications that the studies on adherence and its inhibition may well have a practical outcome which in the long run may be used clinically. This may be a longer run for some workers than for others. However, even those wrestling with the use of fimbrial vaccines, and I am thinking particularly about *E. coli* in this regard, seem to suggest that there is a light at the end of the tunnel. Let us hope that the whole effort is not going to end up as no more than an academic exercise. A clinical application rather than a mere clinical implication is the goal we all want to reach. And, talking of putting clinical implications into effect, I see no reason why Dr Sharon shouldn't irrigate bladders with mannose, although, from what we have heard, it seems much more likely that irrigation with the globoside discussed by Dr Svanborg Edén will be effective. And what of changing the pH of the urine? That is a time-honoured approach to treating urinary tract infections. Could it be that alkalinizing the urine has an effect because it tends to prevent the adhesion of bacteria that occurs at a low pH? I may be wrong but I give this as a simple example of the many thoughts the meeting has engendered for me.

As to terminology, I am sure we shouldn't attempt to influence the scientific community at large too much, particularly when there has been some doubt in our own minds about the correct terms to use. However, my impression is that most of us would be happy to speak about 'adherence' or 'adhesion' of a microorganism to a 'receptor' on the eukaryotic cell, there being a reciprocal 'adhesin' on the microorganism. In deference to Duguid we feel that non-flagellar structures which are on the surface of microorganisms and are considered to be involved in adherence should be called 'fimbriae', and that the term 'pili' should be reserved for structures involved in the transfer of genetic information from one organism to another.

I have been to other meetings where there has been little time for discussion or where everyone was engaged in research on exactly the same topic or microorganism. In these circumstances, human nature being what it is, there is a reluctance to give away one's best ideas and discussion becomes stultified. I don't believe anyone felt threatened at this meeting and in consequence there was a free interchange of data and ideas. For me, and I hope for all those present, the meeting was a very stimulating one. I trust that those who read the compilation of our efforts will find that so too.

TABLE 1 Adhesion of various microorganisms to eukaryotic cells

Microorganism	*Host receptor*	*Binding site or adhesin on organism*	*Can adhesion be blocked?*	*Is adhesion important for pathogenicity?*
Myxovirus	Neuraminic acid-containing glycoprotein and possibly glycolipid on cells and in mucus	HA (protein)	By antibody and receptor analogues	Receptor-binding protein (HA) is important in virus penetration
Paramyxovirus		HN (protein)	By antibody and receptor analogues	HN not important but other surface glycoprotein (F) plays major role in pathogenesis because it is required for penetration
Alpha virus	Cell surface glycoproteins	Spike glycoprotein	By antibody	Yes
Chlamydiae				
C. psittaci	Epithelium of respiratory, intestinal and other surfaces of avian and mammalian species Trypsin-sensitive receptors *N*-Acetylglucosamine-containing (one *C. psittaci* strain)	Heat-sensitive binding sites	By antibody and lectin	Important in penetration
C. trachomatis	Epithelium of human (and simian) urogenital tract, rectum, oropharynx and conjunctiva Sialic acid? (ocular *C. trachomatis* strains) *N*-Acetylglucosamine-containing (LGV *C. trachomatis* strain)	Heat-sensitive binding sites	By antibody and lectin	Important in penetration

Mycoplasmas				
M. pneumoniae	Sialic acid moieties and other unidentified receptors on respiratory tract epithelium of humans	Protein(s) on cell membrane possibly concentrated on special tip at end of filaments	Reduced by neuraminidase treatment of host cells, sialoglycoproteins, trypsin treatment of organisms; by metabolic inhibitors of mycoplasmas?	Yes. Non-adherent variants lose pathogenicity
M. gallisepticum	Sialic acid moieties on respiratory tract epithelium of chickens and turkeys	Proteins on the cell surface of blebs at the poles of the fusiform organisms	Reduced by neuraminidase treatment of host cells, sialoglycoproteins, trypsin treatment of organisms	Not determined
Bacteria				
Escherichia coli		All fimbrial adhesins composed of proteins:		
Enterotoxigenic				
(A) Human strains	Human intestinal epithelium	CFA/I fimbriae (O15, O20, O25 and O78 serogroups) More common in LT^+/ST^+ strains than in LT^-/ST^+ strains; rare in LT^+/ST^- strains	?	Probably. Purified fimbriae used as components of vaccines
		CFA/II fimbriae (O6 and O8 serogroups) Common in LT^+/ST^- strains	?	
		Type 1 somatic fimbriae (all serogroups)	By D-mannose and derivatives	
		Others	?	

Microorganism	*Host receptor*	*Binding site or adhesin on organism*	*Can adhesion be blocked?*	*Is adhesion important for pathogenicity?*
(B) Animal strains	Pig intestinal cells	K88 fimbriae (certain serogroups)	By K88 and antiserum to K88	Yes. Purified fimbrial vaccines protect
	Microvillus membrane of brush borders	N-terminal sequence of hydrophilic amino acids	Partial blocking by K99 and 987 fimbrial preparations, possibly by steric hindrance	
	Attachment is mannose-resistant	C-terminal sequence of hydrophobic amino acids		
	30-50% inhibition by stachyose and galactan			
	Glycolipid receptor site?			
	Attachment of 20% of bacteria to rabbit intestinal cells, no attachment to calf, sheep or guinea-pig			
	Pig, sheep and calf intestinal cells	K99 fimbriae (certain serogroups)	By antibody and homologous purified fimbriae	
	Pig intestinal cells	987 P fimbriae (certain serogroups)		
		Others (serogroups unknown)		

Escherichia coli (contd)				
Human urinary tract infection	(A) Globoseries glycolipids containing the sugar sequence GalNAcβ1→3 Galα1→4Gal in human urinary tract epithelial cells, human erythrocytes	Protein? Associated with fimbriae on certain strains	(1) By glycolipids isolated from human uroepithelial cells (2) By globoseries glycolipids isolated from human erythrocytes and other sources (3) By antibodies to isolated fimbriae (4) Not by D-mannose	Probably
	(B) Unidentified mannose-containing structure in urinary slime (Tamm-Horsfall glycoprotein?)	Protein. Type 1 fimbriae	(1) D-Mannose blocks haemagglutination (2) Tamm-Horsfall glycoprotein may block attachment to uroepithelial cells of strains belonging both to A and B	Not determined
Neisseria gonorrhoeae	Mucosal epithelium of human (and chimpanzee) urogenital tract, rectum, oropharynx and conjunctiva Chemical nature?	Pili (fimbriae) Glycoprotein? Glycolipid? Cell wall components Proteins? Sugar residues?	By antibody to pili	Yes. Not pathogenic for non-human species and no adhesion to these tissues. Also, less virulent type 4 organisms adhere less well than virulent type 1 organisms
Streptococcus spp.				
Group A streptococci	Epithelium of naso-oropharynx and other surfaces Chemical nature? (glycoprotein?)	Lipoteichoic acid (LTA) Glucan polymers on surface of some streptococci (*S. mutans, S. sanguis* and *S. bovis*) possibly involved in adherence to damaged endothelium	By LTA and by antibody	Yes

Microorganism	*Host receptor*	*Binding site or adhesin on organism*	*Can adhesion be blocked?*	*Is adhesion important for pathogenicity?*
Vibrio cholerae	Erythrocytes and brush border membranes of human and animal small intestinal epithelial cells	Unknown. Related to flagella synthesis?	By L-fucose and D-mannose	Not yet demonstrated directly
	Erythrocytes	Soluble haemagglutinin	By culture supernatant fluids and partially purified haemagglutinin. Not by fucose or mannose	Inhibition of adhesion of *V. cholerae* by soluble haemagglutinin and prevention of disease shown in infant rabbits
	Intestinal mucus	None. Invasion of mucus gel requires motility and chemotaxis	By inhibiting motility or by blocking chemotaxis, e.g. by local antibodies	In experimental cholera in several animal models
Protozoa				
Plasmodium spp.	Erythrocyte surface membrane components: likely to be more than one Chemical nature unknown	Surface of malaria merozoites for initial attachment Merozoite apex for junction formation Chemical nature unknown but trypsin-sensitive for *P. knowlesi* merozoites	Invasion of human erythrocytes by *P. falciparum* may be specifically inhibited by *N*-acetyl-D-glucosamine	Yes. An essential step in invasion of host cell by obligate intraerythrocytic parasite

Index of contributors

Entries in **bold** *type refer to papers; other entries refer to discussion contributions*

Indexes compiled by William Hill

Subject index